Lecture Notes in Artificial Intelligence 6992

Subseries of Lecture Notes in Computer Science

Ronen I. Brafman Fred S. Roberts
Alexis Tsoukiàs (Eds.)

Algorithmic Decision Theory

Second International Conference, ADT 2011
Piscataway, NJ, USA, October 26-28, 2011
Proceedings

Springer

Series Editors

Volume Editors

Ronen I. Brafman
Ben-Gurion University of the Negev
Beer-Sheva, Israel
E-mail: brafman@cs.bgu.ac.il

Fred S. Roberts
Rutgers University, DIMACS
Piscataway, NJ, USA
E-mail: froberts@dimacs.rutgers.edu

Alexis Tsoukiàs
Université Paris Dauphine, CNRS - LAMSADE
Paris, France
E-mail: tsoukias@lamsade.dauphine.fr

ISSN 0302-9743 e-ISSN 1611-3349
ISBN 978-3-642-24872-6 ISBN 978-3-642-24873-3 (eBook)
DOI 10.1007/978-3-642-24873-3
Springer Heidelberg Dordrecht London New York

Library of Congress Control Number: 2011938800

CR Subject Classification (1998): I.2, H.3, F.1, H.4, G.1.6, F.4.1-2, C.2

LNCS Sublibrary: SL 7 – Artificial Intelligence

Typesetting: Camera-ready by author, data conversion by Scientific Publishing Services, Chennai, India

Printed on acid-free paper

Springer is part of Springer Science+Business Media (www.springer.com)

Preface

Algorithmic Decision Theory (ADT) is a new interdisciplinary research area aiming at bringing together researchers from different fields such as decision theory, discrete mathematics, theoretical computer science, economics, and artificial intelligence, in order to improve decision support in the presence of massive data bases, combinatorial structures, partial and/or uncertain information and distributed, possibly interoperating decision makers. Such problems arise in real-world decision making in areas such as humanitarian logistics, epidemiology, environmental protection, risk assessment and management, e-government, electronic commerce, protection against natural disasters, and recommender systems.

In 2007, the EU-funded COST Action IC0602 on Algorithmic Decision Theory was started, networking a large number of researchers and research laboratories around Europe (and beyond). For more details see www.algodec.org. In October 2009 the First International Conference on Algorithmic Decision Theory was organized in Venice (Italy) (see www.adt2009.org) with considerable success (the proceedings appeared as LNAI 5783). The success of both the COST Action (now ended) and the conference led to several new initiatives, including the DIMACS 2010-2013 4-year Special Focus on Algorithmic Decision Theory supported by the U.S. National Science Foundation (NSF) (http://dimacs.rutgers.edu/SpecialYears/2010ADT/) and the GDRI ALGODEC (2011-2014) funded by several research institutions of five countries (Belgium, France, Luxembourg, Spain and the USA), including the Centre National de la Recherche Scientifique, France (CNRS) and the NSF.

These initiatives led in turn to the decision to organize the Second International Conference on Algorithmic Decision Theory (ADT 2011) at DIMACS, Rutgers University, October 26–28, 2011 (see www.adt2011.org). This volume contains the papers presented at ADT 2011. The conference received 50 submissions. Each submission was reviewed by at least 2 Program Committee members, and the Program Committee decided to accept 24 papers. There are two kinds of contributed papers, technical research papers and research challenge papers that lay out research questions in areas relevant to ADT. The topics of these contributed papers range from computational social choice to preference modeling, from uncertainty to preference learning, from multi-criteria decision making to game theory. In addition to the contributed papers, the conference had three kinds of invited talks: research talks by Michael Kearns, Don Kleinmuntz, and Rob Schapire; research challenge talks by Carlos Guestrin, Milind Tambe, and Marc Pirlot; and two tutorials: a tutorial on preference learning by Eyke Hullermeier and a tutorial on utility elicitation by Patrice Perny. We believe that colleagues will find this collection of papers exciting and useful for the advancement of the state of the art in ADT and in their respective disciplines.

We would like to take this opportunity to thank all authors who submitted papers to this conference, as well as all the Program Committee members and external reviewers for their hard work. ADT 2011 was made possible thanks to the support of the DIMACS Special Focus on Algorithmic Decision Theory, the GDRI ALGODEC, the EURO (Association of European Operational Research Societies), the LAMSADE at the University of Paris Dauphine, DIMACS, the CNRS, and NSF.

We would also like to acknowledge the support of Easychair in the preparation of the proceedings.

October 2011

Ronen Brafman
Fred Roberts
Alexis Tsoukias

Organization

Program Committee

David Banks	Duke University
Cliff Behrens	Telcordia Technologies, Inc.
Bob Bell	AT&T Labs-Research
Craig Boutilier	University of Toronto
Ronen Brafman	Ben-Gurion University of the Negev
Gerd Brewka	Leipzig University
Ching-Hua Chen-Ritzo	IBM T.J. Watson Research Center
Jan Chomicki	University at Buffalo
Vincent Connitzer	Duke University
Carmel Domshlak	Technion - Israel Institute of Technology
Ulle Endriss	ILLC, University of Amsterdam
Joe Halpern	Cornell University
Ulrich Junker	ILOG, An IBM Company
Werner Kiessling	Augsburg University
Jerome Lang	LAMSADE
Michael Littman	Rutgers University
David Madigan	Columbia University
Janusz Marecki	IBM T.J. Watson Research Center
Barry O'Sullivan	4C, University College Cork, Ireland
Sasa Pekec	Duke University
Patrice Perny	LIP6 - University of Paris 6
Marc Pirlot	University of Mons
Eleni Pratsini	IBM Zurich Research Lab
Bonnie Ray	IBM T.J. Watson Research Center
Fred Roberts	Rutgers University
Francesca Rossi	University of Padova
Andrzej Ruszczynski	Rutgers University
Roman Slowinski	Poznan University of Technology
Milind Tambe	University of Southern California
Alexis Tsoukias	CNRS - LAMSADE
Toby Walsh	NICTA and UNSW
Mike Wellman	University of Michigan
Nic Wilson	4C, University College Cork, Ireland
Laura Wynter	IBM T.J. Watson Research Center

Additional Reviewers

Brown, Matthew
He, Qing
Kamarianakis, Yiannis
Kawas, Ban
Kwak, Jun-Young
Lu, Tyler
Narodytska, Nina
Nonner, Tim
Spanjaard, Olivier
Szabo, Jacint
Wang, Xiaoting
Zhang, Xi

Sponsors

Table of Contents

How Hard Is it to Bribe the Judges? A Study of the Complexity of Bribery in Judgment Aggregation*

Dorothea Baumeister[1], Gábor Erdélyi[2], and Jörg Rothe[1]

[1] Institut für Informatik, Universität Düsseldorf, 40225 Düsseldorf, Germany
[2] SPMS, Nanyang Technological University, Singapore 637371

Abstract. Endriss et al. [1,2] initiated the complexity-theoretic study of problems related to judgment aggregation. We extend their results for manipulating two specific judgment aggregation procedures to a whole class of such procedures, and we obtain stronger results by considering not only the classical complexity (NP-hardness) but the parameterized complexity (W[2]-hardness) of these problems with respect to natural parameters. Furthermore, we introduce and study the closely related issue of bribery in judgment aggregation, inspired by work on bribery in voting (see, e.g., [3,4,5]). In manipulation scenarios one of the judges seeks to influence the outcome of the judgment aggregation procedure used by reporting an insincere judgment set. In bribery scenarios, however, an external actor, the briber, seeks to influence the outcome of the judgment aggregation procedure used by bribing some of the judges without exceeding his or her budget. We study three variants of bribery and show W[2]-hardness of the corresponding problems for natural parameters and for one specific judgment aggregation procedure. We also show that in certain special cases one can determine in polynomial time whether there is a successful bribery action.

1 Introduction

In judgment aggregation (see, e.g., [6,7]), the judges have to provide their judgments of a given set of possibly interconnected propositions, and if the simple majority rule is used to aggregate the individual judgments, the famous doctrinal paradox may occur (see [8] for the original formulation and [9] for a generalization). The study of different ways of influencing a judgment aggregation process is important, since the aggregation of different yes/no opinions about possibly interconnected propositions is often used in practice. To avoid the doctrinal paradox and, in general, inconsistencies in the aggregated judgment set, it is common to use a premise-based approach as we do here. In this approach, the individual judgments are given only over the premises, and the outcome for the conclusion is derived from the outcome for the premises.

A simple example for such a premise-based judgment aggregation procedure under the majority rule is given in Table 1. In this example, which is due to Bovens and Rabinowicz [10] (see also [11]), the three judges of a tenure committee have to decide whether a candidate deserves tenure, based on their judgments of two issues: first,

* This work was supported in part by DFG grant RO 1202/12-1 and the European Science Foundation's EUROCORES program LogICCC. The second author was supported by National Research Foundation (Singapore) under grant NRF-RF 2009-08.

R.I. Brafman, F. Roberts, and A. Tsoukiàs (Eds.): ADT 2011, LNAI 6992, pp. 1–15, 2011.

whether the candidate is good enough in research and, second, whether the candidate is good enough in teaching. The candidate should get tenure if and only if both requirements are satisfactorily fulfilled, which gives the decision of each individual judge in the right column of the table. To aggregate their individual judgments by the majority rule, both of the requirements (teaching and research) are evaluated by "yes" if and only if a strict majority of judges says "yes." The result for the conclusion (whether or not the candidate deserves tenure) is then derived logically from the result of the premises. Note that this premise-based judgment procedure preserves consistency and thus circumvents the doctrinal paradox (which would occur if also the aggregated conclusion were obtained by applying the majority rule to the individual conclusions, leading to the contradiction "(*yes* and *yes*) implies *no*").

Table 1. Example illustrating the premise-based procedure for the majority rule [10,11]

	teaching	research		tenure
judge 1	yes	yes		yes
judge 2	yes	no		no
judge 3	no	yes		no
majority	yes	yes	$\Rightarrow$	yes

On the basis of the above example, List [11] concludes that in a premise-based procedure the judges might have an incentive to report insincere judgments. Suppose that in the above example all judges are absolutely sure that they are right, so they all want the aggregated outcome to be identical to their own conclusions. In this case, judge 3 knows that insincerely changing his or her judgment on the candidate's research capabilities from "yes" to "no" would aggregate with the other individual judgments on this issue to a "no" and thus would deny the candidate tenure. For the same reason, judge 2 might have an incentive to give an insincere judgment of the teaching question. This is a classical *manipulation scenario*, which has been studied in depth in the context of voting (see, e.g., the surveys by Conitzer [12] and Faliszewski et al. [13,14] and the references cited therein). Strategic judging (i.e., changing one's individual judgments for the purpose of manipulating the collective outcome) was previously considered by List [11] and by Dietrich and List [15]. Endriss et al. [2] were the first to study the computational aspects of manipulation for judgment aggregation scenarios.

Returning to the above example, suppose that the judgments of judges 2 and 3 in Table 1 were "no" for both premises. Then the candidate (who, of course, would like to get tenure by any means necessary) might try to make some deals with some of the judges (for example, offering to apply for joint research grants with judge 3, and offering to take some of the teaching load off judge 2's shoulders, or just simply bribe the judges with money not exceeding his or her budget) in order to reach a positive evaluation. This is a classical *bribery scenario* which has been studied in depth in the context of voting (first by Faliszewski et al. [3], see also, e.g., [4,5]) and in the context of optimal lobbying (first by Christian et al. [16], see also [17] and Section 4 for more details). Manipulation, bribery, and lobbying are usually considered to be undesirable, and most of the recent literature on these topics is devoted to exploring the barriers to prevent such actions in terms of the computational complexity of the corresponding decision problems.

We extend the results obtained by Endriss et al. [2] on the complexity of manipulation in judgment aggregation from two specific judgment aggregation procedures to a whole class of such procedures. We study the corresponding manipulation problems not only in terms of their classical complexity but in terms of their parameterized complexity with respect to two natural parameters, one being the total number of judges and the other one being the maximum number of changes in the premises needed in the manipulator's judgment set. The W[2]-hardness results we obtain in particular imply the NP-hardness results Endriss et al. [2] obtained for the unparameterized problem.

Finally, inspired by bribery in voting [3], we introduce the concept of bribery in judgment aggregation. We consider three types of bribery (exact bribery, bribery, and microbribery) and define and motivate the corresponding bribery problems for judgment aggregation, building on the related but simpler model of optimal lobbying (see [16,17]). We show that, for one specific judgment aggregation procedure, each of the three types of bribery is W[2]-hard with respect to natural parameters; again, note that NP-completeness follows for the corresponding unparameterized problems. One natural parameter we study here is again the total number of judges. Showing W[2]-hardness for this parameter implies that the problem remains hard even if the number of judges is bounded by a constant. As this is often the case in judgment aggregation, it is natural to study this parameter. By contrast, we also show that in certain cases one can determine in polynomial time whether there exists a successful bribery action.

Both manipulation and bribery were first defined and studied for preference aggregation, especially in voting scenarios. By the above examples we have argued that it makes sense to study these issues also in the context of judgment aggregation. There is, however, one major difference between the aggregation of preferences via voting systems and judgment aggregation. Both fields are closely related but consider different settings (for further details, see [7,18]). In voting, the individuals report their subjective personal preference over some given alternatives. For example, one voter may prefer alternative a to alternative b, and another voter may prefer b to a. This does not contradict, and even if both voters may not understand the other voter's preferences on a and b, they should accept them. In judgment aggregation, however, the judges report their individual judgment of some given proposition φ. If there are two judges, one reporting "φ is *true*" and the other reporting "φ is *false*," they have contradicting individual judgments regarding φ. These two judges with opposing judgments for the same proposition will simply believe the other one is wrong. In certain cases it might even be possible to objectively determine the truth value of the proposition and decide who of the judges is right and who is wrong. This would be impossible to say for an individual preference.

2 Preliminaries

The formal definition of the judgment aggregation framework follows the work of Endriss et al. [2]. The set of all propositional variables is denoted by PS, and the set of propositional formulas built from PS is denoted by $\mathscr{L}_{PS}$. As connectives in propositional formulas, we allow disjunction ($\vee$), conjunction ($\wedge$), implication ($\rightarrow$), and equivalence ($\leftrightarrow$) in their usual meaning, and the two boolean constants 1 and 0 representing "true" and "false," respectively. Since double negations are undesirable, let $\sim\alpha$ denote the complement of α. This means that if α is not negated then $\sim\alpha = \neg\alpha$, and if $\alpha = \neg\beta$ then $\sim\alpha = \beta$. The set of formulas to be judged by the judges is called the *agenda*. Formally, the agenda is a finite, nonempty subset Φ of $\mathscr{L}_{PS}$. As mentioned above, the agenda does not contain doubly negated formulas, and it also holds that $\sim\alpha \in \Phi$ for all $\alpha \in \Phi$, that is, Φ is required to be closed under complementation. The judgment provided by a single judge is called his or her *individual judgment set* and corresponds to the propositions in the agenda accepted by this judge. The set of propositions accepted by all judges is called their *collective judgment set*. An individual or collective judgment set J on an agenda Φ is a subset $J \subseteq \Phi$.

We consider three basic properties of judgment sets, completeness, complement-freeness, and consistency. A judgment set J is said to be *complete* if it contains α or $\sim\alpha$ for each $\alpha \in \Phi$. We say J is *complement-free* if there is no $\alpha \in J$ with $\sim\alpha \in J$. Finally, J is *consistent* if there is an assignment that satisfies all formulas in J. We denote the set of all complete and consistent subsets of Φ by $\mathscr{J}(\Phi)$. Obviously, all sets in $\mathscr{J}(\Phi)$ are also complement-free.

We let $N = \{1,\dots,n\}$ denote the set of judges taking part in a judgment aggregation scenario, and we will always assume that there are at least two judges, so $n \geq 2$. The individual judgment set of judge $i \in N$ is denoted by J_i, and the profile of all n individual judgment sets is denoted by $\mathbf{J} = (J_1,\dots,J_n)$.

To obtain a collective judgment set from a given profile $\mathbf{J} \in \mathscr{J}(\Phi)^n$, an aggregation procedure F is needed. This is a function $F : \mathscr{J}(\Phi)^n \rightarrow 2^{\Phi}$, mapping a profile of n complete and consistent judgment sets to a subset of the agenda Φ, the collective judgment set. We consider the same three basic properties for judgment aggregation procedures as for judgment sets. A judgment aggregation procedure F is said to be *complete/complement-free/consistent* if $F(\mathbf{J})$ is complete/complement-free/consistent for all profiles $\mathbf{J} \in \mathscr{J}(\Phi)^n$. One particular judgment aggregation procedure studied by Endriss et al. [2] is the *premise-based procedure*.

Definition 1 (Premise-based Procedure [2]). *Let the agenda Φ be divided into two disjoint sets, $\Phi = \Phi_p \uplus \Phi_c$, where Φ_p is the set of premises and Φ_c is the set of conclusions, and both Φ_p and Φ_c are closed under complementation. The* premise-based procedure *is a function $PBP : \mathscr{J}(\Phi)^n \rightarrow 2^{\Phi}$ mapping, for $\Phi = \Phi_p \uplus \Phi_c$, each profile $\mathbf{J} = (J_1,\dots,J_n)$ to the following judgment set:*

$$PBP(\mathbf{J}) = \triangle \cup \{\varphi \in \Phi_c \mid \triangle \models \varphi\}$$

with $\triangle = \{\varphi \in \Phi_p \mid \|\{i \mid \varphi \in J_i\}\| > n/2\}$, where $\|S\|$ denotes the cardinality of set S and $\models$ denotes the satisfaction relation.

According to this definition, the majority procedure is applied only to the premises of the agenda, and the collective outcome for the conclusions is derived from the collective outcome of the premises. However, this is not sufficient to obtain a complete and consistent procedure. To achieve this, it is furthermore required that the agenda is closed under propositional variables (i.e., every variable that occurs in a formula of Φ is contained in Φ), that the set of premises is the set of all literals in the agenda, and that the number of judges is odd. Endriss et al. [2] argue that this definition is appropriate, since the problem of determining whether an agenda guarantees a complete and consistent outcome for the majority procedure is an intractable problem.

We extend this approach to the class of *uniform quota rules* as defined by Dietrich and List [19]. We allow an arbitrary quota and do not restrict our scenarios to an odd number of judges.

Definition 2 (Premise-based Quota Rule). *Let the agenda Φ be divided into two disjoint sets, $\Phi = \Phi_p \uplus \Phi_c$, where Φ_p is the set of premises and Φ_c is the set of conclusions, and both Φ_p and Φ_c are closed under complementation. Divide the set of premises Φ_p into two disjoint subsets, Φ_1 and Φ_2, such that for each $\varphi \in \Phi_p$, either $\varphi \in \Phi_1$ and $\sim\varphi \in \Phi_2$ or $\varphi \in \Phi_2$ and $\sim\varphi \in \Phi_1$. Define a quota $q_\varphi \in \mathbb{Q}$ with $0 \leq q_\varphi < 1$ for every $\varphi \in \Phi_1$. The quota for every $\varphi \in \Phi_2$ is then defined as $q'_\varphi = 1 - q_\varphi$. The* premise-based quota rule *is a function $PQR : \mathscr{J}(\Phi)^n \to 2^\Phi$ mapping, for $\Phi = \Phi_p \uplus \Phi_c$, each profile $\mathbf{J} = (J_1, \ldots, J_n)$ to the following judgment set:*

$$PQR(\mathbf{J}) = \triangle_q \cup \{\varphi \in \Phi_c \mid \triangle_q \models \varphi\},$$

where

$$\triangle_q = \{\varphi \in \Phi_1 \mid \|\{i \mid \varphi \in J_i\}\| > n \cdot q_\varphi\} \cup \{\varphi \in \Phi_2 \mid \|\{i \mid \varphi \in J_i\}\| > \lceil n \cdot q'_\varphi - 1 \rceil\}.$$

To obtain complete and consistent collective judgment sets, we again require that the agenda Φ is closed under propositional variables, and that Φ_p consists of all literals. The number of affirmations needed to be in the collective judgment set may differ for the variables in Φ_1 and in Φ_2. For $\varphi \in \Phi_1$, at least $\lfloor n \cdot q_\varphi + 1 \rfloor$ affirmations from the judges are needed, and for $\varphi \in \Phi_2$, $\lceil n \cdot q'_\varphi \rceil$ affirmations are needed. Clearly, since $\lfloor n \cdot q_\varphi + 1 \rfloor + \lceil n \cdot q'_\varphi \rceil = n + 1$, it is ensured that for every $\varphi \in \Phi$, either $\varphi \in PQR(\mathbf{J})$ or $\sim\varphi \in PQR(\mathbf{J})$. Observe that the quota $q_\varphi = 1$ for a literal $\varphi \in \Phi_1$ is not considered here, since then $n+1$ affirmations were needed for $\varphi \in \Phi_1$ to be in the collective judgment set, which is not possible. Hence, the outcome does not depend on the individual judgment sets. By contrast, considering $q_\varphi = 0$ leads to the case that $\varphi \in \Phi_1$ needs at least one affirmation, and $\sim\varphi \in \Phi_2$ needs n affirmations, which may be a reasonable choice.

If the quota q_φ is identical for all literals in Φ_1 and hence also the quota q'_φ for all literals in Φ_2, we obtain the special case of *uniform premise-based quota rules*. The quotas will then be q for all $\varphi \in \Phi_1$ and q' for all $\varphi \in \Phi_2$. In this paper, we focus on this class of rules, and denote it by $UPQR_q$. For the case of $q = 1/2$ and an odd number of judges, we obtain exactly the premise-based procedure defined by Endriss et al. [2] (see Definition 1).

We assume that the reader is familiar with the basic concepts of complexity theory and with complexity classes such as P and NP; see, e.g., [20]. Downey and Fellows [21] introduced *parameterized* complexity theory; in their framework it is possible to do a more fine-grained multi-dimensional complexity analysis. In particular, NP-complete problems may be easy (i.e., fixed-parameter tractable) with respect to certain parameters confining the seemingly unavoidable combinatorial explosion. If this parameter is reasonably small, a fixed-parameter tractable problem can be solved efficiently in practice, despite its NP-hardness. Formally, a *parameterized decision problem* is a set $L \subseteq \Sigma^* \times N$, and we say it is *fixed-parameter tractable* (FPT) if there is a constant c such that for each input (x,k) of size $n = |(x,k)|$ we can determine in time $O(f(k) \cdot n^c)$ whether (x,k) is in L, where f is a function depending only on the parameter k. The main hierarchy of parameterized complexity classes is:

$$\mathrm{FPT} = \mathrm{W}[0] \subseteq \mathrm{W}[1] \subseteq \mathrm{W}[2] \subseteq \cdots \subseteq \mathrm{W}[\ell] \subseteq \mathrm{XP}.$$

In our results, we will focus on only the class W[2], which refers to problems that are considered to be fixed-parameter intractable. In order to show that a parameterized problem is W[2]-hard, we will give a parameterized reduction from the W[2]-complete problem k-DOMINATING SET (see [21]). We say that a parameterized problem A *parameterized reduces* to a parameterized problem B if each instance (x,k) of A can be transformed in time $\mathcal{O}(g(k) \cdot |x|^c)$ (for some function g and some constant c) into an instance (x',k') of B such that $(x,k) \in A$ if and only if $(x',k') \in B$, where $k' = g(k)$.

3 Problem Definitions

Bribery problems in voting theory, as introduced by Faliszewski et al. [3] (see also, e.g., [4,5]), model scenarios in which an external actor seeks to bribe some of the voters to change their votes such that a distinguished candidate becomes the winner of the election. In judgment aggregation it is not the case that one single candidate wins, but there is a decision for every formula in the agenda. So the external actor might seek to obtain exactly his or her desired collective outcome by bribing the judges, or he or she might be interested only in the desired outcome of some formulas in Φ. The exact bribery problem is then defined as follows for a given aggregation procedure F.

EXACT-F-BRIBERY	
Given:	An agenda Φ, a profile $\mathbf{T} \in \mathcal{J}(\Phi)^n$, a consistent and complement-free judgment set J (not necessarily complete) desired by the briber, and a positive integer k.
Question:	Is it possible to change up to k individual judgment sets in $\mathbf{T}$ such that for the resulting new profile $\mathbf{T}'$ it holds that $J \subseteq F(\mathbf{T}')$?

Note that if J is a complete judgment set then the question is whether $J = F(\mathbf{T}')$.

Since in the case of judgment aggregation there is no winner, we also adopt the approach Endriss et al. [2] used to define the manipulation problem in judgment aggregation. In their definition, an outcome (i.e., a collective judgment set) is more desirable for the manipulator if its Hamming distance to the manipulator's desired judgment set is smaller, where for an agenda Φ the Hamming distance $H(J,J')$ between two complete and consistent judgment sets $J,J' \in \mathcal{J}(\Phi)$ is defined as the number of positive formulas in Φ on which J and J' differ. The formal definition of the manipulation problem in judgment aggregation is as follows, for a given aggregation procedure F.

F-MANIPULATION

Given: An agenda Φ, a profile $\mathbf{T} \in \mathscr{J}(\Phi)^{n-1}$, and a consistent and complete judgment set J desired by the manipulator.

Question: Does there exist a judgment set $J' \in \mathscr{J}(\Phi)$ such that $H(J,F(\mathbf{T},J')) < H(J,F(\mathbf{T},J))$?

Now, we can give the formal definition of bribery in judgment aggregation, where the briber seeks to obtain a collective judgment set having a smaller Hamming distance to the desired judgment set, then the original outcome has. In bribery scenarios, we extend the above approach of Endriss et al. [2] by allowing that the desired outcome for the briber may be an incomplete (albeit consistent and complement-free) judgment set. This reflects a scenario where the briber may be interested only in some part of the agenda. The definition of Hamming distance is extended accordingly as follows. Let Φ be an agenda, $J \in \mathscr{J}(\Phi)$ be a complete and consistent judgment set, and $J' \subseteq \Phi$ be a consistent and complement-free judgment set. The *Hamming distance* $H(J,J')$ *between* J *and* J' is defined as the number of formulas from J' on which J does not agree:

$$H(J,J') = \|\{\varphi \mid \varphi \in J' \wedge \varphi \notin J\}\|.$$

Observe that if J' is also complete, this extended notion of Hamming distance coincides with the notion Endriss et al. [2] use.

F-BRIBERY

Given: An agenda Φ, a profile $\mathbf{T} \in \mathscr{J}(\Phi)^n$, a consistent and complement-free judgment set J (not necessarily complete) desired by the briber, and a positive integer k.

Question: Is it possible to change up to k individual judgment sets in $\mathbf{T}$ such that for the resulting new profile $\mathbf{T}'$ it holds that $H(F(\mathbf{T}'),J) < H(F(\mathbf{T}),J)$?

Faliszewski et al. [5] introduced microbribery for voting systems. We adopt their notion so as to apply to judgment aggregation. In microbribery for judgment aggregation, if the briber's budget is k, he or she is not allowed to change up to k entire judgment sets but instead can change up to k premise entries in the given profile (the conclusions change automatically if necessary).

F-MICROBRIBERY

Given: An agenda Φ, a profile $\mathbf{T} \in \mathscr{J}(\Phi)^n$, a consistent and complement-free judgment set J (not necessarily complete) desired by the briber, and a positive integer k.

Question: Is it possible to change up to k entries among the premises in the individual judgment sets in $\mathbf{T}$ such that for the resulting profile $\mathbf{T}'$ it holds that $H(F(\mathbf{T}'),J) < H(F(\mathbf{T}),J)$?

EXACT-F-MICROBRIBERY is defined analogously to the corresponding bribery problem with the difference that the briber is allowed to change only up to k entries in $\mathbf{T}$ rather than to change k complete individual judgment sets.

In our proofs we will make use of the following two problems. First, we will use DOMINATING SET, a classical problem from graph theory. Given a graph $G = (V,E)$, a *dominating set* is a subset $V' \subseteq V$ such that for each $v \in V \setminus V'$ there is an edge $\{v, v'\}$ in E with $v' \in V'$. The size of a dominating set V' is the number $\|V'\|$ of its vertices.

DOMINATING SET

Given: A graph $G = (V,E)$, with the set V of vertices and the set E of edges, and a positive integer $k \leq \|V\|$.

Question: Does G have a dominating set of size at most k?

DOMINATING SET is NP-complete (see [22]) and, when parameterized by the upper bound k on the size of the dominating set, its parameterized variant (denoted by k-DOMINATING SET, to be explicit) is W[2]-complete [21].

Second, we will use the following problem:

OPTIMAL LOBBYING

Given: An $m \times n$ 0-1 matrix L (whose rows represent the voters, whose columns represent the referenda, and whose 0-1 entries represent No/Yes votes), a positive integer $k \leq m$, and a target vector $x \in \{0,1\}^n$.

Question: Is there a choice of k rows in L such that by changing the entries of these rows the resulting matrix has the property that, for each j, $1 \leq j \leq n$, the jth column has a strict majority of ones (respectively, zeros) if and only if the jth entry of the target vector x of The Lobby is one (respectively, zero)?

OPTIMAL LOBBYING has been introduced and, parameterized by the number k of rows The Lobby can change, shown to be W[2]-complete by Christian et al. [16] (see also [17] for a more general framework and more W[2]-hardness results).

Note that a multiple referendum as in OPTIMAL LOBBYING can be seen as the special case of a judgment aggregation scenario where the agenda is closed under complementation and propositional variables and contains only premises and where the majority rule is used for aggregation. For illustration, consider the following simple example of a multiple referendum. Suppose the citizens of a town are asked to decide by a referendum whether two projects, A and B (e.g., a new hospital and a new bridge), are to be realized. Suppose the building contractor (who, of course, is interested in being awarded a contract for both projects) sets some money aside to attempt to influence the outcome of the referenda, by bribing some of the citizens without exceeding this budget. Observe that an EXACT-*PBP*-BRIBERY instance with only premises in the agenda and with a complete desired judgment set J is nothing other than an OPTIMAL LOBBYING instance, where J corresponds to The Lobby's target vector.[1] Requiring the citizens to give their opinion only for the premises A and B of the referendum and not for the conclusion (whether both projects are to be realized) again avoids the doctrinal paradox.

[1] Although exact bribery in judgment aggregation thus generalizes lobbying in the sense of Christian et al. [16] (which is different from bribery in voting, as defined by Faliszewski et al. [3]), we will use the term "bribery" rather than "lobbying" in the context of judgment aggregation.

As mentioned above, studying the parameterized complexity for the parameter "total number of judges" is very natural. The second parameter we have considered for the manipulation problem in Theorem 1 is the "maximum number of changes in the premises needed in the manipulator's judgment set." Hence this theorem shows that the problem remains hard even if the number of premises the manipulator can change is bounded by a fixed constant. This is also very natural, since the manipulator may wish to report a judgment set that is as close as possible to his or her sincere judgment set, because for a completely different judgment set it might be discovered too easily that he was judging strategically.

In contrast to the hardness results stated in Theorem 1, the following proposition shows that, depending on the agenda, there are cases in which $UPQR_q$-MANIPULATION is solvable in polynomial time.

Proposition 1. *If the agenda contains only premises then* $UPQR_q$-MANIPULATION *is in* P.

Proof. Assume that the agenda Φ contains only premises. Then every variable is considered independently. Let n be the number of judges. If φ is contained in the judgment set J of the manipulator, and φ does not have $\lfloor n \cdot q + 1 \rfloor$ (respectively, $\lceil n(1-q) \rceil$) affirmations without considering J, it cannot reach the required number of affirmations if the manipulator switches from φ to $\sim\varphi$ in his or her judgment set. ❑

The W[2]-hardness result for $UPQR_q$-MANIPULATION, parameterized by the number of judges, stated in Theorem 1 implies that there is little hope to find a polynomial-time algorithm for the general problem even when the number of judges participating is fixed. However, Proposition 1 tells us that if the agenda is simple and contains no conclusions, the problem can be solved efficiently even when the number of judges participating is not fixed.

4.2 Bribery in Judgment Aggregation

In this section we will study the complexity of several bribery problems for the premise-based procedure *PBP*, i.e., $UPQR_{1/2}$ for an odd number of judges. We will again establish even W[2]-hardness results for two natural parameters for these bribery problems.

Theorem 2. *PBP*-BRIBERY *is* W[2]*-hard when parameterized either by the total number of judges, or by the number of judges that can be bribed.*

Proof. We will show W[2]-hardness by a slightly modified construction from Theorem 1. We start by considering the case, where the briber is allowed to bribe exactly one judge. The notation and the agenda from that proof remain unchanged, but the individual judgment sets are slightly different. The first two judges remain unchanged, but the third judge has the same judgment set as the second one, and the desired judgment set J is equal to J_3. Since the quota is $1/2$, two affirmations are needed to be in the collective judgment set. Again the briber cannot benefit from bribing one judge to switch from $\neg y$ to y in his or her individual judgment set. Hence the change must be in the set of variables $\{v_1, \dots, v_n\}$ from the second or the third judge. By a similar argument as in the proof of Theorem 1, there is a successful bribery action if and only if there is a dominating set of size at most k for the given graph.

Now we consider the case that the briber is allowed to bribe more than one judge. If the briber is allowed to bribe k judges, we construct an instance with $2k+1$ judges, where one judgement set is equal to J_1 and the remaining $2k$ individual judgment sets are equal to J_2. It is again not possible for the briber to change the entry for y, and the briber must change the entry for any v_i in the judgment sets from k judges to obtain a different collective outcome. This construction works by similar arguments as above. Since the total number of judges and the number of judges that can be bribed depends only on k, W[2]-hardness follows for both parameters. ❑

As in the case of manipulation, the proof of Theorem 2 immediately implies an NP-completeness result for *PBP*-BRIBERY.

Next, we turn to microbribery. Here the briber can change only up to a fixed number of entries in the individual judgment sets. We again start by proving W[2]-hardness for the parameters number of judges and number of microbribes allowed.

Theorem 3. *PBP*-MICROBRIBERY *is* W[2]*-hard when parameterized either by the total number of judges, or by the number of microbribes allowed.*

Proof. The proof that *PBP*-MICROBRIBERY is W[2]-hard is similar to the proof of Theorem 2. The given instance for the k-DOMINATING SET Problem is the graph $G=(V,E)$ and the positive integer k. The agenda Φ is defined as in the proof of Theorem 1. The number of judges is $2k+1$, where the individual judgment sets of k judges are of type J_1 and the remaining $k+1$ individual judgment sets are of type J_2. The desired outcome of the briber is the judgment set J_3. The number of affirmations needed to be in the collective judgment set is at least $k+1$, and the number of entries the briber is allowed to change is at most k. Since none of the judges have y in their individual judgment sets, the briber cannot change the collective outcome for y to 1. Hence all entries that can be changed are for the variables $v_1,\dots,v_n$. Obviously, setting the value for one v_i in one of the judges of type J_2 to 1 causes v_i to be in the collective judgment set and all other changes have no effect on the collective judgment set. By similar arguments as in the proof of Theorem 1, there is a successful microbribery action if and only if the given graph has a dominating set of size at most k. Since both the total number of judges and the number of entries the briber is allowed to change depend only on k, W[2]-hardness follows directly for both parameters. ❑

Again, NP-hardness of *PBP*-MICROBRIBERY follows immediately from that of DOMINATING SET.

Theorem 4. EXACT-*PBP*-BRIBERY *is* W[2]*-hard when parameterized by the number of judges that can be bribed.*

Proof. Observe that an exact bribery instance with only premises in the agenda and with a complete desired judgment set J is exactly the OPTIMAL LOBBYING problem. Since this problem is W[2]-complete for the parameter number of rows that can be changed, EXACT-*PBP*-BRIBERY inherits the W[2]-hardness lower bound, where the parameter is the number of judges that can be bribed. ❑

Note that W[2]-hardness with respect to any parameter directly implies NP-hardness for the corresponding unparameterized problem, so EXACT-*PBP*-BRIBERY is also NP-complete (all unparameterized problems considered here are easily seen to be in NP).

Theorem 5. EXACT-*PBP*-MICROBRIBERY *is* W[2]*-hard when parameterized either by the number of judges, or by the number of microbribes.*

Proof. Consider the construction in the proof of Theorem 3, and change the agenda such that there are only n^2-2 (instead of n^2-k-2) syntactic variations of the formula $v_1 \vee \cdots \vee v_n$ (i.e., this can be seen as giving a weight of n^2-1 to this formula), and that the desired judgment set J is incomplete and contains all conclusions. By similar arguments as above, a successful microbribery of k entries is possible if and only if there is a dominating set for G of size at most k. ❑

As for the manipulation problem, we studied in Theorems 2 through 5 the bribery problems for the natural parameter "total number of judges." It turned out that for that parameter BRIBERY, MICROBRIBERY, and their exact variants are W[2]-hard for the premise-based procedure for the majority rule. Hence these four problems remain hard even if the total number of judges is fixed. Furthermore we considered the parameter "number of judges allowed to bribe" for *PBP*-BRIBERY and its exact variant and the parameter "number of microbribes allowed" for *PBP*-MICROBRIBERY and its exact variant. Both parameters concern the budget of the briber. Since the briber aims at spending as little money as possible, it is also natural to consider this parameter. But again W[2]-hardness was shown in all cases, which means that bounding the budget by a fixed constant does not help to solve the problem easily (i.e., it is unlikely to be fixed-parameter tractable).

Although the exact microbribery problem is computationally hard in general for the aggregation procedure *PBP*, there are some interesting naturally restricted instances where it is computationally easy.

Theorem 6. *If the desired judgment set J is complete or if the desired judgment set is incomplete but contains all of the premises or only premises, then* EXACT-*PBP*-MICROBRIBERY *is in* P.

Proof. We give only an informal description of the algorithm that computes a successful microbribery.

Input: Our algorithm takes as an input a complete profile **T**, a consistent judgment set J, and a positive integer k.

Step 1: For each premise present in J, compute the minimum number of entries that have to be flipped in order to make the collective judgment on that premise equal to the desired judgment set's entry on that premise. Note that this can be done in linear time, since it is a simple counting. Let d_i denote the number of entries needed to flip for premise i.

Step 2: Check if $\sum_i d_i \leq k$.

Output: If $\sum_i d_i \leq k$, output the entries which have to be flipped and halt. Otherwise, output "bribery impossible" and halt.

Clearly, this algorithm works in polynomial time. The output is correct, since if we need less than k flips in the premises, the premises are evaluated exactly as they are in J, and the conclusions follow automatically, since we are using a premise-based procedure. ❑

5 Conclusions

Following up a line of research initiated by Endriss et al. [1,2], we have studied the computational complexity of problems related to manipulation and bribery in judgment aggregation. In particular, the complexity of bribery—though deeply investigated in the context of voting [3,4,5]—has not been studied before in the context of judgment aggregation. For three natural scenarios modelling different ways of bribery, we have shown that the corresponding problems are computationally hard even with respect to their parameterized complexity (namely, W[2]-hard) for natural parametrizations. In addition, extending the results of Endriss et al. [2] on the (classical) complexity of manipulation in judgment aggregation, we have obtained W[2]-hardness for the class of uniform premise-based quota rules, for each reasonable quota. From all W[2]-hardness results we immediately obtain the corresponding NP-hardness results, and since all problems considered are easily seen to be in NP, we have NP-completeness results. It remains open, however, whether one can also obtain matching upper bounds in terms of parameterized complexity. We suspect that all W[2]-hardness results in this paper in fact can be strengthened to W[2]-completeness results.

Faliszewski et al. [3] introduced and studied also the "priced" and "weighted" versions of bribery in voting. These notions can be reasonably applied to bribery in judgment aggregation: The "priced" variant means that judges may request different amounts of money to be willing to change their judgments according to the briber's will, and the "weighted" variant means that the judgments of some judges may be heavier than those of others. Although we have not defined this in a formal setting here, note that our hardness results carry over to more general problem variants as well. A more interesting task for future research is to try to complement our parameterized worst-case hardness results by studying the typical-case behavior for these problems, as is currently done intensely in the context of voting. Another interesting task is to study these problems for other natural parameters and for other natural judgment aggregation procedures.

Acknowledgments. We thank the anonymous reviewers for their helpful reviews and literature pointers.

References

1. Endriss, U., Grandi, U., Porello, D.: Complexity of judgment aggregation: Safety of the agenda. In: Proceedings of the 9th International Joint Conference on Autonomous Agents and Multiagent Systems, IFAAMAS, pp. 359–366 (May 2010)
2. Endriss, U., Grandi, U., Porello, D.: Complexity of winner determination and strategic manipulation in judgment aggregation. In: Conitzer, V., Rothe, J. (eds.) Proceedings of the 3rd International Workshop on Computational Social Choice, Universität Düsseldorf, pp. 139–150 (September 2010)
3. Faliszewski, P., Hemaspaandra, E., Hemaspaandra, L.: How hard is bribery in elections? Journal of Artificial Intelligence Research 35, 485–532 (2009)
4. Elkind, E., Faliszewski, P., Slinko, A.: Swap bribery. In: Mavronicolas, M., Papadopoulou, V.G. (eds.) SAGT 2009. LNCS, vol. 5814, pp. 299–310. Springer, Heidelberg (2009)

5. Faliszewski, P., Hemaspaandra, E., Hemaspaandra, L., Rothe, J.: Llull and Copeland voting computationally resist bribery and constructive control. Journal of Artificial Intelligence Research 35, 275–341 (2009)
6. List, C., Pettit, P.: Aggregating sets of judgments: An impossibility result. Economics and Philosophy 18(1), 89–110 (2002)
7. List, C., Pettit, P.: Aggregating sets of judgments: Two impossibility results compared. Synthese 140(1-2), 207–235 (2004)
8. Kornhauser, L.A., Sager, L.G.: Unpacking the court. Yale Law Journal 96(1), 82–117 (1986)
9. Pettit, P.: Deliberative democracy and the discursive dilemma. Philosophical Issues 11, 268–299 (2001)
10. Bovens, L., Rabinowicz, W.: Democratic answers to complex questions – an epistemic perspective. Synthese 150(1), 131–153 (2006)
11. List, C.: The discursive dilemma and public reason. Ethics 116(2), 362–402 (2006)
12. Conitzer, V.: Making decisions based on the preferences of multiple agents. Communications of the ACM 53(3), 84–94 (2010)
13. Faliszewski, P., Hemaspaandra, E., Hemaspaandra, L.: Using complexity to protect elections. Communications of the ACM 53(11), 74–82 (2010)
14. Faliszewski, P., Procaccia, A.: AI's war on manipulation: Are we winning? AI Magazine 31(4), 53–64 (2010)
15. Dietrich, F., List, C.: Strategy-proof judgment aggregation. Economics and Philosophy 23(3), 269–300 (2007)
16. Christian, R., Fellows, M., Rosamond, F., Slinko, A.: On complexity of lobbying in multiple referenda. Review of Economic Design 11(3), 217–224 (2007)
17. Erdélyi, G., Fernau, H., Goldsmith, J., Mattei, N., Raible, D., Rothe, J.: The complexity of probabilistic lobbying. In: Rossi, F., Tsoukias, A. (eds.) ADT 2009. LNCS, vol. 5783, pp. 86–97. Springer, Heidelberg (2009)
18. Dietrich, F., List, C.: Arrow's theorem in judgment aggregation. Social Choice and Welfare 29(1), 19–33 (2007)
19. Dietrich, F., List, C.: Judgment aggregation by quota rules: Majority voting generalized. Journal of Theoretical Politics 19(4), 391–424 (2007)
20. Papadimitriou, C.: Computational Complexity, 2nd edn. Addison-Wesley, Reading (1995) Reprinted with corrections
21. Downey, R., Fellows, M.: Parameterized Complexity. Springer, Heidelberg (1999)
22. Garey, M., Johnson, D.: Computers and Intractability: A Guide to the Theory of NP-Completeness. W. H. Freeman and Company, New York (1979)

A Translation Based Approach to Probabilistic Conformant Planning

Ronen I. Brafman and Ran Taig

Department of Computer Science, Ben-Gurion University of the Negev,
Beer-Sheva, Israel
{brafman,taig}@cs.bgu.ac.il

Abstract. In conformant probabilistic planning (CPP), we are given a set of actions with stochastic effects, a distribution over initial states, a goal condition, and a value $0 < p \leq 1$. Our task is to find a plan π such that the probability that the goal condition holds following the execution of π in the initial state is at least p. In this paper we focus on the problem of CPP with deterministic actions. Motivated by the success of the translation-based approach of Palacious and Geffner [6], we show how deterministic CPP can be reduced to a metric-planning problem. Given a CPP, our planner generates a metric planning problem that contains additional variables. These variables represent the probability of certain facts. Standard actions are modified to update these values so that this semantics of the value of variables is maintained. An empirical evaluation of our planner, comparing it to the best current CPP solver, Probabilistic-FF, shows that it is a promising approach.

1 Introduction

An important trend in research on planning under uncertainty is the emergence of planners that utilize an underlying classical, deterministic planner. Two highly influential examples are the replanning approach [7] in which an underlying classical planner is used to solve MDPs by repeatedly generating plans for a determinized version of the domain, and the translation-based approach for conformant planning [6] and contingent planning [1], where a problem featuring uncertainty about the initial state is transformed into a classical problem on a richer domain. Both approaches have drawbacks: replanning can yield bad results given dead-ends and low-valued, less likely states. The translation-based approach can blow-up in size given complex initial belief states and actions. In both cases, however, there are efforts to improve these methods, and the reliance on fast, off-the-shelf, classical planners seems to be very useful.

This paper continues this trend, leveraging the translation-based approach of Palacious and Geffner [6] to handle a quantitative version of conformant planning, in which there is a probability distribution over the initial state of the world, although actions remain deterministic. The task now is to attain the goal condition with certain probability, rather than with certainty. More generally, conformant probabilistic planning (CPP) allows for stochastic actions, but as in earlier work, we will focus on the simpler case of deterministic actions. Our algorithm takes a deterministic CPP, and generate a metric-planning problem, which we give as input to the *Metric-FF* planner [3]. The classical

R.I. Brafman, F. Roberts, and A. Tsoukiàs (Eds.): ADT 2011, LNAI 6992, pp. 16–27, 2011.

problem we generate contains boolean propositions of the form q/t, which intuitively denote the fact that q is true now, given that the initial state satisfied t, as well as a numeric functions of the form $Pr(q)$ which maintain the probability that q holds currently. The original set of actions is transformed in order to maintain the semantics of these variables. Finally, a goal such as "make q true with probability at least Θ," is now captured by setting the numeric goal for the metric-planning problem: "$Pr(q) > \Theta$."

We compare our planner empirically against *PFF* [2], which is the state of the art in CPP. Although this is a preliminary evaluation, it is quite promising. It shows that on various domains our planner is faster than *PFF*. However, there are some domains and problems that are still challenging to our planner, partly due to shortcomings of the underlying metric planner (it's restricted language) or large conformant width of the problem.

In the following section we provide some needed background on CPP and *PFF*. Then, we explain our compilation scheme, showing its correctness. Then, we discuss our system and its empirical performance, evaluating it against *PFF* on standard CPP domains. Finally, we discuss some extensions.

2 Background

2.1 Conformant Probabilistic Planning

The probabilistic planning framework we consider adds probabilistic uncertainty to a subset of the classical ADL language, namely (sequential) STRIPS with conditional effects. Such STRIPS planning tasks are described over a set of propositions $\mathcal{P}$ as triples (A, I, G), corresponding to the *action set*, *initial world state*, and *goals*. I and G are sets of propositions, where I describes a concrete initial state w_I, while G describes the set of goal states $w \supseteq G$. Actions a are pairs $(pre(a), E(a))$ of the *precondition* and the *(conditional) effects*. A conditional effect e is a triple $(con(e), add(e), del(e))$ of (possibly empty) proposition sets, corresponding to the effect's *condition*, *add*, and *delete* lists, respectively. The precondition $pre(a)$ is also a proposition set, and an action a is *applicable* in a world state w if $w \supseteq pre(a)$. If a is not applicable in w, then the result of applying a to w is undefined. If a is applicable in w, then all conditional effects $e \in E(a)$ with $w \supseteq con(e)$ occur. Occurrence of a conditional effect e in w results in the world state $w \cup add(e) \setminus del(e)$, which we denote by $a(w)$. We will use $\bar{a}(w)$ to denote the state resulting from the the sequence of actions $\bar{a}$ in world state w.

If an action a is applied to w, and there is a proposition q such that $q \in add(e) \cap del(e')$ for (possibly the same) occurring $e, e' \in E(a)$, then the result of applying a in w is undefined. Thus, we require the actions to be not self-contradictory, that is, for each $a \in A$, and every $e, e' \in E(a)$, if there exists a world state $w \supseteq con(e) \cup con(e')$, then $add(e) \cap del(e') = \emptyset$. Finally, an action sequence $\overline{a}$ is a *plan* if the world state that results from iterative execution of $\overline{a}(w_I) \supseteq G$.

Our probabilistic planning setting extends the above with probabilistic uncertainty about the initial state. In its most general form, CPP covers stochastic actions as well, but we leave this to future work. Conformant probabilistic planning tasks are quadruples (A, b_I, G, θ), corresponding to the *action set*, *initial belief state*, *goals*, and *acceptable goal satisfaction probability*. As before, G is a set of propositions. The initial state is no longer assumed to be known precisely. Instead, we are given a probability distribution over the world states, b_I, where $b_I(w)$ describes the likelihood of w being the initial world state.

There is no change in the definition of actions and their applications in states of the world. But since we now work with belief states, actions can also be viewed as transforming one belief state to another. The likelihood $[b,a](w')$ of a world state w' in the belief state $[b,a]$, resulting from applying action a in belief state b, is given by

$$[b,a](w') = \sum_{a(w)=w'} b(w) \tag{2.1}$$

We will also use the notation $[b,a](\varphi)$ to denote $\sum_{a(w)=w', w' \models \varphi} b(w)$, and we somewhat abuse notation and write $[b,a] \models \varphi$ for the case where $[b,a](\varphi) = 1$.

For any action sequence $\overline{a} \in A^*$, and any belief state b, the new belief state $[b, \overline{a}]$ resulting from applying $\overline{a}$ at b is given by

$$[b,\overline{a}] = \begin{cases} b, & \overline{a} = \langle\rangle \\ [b,a], & \overline{a} = \langle a \rangle, a \in A \\ [[b,a], \overline{a}'], & \overline{a} = \langle a \rangle \cdot \overline{a}', a \in A, \overline{a}' \neq \emptyset \end{cases} . \tag{2.2}$$

In such setting, achieving G with certainty is typically unrealistic. Hence, θ specifies the required *lower bound* on the probability of achieving G. A sequence of actions $\overline{a}$ is called a *plan* if we have $b_{\overline{a}}(G) \geq \theta$ for the belief state $b_{\overline{a}} = [b_I, \overline{a}]$. Because our actions are deterministic, this is essentially saying that $\overline{a}$ is a *plan* if $Pr(\{w : \overline{a}(w) \models G\}) \geq \theta$, i.e,. the weight of the initial states from which the plan reaches the goal is at least θ.

2.2 PFF

The best current probabilistic conformant planner is *Probabilistic FF* (PFF) [2], which we now briefly describe. The basic ideas underlying Probabilistic-FF are:

1. Define time-stamped Bayesian Networks (BN) describing probabilistic belief states.
2. Extend Conformant-FF's belief state CNFs to model these BN.
3. In addition to the SAT reasoning used by Conformant-FF [4] , use weighted model-counting to determine whether the probability of the (unknown) goals in a belief state is high enough.
4. Introduce approximate probabilistic reasoning into Conformant-FF's heuristic function.

In more detail, given a probabilistic planning task (A, b_I, G, θ), a belief state $b_{\overline{a}}$ corresponding to some applicable in b_I m-step action sequence $\overline{a}$, and a proposition $q \in \mathcal{P}$, we say that q is *known* in $b_{\overline{a}}$ if $b_{\overline{a}}(q) = 1$, *negatively known* in $b_{\overline{a}}$ if $b_{\overline{a}}(q) = 0$, and *unknown* in $b_{\overline{a}}$, otherwise. We begin with determining whether each q is known, negatively known, or unknown at time m. Re-using the Conformant-FF machinery, this classification requires up to two SAT tests of $\phi(b_{\overline{a}}) \wedge \neg q(m)$ and $\phi(b_{\overline{a}}) \wedge q(m)$, respectively. The information provided by this classification is used threefold. First, if a subgoal $g \in G$ is negatively known at time m, then we have $b_{\overline{a}}(G) = 0$. On the other extreme, if all the subgoals of G are known at time m, then we have $b_{\overline{a}}(G) = 1$. Finally, if some subgoals of G are known and the rest are unknown at time m, then *PFF* evaluates the belief state $b_{\overline{a}}$ by testing whether

$$b_{\overline{a}}(G) = \mathsf{WMC}\,(\phi(b_{\overline{a}}) \wedge G(m)) \geq \theta. \tag{2.3}$$

(where WMC stands for weighted-model counting).

After evaluating the considered action sequence $\overline{a}$, if $b_{\overline{a}}(G) \geq \theta$, then *PFF* has found a plan. Otherwise, the forward search continues, and the actions that are applicable in $b_{\overline{a}}$ (and thus used to generate the successor belief states) are actions whose preconditions are all known in $b_{\overline{a}}$.

2.3 Metric Planning and Metric-FF

Metric planning extends standard classical planning with numerical variables and numerical constraints. Actions can have such constraints as their preconditions, as well as numeric effects. More specifically, arithmetic expression are defined using the operators: $+, -, *, /$, and allow the formation of numeric constraints of the form (e, c, e') where e and e' are numeric expressions and $comp \in \{>, \geq, =, \leq, <\}$. A numeric effect is a triple (v^i, ass, e) where $v^i \in V$,$ass \in \{:=, + =, - =, * =, / =\}$ and e is a numeric expression.

Formally, a numeric planning task is a tuple (P, V, A, I, G), where P is a set of propositions and V is a set of numeric variables that can take on rational values. As usual, A is a set of actions, I is the initial state, and G is the goal condition, but their form is somewhat different, as we now explain. Action conditions now take the form of a pair $(p(con), v(con))$ s.t. $p(con) \subseteq P$ and $v(con)$ is a set of numeric constraints. An effect is simply a triple $(p(eff)^+, p(eff)^-, v(eff))$ s.t. $p(eff)^+, p(eff)^- \subseteq P$ (The ordinary add and delete lists) and $v(eff)$ is a set of numeric effects s.t $i \neq j$ for all $(v^i, ass, e), (v^j, ass, e') \in v(eff)$. An action $a \in A$ is then a pair $(pre(a), eff(a))$ where $pre(a)$ is a condition and $eff(a)$ is an effect. Conditional effects are defined in the same manner. I is the usual initial state with the addition of the the vector $v(I)$ marking the initial values of the numeric variables, and G is simply a conjunction of conditions.

A state s for a metric planning task is now a pair $s = (p(s), v(s))$ where $p(s) \subseteq P$ is a set of propositions that are true in s and $v(s) = (v^1(s), v^2(s)..., v^n(s)) \in \mathbb{Q}^n$ is a vector of rational numbers s.t $v^i(s)$ is the numeric value of v^i in the state s. The value of an expression e in as state s is the rational number that the expression simplifies to when replacing all numeric variables with their respective value, $v(s)$, or it is undefined if division by 0 occurs. A constraint $(e, comp, e')$ holds in a state s if e and e' are defined in s and stand in the relation $comp$. A condition holds in s if all the constraints in this condition hold in s. The value of a numeric effect (v^i, ass, e) of an action executed in s is the outcome of of modifying the value of v^i in s with the value of exp in s, using the operator ass. For a numeric effect to be applicable in s it must be defined in s.

The *Metric-FF* [3] planner is a natural extension of the forward heuristic-search planner, FF [5] with a heuristic that extends the idea of delete relaxation to numeric variables so as to maintain the monotonic change characteristic of delete-relaxation planning. The basic idea is to look at a restricted language which allows the use of the operators: $+=, -=$ and only and the comparators $>, \geq$ where the right hand side of the comparators are positive rational numbers. The problem is pre-processed to this format, and from that point on, the heuristic computation ignores delete-lists and numeric effects that use "$-=$" as their assignment operator.

2.4 The Translation Approach

We present here a modified version of the translation-based method of [6], adapted to our settings. The essential idea behind the translation approach to conformant planning implemented in the T_0 planner is to reason by cases. The different cases correspond to different conditions on the initial state, or, equivalently, different sets of initial states. These sets of states, or conditions are captured by tags. That is, a tag is identified with a subset of b_I.

With every proposition p, we associate a set of tags T_p. We require that this set be *deterministic* and *complete*. We say that T_p is *deterministic* if for every $t \in T_p$ and any sequence of actions $\bar{a}$, the value of p is uniquely determined by t, the initial belief state b_I and $\bar{a}$. We say that T_p is *complete* w.r.t. an initial belief state b_I if $b_I \subseteq \bigcup_{t \in T_p}$. That is, it covers all possible relevant cases.

Once we determine what tags are required for a proposition p, (see below) we augment the set of propositions with new propositions of the form p/t, where t is one of the possible tags for p. p/t holds the current value of p given that the initial state satisfies the condition t. The value of each proposition of the form p/t is known initially – it reflects the value of p in the initial states represented by t, and since we focus on deterministic tags only, then $p/t \vee \neg p/t$ is a tautology throughout. Our notation p/t differs a bit from the Kp/t notation of Palacious and Geffner. The latter is used to stress the fact that these propositions are actually representing knowledge about the belief state. However, because of our assumption that tags are deterministic, we have that $\neg Kp \rightarrow K\neg p$. To stress this and remove the redundancy, we use a single proposition p/t instead of two propositions $Kp/t, K\neg p/t$.

The actions are transformed accordingly to maintain our state of knowledge. Given the manner tags were selected, we always know how an action would alter the value of some proposition given any of its tags. Thus, we augment the description of actions to reflect this. If the actions are deterministic (which we assume in this paper), then the change to our state of knowledge is also deterministic, and we can reflect it by altering the action description appropriately.

In addition to the propositions p/t, we also maintain numeric variables of the form Pr_p, which denote the probability that p is true. These correspond to the variables Kp used in the conformant case. Their use is explained later.

Ignoring the numeric variables for the moment, the resulting problem is a classical planning problem defined on a larger set of variables. The size of this set depends on the original set of variables and the number of tags we need to add. Hence, an efficient tag generation process is important. A trivial set of tags is one that contains one tag per each possible initial state. Clearly, if we know the initial state of the world, then we know the value of all variables following the execution of any set of actions. However, we can often do much better, as the value of each proposition at the current state depends only on a small number of propositions in the initial state. This allows us to use many fewer tags (=cases). In fact, the current value of different propositions depend on different aspects of the initial state. Thus, in practice, we select different tags for each proposition. We generate the tags for p by finding which literals are relevant to its value using the following recursive definition:

- p is relevant to p
- If q appears (possibly negated) in an effect condition c for action A such that $c \rightarrow r$ and r contains p or $\neg p$ then q is relevant to p
- If r is relevant to q and q is relevant to p then r is relevant to p

Let C_p denote the set containing all the propositions relevant to p. In principle, if we have a tag for every possible assignment to C_p, we would have a fine-grained enough partition of the initial states to sets in which p will always have the same value. However, we can do better. A first reduction in the number of tags is trivial: we can ignore any assignment to C_p which is not satisfied by some possible initial state. A second reduction is related to dependence between variable values in the initial state. Imagine that $r, s \in C_p$, but that in all possible initial states $r \leftrightarrow s$. Thus, we can actually remove one of these variables from the set C_p. More complex forms of dependencies can be discovered and utilized to reduce the tag set. For example, suppose that we know that only one of $x_1, \ldots, x_k$ can be true initially. And suppose that the value of p depends only on which one of these variables is true. Thus, we can have $\{x_1, \ldots, x_k\}$ as tags, denoting, respectively, the state in which x_1 is initially true (and all else are false), the state in which x_2 is true, etc. See [6] for more details on how the tags can be computed efficiently, and for the definition of the notion of the conformant width of the problem.

3 Compiling CPP into Metric Planning

As explained we create a metric planning problem which is then given to *Metric-FF* and the plan created is returned as a plan for the CPP given as input.

3.1 The Metric Planning Problem

Let $P = (V, A, b_I, G, \theta)$ be the CPP given as input. Recall that T_p is the set of tags for p. We use T to denote the entire set of tags (i.e., $\cup T_p$). We generate a metric-planning problem $\widehat{P} = (\widehat{V}, \widehat{F}, \hat{A}, \widehat{I}, \widehat{G})$ as follows:

Propositions: $\widehat{V} = \{p/t \mid p \in V, t \in T_p\}$.

Functions: $\widehat{F} = \{Pr_p \mid p \in V\} \cup \{Pr_{goal}\}$. That is functions that keep the current probability of each original proposition. We sometimes abuse notation and write $Pr_{\neg p}$ instead of $1 - Pr_p$. Finally Pr_{goal} denotes the probability that the goal is true.

Numerical Constants: We use a group $\widehat{c}$ of constants to save the initial probability of each tag $t \in T$. Then, $\widehat{c} = \{b_I(t) \mid t \in T\}$. Note that these can be computed from the initial state description.

Initial State:

- $\widehat{I} = \{l/t \mid l \text{ is a literal, and } t, I \models l, \}$
- $Pr_p = b_I(\{s|s \models p\}$, i.e., the initial probability that p holds.
- $Pr_{goal} = b_I(\{s|s \models G\})$. Again, this can be computed directly from the initial state description.

Goal: $\widehat{G} = \{Pr_{goal} \geq \theta\}$.

Actions: First, for every action $a \in \hat{A}$, we make all its effects conditionals. Thus, if e is an effect of a, we now treat it as a conditional effect of the form $\emptyset \rightarrow \{e\}$.

For every action $a \in A$, $\hat{A}$ contains an action $\widehat{a}$ defined as follows:

- $pre(\widehat{a}) = \{P_l = 1 \mid l \in pre(a)\}$. This reflects the need to make sure actions in the plan are always applicable: The probability of the preconditions is 1 only if they hold given all possible initial states.[1]
- For every conditional effect $(con \rightarrow eff) \in E(a)$, $\widehat{a}$ contains the following conditional effects for each $e \in eff$ and for every $t \in T$:
  - $\{c/t \mid c \in con \cup \{\neg e\}\} \rightarrow \{e/t, Pr_e = Pr_e + b_I(t)\}$.
    That is, if we know all conditions of the conditional effects are true before applying the action given t is true initially then we can conclude that the effect takes place so we now know that e is true under the same assumption. This information is captured by adding e/t. Note that we care only about conditional effects that actually change the state of the world. Hence, we require that the effect not hold prior to the execution of the action. In that case, the new probability of e is the old probability of e plus the probability of the case (as captured by the tag t) we are considering now.
  - If $e \in G$ we also add the following effect to the last condition:
    $Pr_{goal} = Pr_{goal} + (b_I(t) \times \prod_{e' \in G \setminus \{e\}} Pr_{e'})$
    If $\neg e \in G$ we add the following effect to the last condition:
    $Pr_{goal} = Pr_{goal} - (b_I(t) \times \prod_{e' \in G \setminus \{e\}} Pr_{e'})$
    If $e \in G$ then our knowledge of the probability of the goal was changed by the action so that now: $Pr_{goal}^{new} = \prod_{e \in G} Pr_e$. Note that here we assume that the probability of the different sub-goals is independent.[2] Given the increase in the probability of e and the independence assumption, the the new goal probability is : $\prod_{e' \in G \setminus \{e\}} Pr_{e'} \times (Pr_e + b_I(t)) = Pr_{goal}^{old} + (b_I(t) \times \prod_{e' \in G \setminus \{e\}} Pr_{e'})$. The same rational guides us when the action reduces the probability of some sub-goal.

[1] We follow the convention of earlier planners here. In fact, we see no reason to require that actions be always applicable, as long as the goal is achieved with the desired probability.

[2] We can handle the case of dependent goals, but that requires adding more tags, i.e., by adding tags that determinize the goal.

4 Accuracy of Probabilistic Calculations

The soundness of our algorithm rests on the accuracy of our probabilistic estimate of the value of P_{goal}. We now prove that this value is correct under the assumption that the set of tags is *deterministic*, *complete*, and *d*isjoint. We defined the notion of deterministic and complete tags earlier. We say that a set of tags T_p is **disjoint** if $\forall t_i, t_j \in T_p$ s.t. $i \neq j$ and for every possible initial state s_I: $t_i \models s_I \Rightarrow t_j \not\models s_I$.

Lemma 1. *Let $\bar{a}$ be a sequence of actions from A, let $\hat{a}$ be the corresponding sequence of actions from $\hat{A}$, and let $t \in T_p$ be a deterministic tag. Then, $[b_I, \bar{a}] \models p/t$ iff for every initially possible world state $w \in t$ we have that $(\hat{a})(w) \models p$.*

This lemma follows from our construction of the new actions, together with the fact that the tags are deterministic, i.e., the value of p in $(\hat{a})(w)$ for all initially possible world states $w \in t$ is the same.

Lemma 2. *Let $p \in V$, and assume that T_p is* deterministic, complete, *and* disjoint*. Let $\bar{a}$ be a sequence of actions in A. Let $\hat{a}$ be the corresponding sequence in $\hat{A}$. Then, $\hat{a}(Pr_p) = [b_I, \bar{a}]\,(p)$. That is, at this stage, Pr_p equals the probability of p following the execution of $\bar{a}$.*

Proof. By induction on the length of $\bar{a}$. For $|\bar{a}| = 0$, this is immediate from the initialization of Pr_p. Assume the correctness of this lemma given a sequence $\bar{a'}$ of length k, and let $\bar{a} = \bar{a'}a$ be a sequence of length $k+1$. By definition, $[b_I, \bar{a}]\,(p)$ is the sum of the probability of all possible worlds in $[b_I, \bar{a}]$ in which p holds, which is identical to the sum of the probability of all possible worlds in $w \in b_I$ such that p holds after executing $\bar{a}$ in w. Because T_p is complete, disjoint, and deterministic, this is identical to the sum of probability of the tags $t \in T_p$ such that p holds after executing $\bar{a}$ in all $w \in T$. Thus, it suffices to show that $\hat{a}(Pr_p)$ contains the sum of probability of these tags. According to Lemma 1, p holds after executing $\bar{a}$ in all $w \in T$ iff p/t holds after executing $\hat{a}$. Assuming that $\hat{a'}(Pr_p)$ was correct, i.e., it summed the right set of tags, if we add to it the weight of any new tag for which p/t holds and remove the weight of any tag for which p/t held after $\bar{a'}$ but p/t does not hold after $\bar{a}$, then (due to disjointness of tags) Pr_p will still maintain the correct sum of tag weights. By construction, we add the weight of t to Pr_p only if there is a real change in the value of p given t. □

Corollary 1. *The plan returned is a legal plan for P.*

Proof. Each action pre-condition L is replaced by the pre-condition $P_L = 1$, from lemma 1 we learn that this property holds if and only if L is known with full certainty and the action can be applied.

Corollary 2. *Assuming that sub-goals are probabilistically independent, then, in each stage of the planning process Pr_{goal} holds the accurate probability of the goal state.*

Proof. If $G = \{L\}$ then it's immediate from lemma 2. Otherwise, from lemma 2 it follows that this holds true for every sub-goal. Thus, the probability of the goal is the product of the probability of the sub-goals. The proof follows by induction from the fact that we initialize Pr_{goal} correctly, and from the updates performed following each action. Specifically, suppose that the probability of subgoal g increased following the last action. The new goal probability is : $\prod_{g' \in G \setminus \{g\}} Pr_{g'} \times (Pr_g + b_I(t)) = Pr_{goal}^{old} + (b_I(t) \times \prod_{g' \in G \setminus \{g\}} Pr_{g'})$. By construction, one effect of the corresponding action in $\hat{A}$ is $Pr_{goal} = Pr_{goal} + (b_I(t) \times \prod_{g' \in G \setminus \{g\}} Pr_{e'})$. This maintains the correct value. A similar update occurs in the case of a reduction. Since updates are done sequentially, the value remains correct even if an action affects multiple goals.

These results assume that the set of tags is complete, deterministic, and disjoint. The discussion in Section 2.4 explains the tag generation process, and it is easy to see that the set of tags generated in this way is indeed complete, deterministic, and disjoint. See [6] for a more sophisticated algorithm.

5 Example

We illustrate the ideas behind our planner using an example adapted from [6]. We need to move an object from an origin to a destination using two actions: $pick(l)$ that picks up an object from a location if the hand is empty and the object is in that location but if the hand is full it drops the object being held in the location. The second action is $drop(l)$, that drops the object at a location if the object is being held. All effects are conditional effects so there are no action preconditions. We assume, for simplicity, there's only a single object. Formally, The actions are as follows:

- $pick(l) : \neg hold, at(l) \rightarrow hold \wedge \neg at(l)$
  $hold \rightarrow \neg hold \wedge at(l)$
- $drop(l) : hold \rightarrow \neg hold \wedge at(l)$

Consider an instance P of the described domain where the hand is initially empty with certainty, and the object is initially at either l_1 or l_2 or l_3, and it needs to be moved to l_4 with a probability of 0.5. That is: $I = \{Pr[\neg hold] = 1, Pr[at(l_1)] = 0.2, Pr[at(l_2)] = 0.4, Pr[at(l_3)] = 0.4, Pr[at(l_4)] = 0\}, G = \{Pr[at(l_4)] \geq 0.5\}$.

A brief look at the domain shows that a plan can achieve the goal by only considering two possible original object locations, unlike in conformant planning where we must consider all three possible initial locations to succeed. The tags sets needed for the input are: $T_L = \{at(l_1), at(l_2), at(l_3)\}$ for $L \in \{hold, at(l_4)\}$. Note that T_L is indeed disjoint,deterministic and complete for L. Based on these tags our algorithm outputs the following Metric-Planning task $\hat{P} = \{\hat{F}, \hat{V}, \hat{A}, \hat{I}, \hat{G}\}$ as follows:

$\hat{F} = \{L/t \mid L \in \{at(l), hold\}, l \in \{l_1, l_2, l_3\}\}$.
$\hat{V} = \{Pr_{at(l)} \mid l \in \{l_1, l_2, l_3, l_4\}\} \cup \{Pr_{hold}\}$.
$\hat{I} = \{at(l)/at(l) \mid l \in \{l_1, l_2, l_3\}\} \cup \{Pr_{at(l_1)} = 0.2, Pr_{at(l_2)} = 0.4, Pr_{at(l_3)} = 0.4, Pr_{at(l_4)} = 0, Pr_{hold} = 0, Pr_{\neg hold} = 1, Pr_{\neg at(l_i)} = 1 - Pr_{at(l_i)} (1 \leq i \leq 4)\}$.
$\hat{G} = \{Pr_{at(l_4)} \geq 0.5\}$.

Please note that since the goal is not a conjunction of literals we actually only need to track the probability of $at(l_4)$ to check if we achieved the goal so no special Pr_{goal} numerical variable is needed. Now we modify the original actions, making it update the probabilities during the planning process. This is done as follows:

- Original conditional effect (action $pick(l)$): $\neg hold, at(l) \rightarrow hold \wedge \neg at(l)$.

  Output :
  - $\neg hold, at(l) \rightarrow hold \wedge \neg at(l), Pr_{hold} = 1, Pr_{\neg hold} = 0, Pr_{at(l)} = 0, Pr_{\neg at(l)} = 1;$
  - For each $l' \in \{l_1, l_2, l_3\}$ we add the following: $\neg hold/at(l'), at(l)/at(l') \rightarrow hold \wedge \neg at(l), hold/at(l') \wedge \neg at(l)/at(l'), Pr_{hold}+ = b_I(at(l')), Pr_{\neg hold}- = b_I(at(l')), Pr_{at(l)}- = b_I(at(l')), Pr_{\neg at(l)}+ = b_I(at(l'));$

- Original conditional effect(actions: $pick(l), drop(l)$): $hold \rightarrow \neg hold \wedge at(l)$.

  Output :
  - $hold \rightarrow \neg hold \wedge at(l), Pr_{hold} = 0, Pr_{\neg hold} = 1, Pr_{at(l)} = 1, Pr_{\neg at(l)} = 0;$
  - For each $l' \in \{l_1, l_2, l_3\}$ we add the following: $hold/at(l') \rightarrow \neg hold/at(l') \wedge at(l)/at(l'), Pr_{hold}- = b_I(at(l')), Pr_{\neg hold}+ = b_I(at(l')), Pr_{at(l)}+ = b_I(at(l')), Pr_{\neg at(l)}- = b_I(at(l'));$

It's now easy to observe how the plan $\pi = < pick(l_1), drop(l_4), pick(l_2), drop(l_4) >$ solves both the Metric-Planning Problem and the original CPP - let's examine the values of some of the variables throughout the plan execution process:

- Time 0 : $at(l_1)/at(l_1), at(l_2)/at(l_2), Pr_{at(l_4)} = 0, Pr_{hold} = 0$
- Time 1 : $hold/at(l_1), at(l_2)/at(l_2), Pr_{at(l_4)} = 0, Pr_{hold} = 0.2$
- Time 2 : $at(l_4)/at(l_1), at(l_2)/at(l_2), Pr_{at(l_4)} = 0.2, Pr_{hold} = 0$
- Time 3 : $at(l_4)/at(l_1), hold/at(l_2), Pr_{at(l_4)} = 0.2, Pr_{hold} = 0.4$
- Time 4 : $at(l_4)/at(l_1), at(l_4)/at(l_2), Pr_{at(l_4)} = 0.6, Pr_{hold} = 0$: goal achieved.

6 Empirical Evaluation

We implemented the algorithm as follows. Our input problem is stripped of probabilistic information and transformed into a conformant planning problem. This is fed to the *cf2cs* program, which is a part of *T-0* planner, and computes the set of tags. Using this set of tags, we generate the new metric planning problem. Currently, we have a somewhat inefficient tool for generating the new domains, which actually uses part of the *T-0*'s domain generation code and another tool that augments it with numeric information. This results is a large overhead in many domains, where the translation process takes longer than the planner. In the future, we will construct a dedicated translator, which we believe will result in improved performance. In addition, we are also limited in our ability to support multiple conjunctive goals. *Metric-FF* supports only linear numerical expressions. Our theory requires multi-linear expressions when there are more than two goals (i.e., we must multiply non-constants). Consequently, when there are more than two independent sub-goals, we basically require the achievement of each of them so that the product of their probabilities be sufficient. That is, if $G = g_1 \wedge \cdots \wedge g_m$, and it must be achieved with probability θ, we pose the metric goal: $Pr_{g_1} > \sqrt[m]{\theta} \wedge \cdots \wedge Pr_{g_m} > \sqrt[m]{\theta}$. This is a stronger requirement then $Pr_G > \theta$.

Table 1 below shows the results of our experimental evaluation. We refer to our planner as PTP (for probabilistic translation-based planner).

Table 1. Empirical results for problems with probabilistic initial states. Times t in seconds, plan length l. (P-FF results for *Bomb* are given by the table in [2] due to technical issues preventing us from running it on our system).

		$\theta = 0.25$		$\theta = 0.5$		$\theta = 0.75$		$\theta = 1.0$	
Instance	#actions/#facts/#states	*t/l*		*t/l*		*t/l*		*t/l*	
		P-FF	PTP	P-FF	PTP	P-FF	PTP	P-FF	PTP
Safe-uni-70	70/71/140	2.65 /18	0.87/18	5.81/35	0.85/35	10.1/53	0.9/53	5.1/70	0.88/70
Safe-cub-70	70/70/138	0.88/5	0.9/5	1.7/12	0.94/12	3.24/21	0.95/21	4.80/69	0.96/69
Cube-uni-15	6/90/3375	4.25/26	2.4/33	6.35/34	2.49/45	9.20/38	2.65/50	31.2/42	2.65/50
Cube-cub-11	6/90/3375	0.3/5	1.17/12	0.9/9	1.31/15	1.43/13	1.41/21	28.07/31	3.65 /36
Bomb-50-50	2550/200/> 2^{100}	0.01/0	0.01/0	0.10/16	3.51/50	0.25/36	3.51/50	0.14/51/50	3.51/50
Bomb-50-10	510/120/> 2^{60}	0.01/0	0.01/0	0.89/22	1.41/90	4.04/62	1.41/90	1.74/90	1.46/90
Bomb-50-5	255/110/> 2^{55}	0.01/0	0.01/0	1.70/27	1.32/95	4.80/67	1.32/95	2.17/95	1.32/95
Bomb-50-1	51/102/> 2^{51}	0.01/0	0.01/0	2.12/31	0.64/99	6.19/71	0.64/99	2.58/99	0.64/99
Log-2	3440/1040/> 20^{10}	0.90/54	–	1.07/62	–	1.69/69	–	1.84/78	–
Log-3	3690/1260 /> 30^{10}	2.85/64	–	8.80/98	–	4.60/99	–	4.14/105	–
Log-4	3960/1480/> 40^{10}	2.46/75	–	8.77/81	–	6.20/95	–	8.26/107	–

The results reported are on benchmarks tested by *PFF*. On the *safe* domain, both on a uniform and cubic distributions *PTP* is faster than *PFF*. In this domain *PTP* enjoys the fact that there is a single goal, so we do not face the limitations of *Metric-FF* discussed above. In $cube - n$ *PTP* is again faster, although it outputs longer plans. This is likely to be a byproduct of formulation of the goal as a conjunction of three probabilistic goals, each of which needs to be achieved with much higher probability. This phenomenon is more dramatic in the experiments on *bomb* where 50 goals needs to be achieved, so actually we need to disarm all bombs in order to reach the goals where, in fact, the desired goal probability can be achieved without disarming all bombs. Still, *PTP* is faster than *PFF* on the harder instances of the problem where only 1 or 5 toilets can be used for disarming all bombs. On the other hand, on the *logistics* domain, *PTP* performs poorly. Although theoretically (in terms of conformant width) the problem does not appear especially challenging, *PTP* cannot solve most *logistics* instances. It appears that *Metric-FF*'s heuristic function provides poor indication of the quality of states in this case. Two additional domains are *rovers* and *grid*. They have large conformant width, and hence exact computation on them requires generating very large domains, which we currently cannot handle. *T-0* is able to deal with these domains by using various simplifications. One of the main challenges for PTP is to adapt some of these simplifications to the probabilistic case.

7 Summary

We described PTP, a novel probabilistic conformant planner based on the translation approach of Pallacious and Geffner [6]. PTP performs well on some domains, whereas in others it faces fundamental problems that require an extension of the theory behind this approach. We intend to extend this theory and devise methods for more efficient translations.

Acknowledgements. The authors were partly supported by ISF Grant 1101/07, the Paul Ivanier Center for Robotics Research and Production Management, and the Lynn and William Frankel Center for Computer Science.

References

1. Albore, A., Palacios, H., Geffner, H.: A translation-based approach to contingent planning. In: IJCAI, pp. 1623–1628 (2009)
2. Domshlak, C., Hoffmann, J.: Probabilistic planning via heuristic forward search and weighted model counting. J. Artif. Intell. Res (JAIR) 30, 565–620 (2007)
3. Hoffmann, J.: The metric-ff planning system: Translating "ignoring delete lists" to numeric state variables. J. Artif. Intell. Res (JAIR) 20, 291–341 (2003)
4. Hoffmann, J., Brafman, R.I.: Conformant planning via heuristic forward search: A new approach. Artif. Intell. 170(6-7), 507–541 (2006)
5. Hoffmann, J., Nebel, B.: The ff planning system: Fast plan generation through heuristic search. J. Artif. Intell. Res (JAIR) 14, 253–302 (2001)
6. Palacios, H., Geffner, H.: Compiling uncertainty away in conformant planning problems with bounded width. J. Artif. Intell. Res (JAIR) 35, 623–675 (2009)
7. Yoon, S.W., Fern, A., Givan, R.: FF-replan: A baseline for probabilistic planning. In: ICAPS, p. 352 (2007)

Committee Selection with a Weight Constraint Based on a Pairwise Dominance Relation*

Charles Delort, Olivier Spanjaard, and Paul Weng

UPMC, LIP6-CNRS, UMR 7606
4 Place Jussieu, F-75005 Paris, France
{charles.delort,olivier.spanjaard,paul.weng}@lip6.fr

Abstract. This paper is devoted to a knapsack problem with a cardinality constraint when dropping the assumption of additive representability [10]. More precisely, we assume that we only have a classification of the items into ordered classes. We aim at generating the set of preferred subsets of items, according to a pairwise dominance relation between subsets that naturally extends the ordering relation over classes [4,16]. We first show that the problem reduces to a multiobjective knapsack problem with cardinality constraint. We then propose two polynomial algorithms to solve it, one based on a multiobjective dynamic programming scheme and the other on a multiobjective branch and bound procedure. We conclude by providing numerical tests to compare both approaches.

Keywords: Committee selection, Ordinal combinatorial optimization, Multiobjective combinatorial optimization, Knapsack with cardinality constraint, Polynomial algorithms.

1 Introduction

Ranking sets of objects based on a ranking relation on objects has been extensively studied in social choice theory within an axiomatic approach [1]. Many *extension rules* have been proposed and axiomatically justified to extend an order relation over a set of objects to an order relation over its power set. This issue is indeed of primary interest in various fields such as choice under uncertainty [12], ranking opportunity sets [3], and of course committee selection [11]. The committee selection problem consists in choosing a subset of invividuals based on an ordering of individuals. Although a lot of works deal with this problem in the economic literature, it has received much less attention from the algorithmic viewpoint. In other words, the computational aspect (i.e., the effective calculability of the preferred committees) is often a secondary issue. This is precisely the issue we study in this paper.

More formally, we investigate the problem of selecting K individuals (or more generally objects) among n with budget B, where the selection of individual i

* This research has been supported by the project ANR-09-BLAN-0361 GUaranteed Efficiency for PAReto optimal solutions Determination (GUEPARD).

R.I. Brafman, F. Roberts, and A. Tsoukiàs (Eds.): ADT 2011, LNAI 6992, pp. 28–41, 2011.

requires a cost w^i. The only preferential information is an assignment of each individual i in a preference class $\gamma^i \in \{1, \ldots, C\}$, with $1 \rhd 2 \rhd \ldots \rhd C$, where $\rhd$ means "is strictly preferred to". For illustration, consider the following example. Assume that an English soccer team wishes to recruit $K = 2$ players with budget $B = 6$. The set $\mathcal{N}$ of available players consists of international players (class 1), Premier League players (class 2) and Division 1 players (class 3). This problem can be modeled as a knapsack problem where one seeks a subset $\mathcal{S} \subseteq \mathcal{N}$ such that $\sum_{i \in \mathcal{S}} w^i \leq 6$ and $|\mathcal{S}| = 2$, but where the objective function is not explicited. Consider now the following instance: $\mathcal{N} = \{1, 2, 3, 4\}$, $w^1 = 5$, $w^2 = 2$, $w^3 = 4$, $w^4 = 1$. Player 1 is international, players 2, 3 are from the Premier League, and player 4 is from the Division 1 championship: $\gamma^1 = 1$, $\gamma^2 = \gamma^3 = 2$, $\gamma^4 = 3$.

When the individuals are evaluated in this way (i.e., on an ordinal scale), arbitrarily assigning numerical values to classes (each class can be viewed as a grade in the scale) introduces a bias in the modeling [2]. For instance, if value 8 is assigned to class 1, value 4 is assigned to class 2 and value 1 to class 3, then the ensuing recruitment choice (the one maximizing the sum of the value according to the budget) is $\{1, 4\}$. By valuing class 2 by 5 instead of 4 (which is still compatible with the ordinal classes), the ensuing recruitment choice becomes $\{2, 3\}$. Thus, one observes that slight changes in the numerical values lead to very different choices. This illustrates the need for algorithms specifically dedicated to combinatorial problems with ordinal measurement.

This problem has been studied in a slightly different setting by Klamler *et al.* [14]. They assume to have a preference relation over the set of individuals, expressed as a reflexive, complete and transitive binary relation. Note that, in our setting with C predefined preference classes, it amounts to set $C = n$ (some preference classes may be empty if there are equivalent individuals). The authors provide linear time algorithms to compute optimal committees according to various extension rules, namely variations of max ordering, leximax and leximin. The max (resp. min) ordering relation consists in ranking committees according to the best (resp. worst) individual they include, while the leximax and leximin relations are enrichments of max and min respectively that consist in breaking ties by going down the ranking (e.g., if the best individuals are indifferent, one compares the second bests, and so on...). Though appealing from the algorithmic viewpoint, these extension rules are nevertheless quite simple from the normative and descriptive viewpoints.

In this paper, we investigate an extension rule that encompasses a much larger set of decision behaviors (at the expense of working with preference classes instead of a complete ranking of individuals). Actually, it leads to identify a set of preferred committees, instead of a single one. Provided the ordinal nature of data, it seems indeed relevant to determine a set of acceptable committees, among which the final choice will be made. In order to extend order relation $\rhd$ over the preference classes (1, 2, 3 in the recruitment example, with $1 \rhd 2 \rhd 3$) to a (reflexive and transitive) preference relation $\succsim$ over the committees, the extension rule we study is the following pairwise dominance relation: a committee $\mathcal{S}$ is preferred to another committee $\mathcal{S}'$ if, to each individual i in $\mathcal{S}'$, one can assign a not-yet-assigned individual i' in $\mathcal{S}$ such that $\gamma^{i'} \trianglerighteq \gamma^i$ (i.e. $\gamma^{i'} \triangleright \gamma^i$ or $\gamma^{i'} = \gamma^i$). For instance, in the previous recruiting example, one has $\{1,3\} \succsim \{2,3\}$ since $\gamma^3 = \gamma^2$ and $\gamma^1 \triangleright \gamma^3$ (note that $\{1,3\}$ is actually not feasible, due to the budget constraint, but it does not matter for our purposes). To our knowledge, this extension rule was proposed by Bossong and Schweigert [4,16]. More recent works with ordinal data also use this rule [5,6,7].

Our first contribution in the present paper is to relate ordinal combinatorial optimization to multiobjective combinatorial optimization, by reducing the determination of the non-dominated solutions in an ordinal problem to the determination of the Pareto set in an appropriately defined corresponding multiobjective problem. We then propose two algorithms to determine a set of optimal committees according to the pairwise dominance relation, one based on a multiobjective dynamic programming scheme and the other one on a multiobjective branch and bound procedure. The complexity of both procedures is polynomial *for a fixed number C of preference classes*. Note that in another context, Della Croce et al. [8] also represented an ordinal optimization problem as a multiobjective problem, but their transformation is different from the one presented here.

The paper is organized as follows. Section 2 relates ordinal optimization to multiobjective optimization. Two polynomial (multiobjective) procedures to solve the commitee selection problem are then presented in Sections 3 and 4. Finally, experimental results are provided in Section 5 to compare both approaches.

2 From Ordinal Combinatorial Optimization to Multiobjective Optimization

Formally, an ordinal combinatorial optimization problem can be defined as follows. Consider a set $\mathcal{N}$ of objects (e.g. items in a knapsack problem, edges in a path or tree problem...). A feasible solution is a subset $\mathcal{S} \subseteq \mathcal{N}$ satisfying a given property (for example, satisfying knapsack constraints). As mentioned in the introduction, for each object $i \in \mathcal{N}$, the only preferential information at our disposal is the preference class $\gamma^i \in \{1, \ldots, C\}$ it belongs to, with $1 \triangleright 2 \triangleright \ldots \triangleright C$. Given an extension rule that lifts preference relation $\triangleright$ to a preference relation $\succsim$ over subsets of $\mathcal{N}$, a feasible solution $\mathcal{S}$ is said to be *preferred* if there exists no feasible solution $\mathcal{S}'$ such that $\mathcal{S}' \succ \mathcal{S}$, where $\succ$ denotes the asymmetric part of $\succsim$. The aim of an ordinal combinatorial optimization problem is then to find a *complete minimal* set of preferred solutions [13]. A set of solutions is said to be *complete* if for any preferred solution, there is a solution in that set that is indifferent to it. A set of solutions is said to be *minimal* if there does not exist a pair $\mathcal{S}, \mathcal{S}'$ of solutions in this set such that $\mathcal{S} \neq \mathcal{S}'$ and $\mathcal{S} \succsim \mathcal{S}'$.

Let us denote by $\max_{\succsim}$ the operation that consists in determining a complete minimal set of preferred solutions according to $\succsim$. The committee selection problem we consider in this paper can then be simply stated as follows:

$$\max_{\succsim}\{\mathcal{S} \subseteq \mathcal{N} \ : \ |\mathcal{S}| = K \text{ and } \sum_{i \in \mathcal{S}} w^i \leq B\}$$

where K is the size of the committee and w^i the cost of selecting individual i.

In the sequel, we consider the following extension rule:

Definition 1. *The* pairwise dominance relation *between subsets of a set $\mathcal{N}$ is defined, for all $\mathcal{S}, \mathcal{S}' \subseteq \mathcal{N}$, by $\mathcal{S} \succsim \mathcal{S}'$ if there exists an injection $\pi : \mathcal{S}' \rightarrow \mathcal{S}$ such that $\forall i \in \mathcal{S}'$, $\gamma^{\pi(i)} \trianglerighteq \gamma^i$.*

Coming back to the example of the introduction, one detects that $\{1,3\} \succsim \{2,3\}$ by setting $\pi(2) = 1$ ($\gamma^1 = 1 \triangleright 2 = \gamma^2$) and $\pi(3) = 3$, or by setting $\pi(2) = 3$ ($\gamma^2 = \gamma^3 = 2$) and $\pi(3) = 1$ ($\gamma^1 = 1 \triangleright 2 = \gamma^3$). Since the opposite relation is not true, one has $\{1,3\} \succ \{2,3\}$.

We are now going to make an original link between ordinal optimization and multiobjective optimization. In this purpose, the following notion will prove useful: for each solution $\mathcal{S}$ and each preference class $c \leq C$, one defines $\mathcal{S}_c = \{i \in \mathcal{S} : \gamma^i \trianglerighteq c\}$. To each solution one associates a *cumulative* vector $(|\mathcal{S}_1|, \ldots, |\mathcal{S}_C|)$. Therefore, one has $|\mathcal{S}_1| \leq |\mathcal{S}_2| \leq \ldots \leq |\mathcal{S}_C|$. Interestingly enough, we now show that comparing solutions according to pairwise dominance amounts to compare those vectors according to weak (Pareto) dominance, which is defined as follows:

Definition 2. *The* weak dominance *relation on C-vectors of $\mathbb{N}^C$ is defined, for all $y, y' \in \mathbb{N}^C$, by $y \succcurlyeq y' \Leftrightarrow [\forall c \in \{1,\ldots,C\}, y_c \geq y'_c)]$. The* dominance *relation $\succ$ is defined as the asymmetric part of $\succcurlyeq$: $y \succ y' \Leftrightarrow [y \succcurlyeq y'$ and $y' \not\succcurlyeq y]$.*

The equivalence result writes formally as follows:

Proposition 1. *For any pair $\mathcal{S}, \mathcal{S}'$ of solutions, we have:*

$$\mathcal{S} \succsim \mathcal{S}' \iff (|\mathcal{S}_1|, \ldots, |\mathcal{S}_C|) \succcurlyeq (|\mathcal{S}'_1|, \ldots, |\mathcal{S}'_C|)$$

Proof. We first prove that $\mathcal{S} \succsim \mathcal{S}' \Rightarrow (|\mathcal{S}_1|, \ldots, |\mathcal{S}_C|) \succcurlyeq (|\mathcal{S}'_1|, \ldots, |\mathcal{S}'_C|)$. Assume there exists an injection $\pi : \mathcal{S}' \rightarrow \mathcal{S}$. Then $|\mathcal{S}_c| \geq |\pi(\mathcal{S}'_c)| = |\mathcal{S}'_c|$ for all c, since $\gamma^{\pi(i)} \trianglerighteq \gamma^i \trianglerighteq c$ for all $i = 1, \ldots, n$. Therefore $(|\mathcal{S}_1|, \ldots, |\mathcal{S}_C|) \succcurlyeq (|\mathcal{S}'_1|, \ldots, |\mathcal{S}'_C|)$ by definition of $\succcurlyeq$.

Conversely, we now show that $(|\mathcal{S}_1|, \ldots, |\mathcal{S}_C|) \succcurlyeq (|\mathcal{S}'_1|, \ldots, |\mathcal{S}'_C|) \Rightarrow \mathcal{S} \succsim \mathcal{S}'$. Assume that $|\mathcal{S}_c| \geq |\mathcal{S}'_c|$ for all c. Since $|\mathcal{S}_1| \geq |\mathcal{S}'_1|$, there exists an injection $\pi_1 : \mathcal{S}'_1 \rightarrow \mathcal{S}_1$. Obviously, $\forall i \in \mathcal{S}'_1, \gamma^{\pi_1(i)} \trianglerighteq \gamma^i$. For any $c > 1$, one can then define by mutual recursion:

- an injection $\pi'_c : \mathcal{S}'_c \backslash \mathcal{S}'_{c-1} \rightarrow \mathcal{S}_c \backslash \pi_{c-1}(\mathcal{S}'_{c-1})$
- an injection $\pi_c : \mathcal{S}'_c \rightarrow \mathcal{S}_c$ by $\pi_c(i) = \pi_{c-1}(i)$ if $i \in \mathcal{S}'_{c-1}$ and $\pi_c(i) = \pi'_c(i)$ otherwise.

Injection π'_c exists for any $c > 1$ because $|\mathcal{S}_c \backslash \pi_{c-1}(\mathcal{S}'_{c-1})| \geq |\mathcal{S}'_c \backslash \mathcal{S}'_{c-1}|$. We have indeed $|\mathcal{S}_c \backslash \pi_{c-1}(\mathcal{S}'_{c-1})| = |\mathcal{S}_c| - |\pi_{c-1}(\mathcal{S}'_{c-1})|$ since $\pi_{c-1}(\mathcal{S}'_{c-1}) \subseteq \mathcal{S}_c$, $|\mathcal{S}_c| - |\pi_{c-1}(\mathcal{S}'_{c-1})| = |\mathcal{S}_c| - |\mathcal{S}'_{c-1}|$ since π_{c-1} is an injection, $|\mathcal{S}_c| - |\mathcal{S}'_{c-1}| \geq |\mathcal{S}'_c \backslash \mathcal{S}'_{c-1}|$ since $|\mathcal{S}_c| \geq |\mathcal{S}'_c|$. Note that by construction, for any c, $\forall i \in \mathcal{S}'_c$, $\gamma^{\pi_c(i)} \trianglerighteq \gamma^i$. For $c = C$ this is precisely the definition, therefore $\mathcal{S} \succsim \mathcal{S}'$. ∎

Coming back again to the example of the introduction, cumulative vector $(1,2,2)$ is associated to $\{1,3\}$, and $(0,2,2)$ to $\{2,3\}$. Note then, that $(1,2,2) \succ (0,2,2)$, consistently with $\{1,3\} \succ \{2,3\}$.

The committee selection problem we consider in this paper can then be formulated as a multiobjective knapsack problem with a cardinality constraint. An instance of this problem consists of a knapsack of integer capacity B, and a set of items $\mathcal{N} = \{1, \ldots, n\}$. Each item i has a weight w^i and a profit $p^i = (p^i_1, \ldots, p^i_C)$, variables w^i, p^i_c ($c \in \{1, \ldots, C\}$) being integers. Without loss of generality, we assume from now on that items in $\mathcal{N}$ are such that $\gamma^1 \trianglerighteq \gamma^2 \trianglerighteq \cdots \trianglerighteq \gamma^n$ and $\forall i, i' \in \mathcal{N}$, $(\gamma^i = \gamma^{i'}$ and $i \leq i') \Rightarrow w^i \leq w^{i'}$ (i.e. the items of $\mathcal{N}$ are indexed in decreasing order of preference classes and in increasing order of weights in case of ties). Otherwise, one can renumber the items.

Consequently, the profit vector of item i is defined by $p^i_c = 0$ for $c < \gamma^i$, and $p^i_c = 1$ for $c \geq \gamma^i$. This way, summing up the profit vectors of the items in a solution $\mathcal{S}$ yields the cumulative vector of $\mathcal{S}$. A solution $\mathcal{S}$ is characterized by a binary n-vector x, where $x^i = 1$ iff $i \in \mathcal{S}$. A solution is feasible if binary vector x satisfies the constraints $\sum_{i=1}^n w^i x^i \leq B$ and $\sum_{i=1}^n x^i = K$. The goal of the problem is to find a complete minimal set of feasible solutions (i.e. one feasible solution by non-dominated cumulative vector), which can be formally stated as follows:

$$
\begin{aligned}
&\text{maximize} \sum_{i=1}^{n} p^i_c x^i \qquad c \in \{1, \ldots, C\}\\
&\text{subject to} \sum_{i=1}^{n} w^i x^i \leq B\\
&\qquad\qquad \textstyle\sum_{i=1}^{n} x^i = K\\
&\qquad\qquad x^i \in \{0,1\} \quad i \in \{1, \ldots, n\}
\end{aligned}
$$

Note that, since vectors p^i are non-decreasing (i.e. $p^i_1 \leq \ldots \leq p^i_C$), the image of all feasible solutions is a subset of $[\![0, K]\!]^C_\uparrow$, which denotes the set of non-decreasing vectors in $[\![0, K]\!]^C = \{0, \ldots, K\}^C$. Furthermore, one has $|\mathcal{S}_C| = K$ for any feasible solution $\mathcal{S}$.

Example 1. *The example of the introduction is formalized as follows:*

$$
\begin{aligned}
&\text{maximize } x_1\\
&\text{maximize } x_1 + x_2 + x_3\\
&\text{maximize } x_1 + x_2 + x_3 + x_4\\
&\text{subject to } 5x_1 + 2x_2 + 4x_3 + x_4 \leq 6\\
&\qquad\qquad x_1 + x_2 + x_3 + x_4 = 2\\
&\qquad\qquad x^i \in \{0,1\} \quad i \in \{1, \ldots, 4\}
\end{aligned}
$$

3 A Multiobjective Dynamic Programming Algorithm

Multiobjective dynamic programming is a well-known approach to solve multiobjective knapsack problems [15]. In this section, we will present an algorithm proposed by Erlebach et al. [9], and apply it to our committee selection problem. The method is a generalization of the dynamic programming approach for the single objective knapsack problem using the following recursion:

$$W[p + p^i, i] = \min\{W[p + p^i, i-1], W[p, i-1] + w^i\} \text{ for } i = 1, \ldots, n$$

where $W[p, i]$ is the minimal weight for a subset of items in $\{1, \ldots, i\}$ with profit p. The recursion is initialized by setting $W[0, 0] = 0$ and $W[p, 0] = B + 1$ for all $p \geq 1$. The formula can be explained as follows. To compute $W[p + p^i, i]$, one compares the minimal weight for a subset of $\{1, \ldots, i\}$ with profit $p + p^i$ that *does not* include item i, and the minimal weight for a subset of $\{1, \ldots, i\}$ with profit $p + p^i$ that *does* include item i.

In a multiobjective setting, the difference lies in the profits, which are now vectors instead of scalars. Nevertheless, the dynamic programming procedure works in a similar way, by using the following recursion:

$$W[(p_1 + p_1^i, \ldots, p_C + p_C^i), i] = \min \begin{cases} W[(p_1 + p_1^i, \ldots, p_C + p_C^i), i - 1] \\ W[(p_1, \ldots, p_C), i - 1] + w^i \end{cases}$$

for $i = 1, \ldots, n$. The recursion is initialized by setting $W[(0, \ldots, 0), 0] = 0$ and $W[p, 0] = B + 1$ for all $p \neq (0, \ldots, 0)$. Once column $W[\cdot, n]$ is computed, the preferred items can then be identified in two steps:

1. one identifies profit vectors p for which $W[p, n] \leq B$;
2. one extracts the non-dominated elements among them.

The corresponding preferred solutions can then be retrieved by using standard bookkeeping techniques.

We adapt this method as follows to fit the committee selection problem, where one has to take into account cardinality constraint $\sum_{i=1}^{n} x^i = K$ and where $(p_1, \ldots, p_C) \in [\![0, K]\!]_{\uparrow}^{C}$. In step 1 above, one identifies profit vectors p for which $W[p, n] \leq B$ and $p_C = K$. This latter condition amounts to check that the cardinality of the corresponding solution is K: all items are indeed of preference class at least C (in other words, $p_C^i = 1$ for $i \in \{1, \ldots, n\}$).

Example 2. *For the instance of Example 1, the dynamic programming procedure can be seen as filling the cells of Table 1.*

Table 1. Dynamic programming table for Example 1. Each cell is computed by using the recursion $W[p + p^i, i] = \min\{W[p + p^i, i - 1], W[p, i - 1] + w^i\}$. For instance, the dark gray cell is computed from the light gray cells.

p \ i	1	2	3	4
$(0, 0, 0)$	0	0	0	0
$(0, 0, 1)$	7	7	7	$\min(7, 0 + 1) = 1$
$(0, 0, 2)$	7	7	7	7
$(0, 1, 1)$	7	$\min(7, 0 + 2) = 2$	2	2
$(0, 1, 2)$	7	7	7	3
$(0, 2, 2)$	7	7	$\min(7, 2 + 4) = 6$	6
$(1, 1, 1)$	$\min(7, 0 + 5) = 5$	5	5	5
$(1, 1, 2)$	7	7	7	$\min(7, 5 + 1) = 6$
$(1, 2, 2)$	7	7	7	7
$(2, 2, 2)$	7	7	7	7

In order to determine the complexity of this procedure, we assume that the number C of preference classes is fixed. At each step of the recursion, the computations required to compute one cell of the dynamic programming table are performed in constant time since it simply consists in a min operation. Furthermore, the number of steps is also polynomial since the number of rows (resp. columns) is within $\Theta(K^C)$ (resp. n). There are indeed as many rows as the number of vectors in $[\![0, K]\!]^C_{\uparrow}$. The cardinality of $[\![0, K]\!]^C_{\uparrow}$ is upper bounded by K^C (the cardinality of $[\![0, K]\!]^C$) and lower bounded by $K^C/C!$ (since there are at most $C!$ distinct vectors in $[\![0, K]\!]^C$ which are permutations of a same vector in $[\![0, K]\!]^C_{\uparrow}$), and therefore the number of rows is within $\Theta(K^C)$. Finally, the identification of preferred items can of course also be done in polynomial time as the number of cells in column $W[\cdot, n]$ is within $\Theta(K^C)$.

To summarize, the time complexity of the procedure is polynomial for a fixed number C of preference classes, and the dynamic programming table has $\Theta(nK^C)$ cells. Regarding the spatial complexity, note that one only needs to keep one column at each step to perform the recursion, and therefore it is in $\Theta(K^C)$.

4 A Multiobjective Branch and Bound Algorithm

4.1 Principle

A classical branch and bound algorithm (BB) explores an enumeration tree whose leaves represent a set of *possibly optimal* solutions (i.e., it is not required that the leaves represent the set of *all feasible* solutions, provided it is guaranteed that at least one optimal solution is present). One can distinguish two main parts in this type of procedure: the *branching part* describing how the set of solutions associated with a node of the tree is separated into subsets, and the *bounding part* describing how the quality of the current subset of solutions is optimistically evaluated. The complete enumeration of the children of a node can be avoided when its optimistic evaluation is worse than the best solution found so far.

A multiobjective BB (MOBB) is an extension of the classical BB. The branching scheme now must be able to enumerate a complete set of feasible solutions in an enumeration tree (i.e., at least one solution must be present in the leaves for each Pareto point in the objective space). In the bounding part, the optimistic evaluation is a vector and the enumeration is stopped when the optimistic evaluation of a node is dominated by an already found non-dominated solution.

4.2 Branching Part

Let us introduce a new notation. For any pair of classes c, c', let $\mathcal{N}_{c,c'} = \{i \in \mathcal{N} : c \trianglerighteq \gamma^i \trianglerighteq c'\}$ be the set of items whose classes are between classes c and c'. Set $\mathcal{N}_{1,c}$ will be denoted by $\mathcal{N}_c$.

Our multiobjective branch and bound approach for the committee selection problem relies on the following property:

Proposition 2. *For any feasible profit vector* $p = (p_1, \ldots, p_C)$*, solution* $x = (x^1, \ldots, x^n)$ *defined by*

- $x^i = 1,\ \forall i = 1, \ldots, p_1$
- $x^i = 0,\ \forall i = p_1 + 1, \ldots, |\mathcal{N}_1|$
- $x^i = 1,\ \forall i = |\mathcal{N}_{c-1}| + 1, \ldots, |\mathcal{N}_{c-1}| + p_c - p_{c-1} \quad \forall c = 2, \ldots, C$
- $x^i = 0,\ \forall i = |\mathcal{N}_{c-1}| + p_c - p_{c-1} + 1, \ldots, |\mathcal{N}_c| \quad \forall c = 2, \ldots, C$

is a minimal weight feasible solution for this profit vector.

Proof. We recall that the lower the index, the better the class, and within each class, the lower the index, the lighter the item. We first show by induction that solution x yields profit vector p. Clearly, solution x admits p_1 items of class 1 and therefore its value on component 1 is p_1. Assume now that the value of x on component $c-1$ is p_{c-1} (induction hypothesis). Then its value on component c is by construction $p_c = p_c - p_{c-1} + p_{c-1}$ since $p_c - p_{c-1}$ items of class c are selected. Solution x yields therefore profit vector $(p_1, \ldots, p_C)$.

By noting that at each step one selects the lightest items of each class, one concludes that x is a minimal weight feasible solution for profit vector p. ■

This observation justifies that we focus on feasible solutions of this type. Now, the branching scheme can be simply explained. Let $P(k, c, p, b)$ denote the subproblem where one wants to select k items whose total weight is less than budget b, where the remaining items are classified in classes $(c, \ldots, C)$ and the profit vector of the already selected items is $p \in [\![0, K]\!]_{\uparrow}^{C}$. The initial problem is then denoted by $P(K, 1, (0, \ldots, 0), B)$. A node in the enumeration tree represents a problem $P(k, c, p, b)$ where $p = (p_1, \ldots, p_C)$ accounts for the items selected in the previous steps. Such a problem can be subdivided into at most $k+1$ subproblems $P(k', c+1, p', b')$ for $k' = 0, \ldots, \min\{k, |\mathcal{N}_{c,c}|\}$, where branching consists in deciding to select exactly $k - k'$ items in class c (the ones with the lowest weights in class c), and p', b' are the updated profit vector and budget to take into account these newly selected items.

Note that in some cases, some subproblems have an empty set of feasible solutions due to the budget constraint, and are therefore discarded. For illustration, the enumeration tree for Example 1 is provided in Figure 1. The vector in a node is the current value of p, and each branch is labelled by the selected items at this step. The dashed node (on the right) is discarded due to the budget constraint, and the gray nodes correspond to non-dominated solutions.

4.3 Bounding Part

For a problem $P(k, c, p, b)$ and a preference class c' such that $c \trianglerighteq c'$, the optimistic evaluation UB of the corresponding node in the enumeration tree is defined by:

$$UB = (m_1, \ldots, m_C) \text{ where } \begin{cases} m_{c'} = p_{c'} & \forall c' = 1, \ldots, c-1 \\ m_{c'} = m_{c,c'} & \forall c' = c, \ldots, C \end{cases} \tag{1}$$

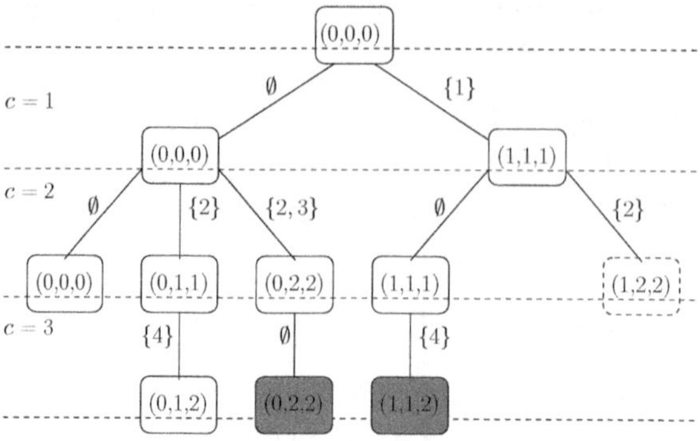

Fig. 1. Enumeration tree for Example 1

and where $m_{c,c'}$ is defined by:

$$m_{c,c'} = \begin{cases} \max \sum_{i \in \mathcal{N}_{c,c'}} x_i \\ \text{s.t.} \sum_{i \in \mathcal{N}_{c,c'}} w_i x_i \leq b \qquad \sum_{i \in \mathcal{N}_{c,c'}} x_i \leq k \\ x_i \in \{0,1\} \quad \forall i \in \mathcal{N}_{c,c'} \end{cases}$$

Note that the above program can be very simply solved by a greedy algorithm.
The following proposition states that UB is indeed an optimistic evaluation:

Proposition 3. *For any $k = 0, \ldots, K$, any $c = 1, \ldots, C$, any vector p of $[\![0,K]\!]_{\uparrow}^{C}$ and any $b = 0, \ldots, B$, the profit vector of any feasible solution in $P(k,c,p,b)$ is weakly dominated by UB.*

Proof. Let p' be the profit vector of a feasible solution in $P(k,c,p,b)$. Let $UB = (m_1, \ldots, m_C)$ be computed as in Eq. 1. For $c' = 1, \ldots, c-1$, by definition, $m_{c'} \geq p'_{c'}$. For $c' = c, \ldots, C$, by definition, $m_{c,c'}$ is the greatest number of items one can pick in $\mathcal{N}_{c,c'}$. Therefore $m_{c'} \geq p'_{c'}$. ■

Example 3. *At the root of the enumeration tree for Example 1, one has $UB = (1,2,2)$. For instance, when considering class 1 and 2, the greatest number of items that can be selected under the constraints is 2 (individuals 2 and 3, with $w^2 + w^3 = 6$), and therefore the second component of UB equals 2.*

4.4 Complexity

The number of nodes in the enumeration tree is clearly upper bounded by $(K+1)^C$, since the tree is of depth C and the number of children of a node is upper bounded by $K+1$. Furthermore, note that each node representing a problem $P(k, C-1, \cdot, \cdot)$ with $k \leq K$ has at most one child: the only decision that can be made is indeed to select $K-k$ items of class C, so that the cardinality constraint holds. The number of nodes in the enumeration tree is therefore in $\mathcal{O}(K^{C-1})$.

As the computation time required for the bounding procedure (at each node) is polynomial provided C is a constant, the complexity of the whole branch and bound algorithm is also polynomial. By comparing the number of cells in the dynamic programming table ($\Theta(nK^C)$) and the number of nodes in the enumeration tree ($\mathcal{O}(K^{C-1})$), it appears that the branch and bound algorithm should perform better. This observation is confirmed experimentally for all problems we tested.

Besides, the spatial complexity of the branch and bound algorithm in the worst case is in $\mathcal{O}(K^{C-1})$. Therefore it is also better than the dynamic programming algorithm from this point of view.

5 Experimental Results

We present here numerical results concerning the multiobjective dynamic programming method, and the branch and bound method. The computer used is an Intel Core 2 duo @3GHz, with 3GB RAM, and the algorithms were coded in C++. We first test our methods on randomly generated instances, and then on a real-world data set (IMDb dataset).

5.1 Randomly Generated Instances

We chose to run our tests on two different types of instances:

- uncorrelated instances (Un): for each item i, γ^i is randomly drawn in $\{1,\ldots,C\}$, and w^i is randomly drawn in $\{1,\ldots,1000\}$.
- correlated instances (Co): for each item i, γ^i is randomly drawn in $\{1,\ldots,C\}$, and w^i is randomly drawn in $\{1+1000*(C-\gamma^i)/C,\ldots,1000*(C-\gamma^i+1)/C\}$. In other words, the better the class, the higher the weight; for instance, if $\gamma^i=3$ (resp. $\gamma^i=2$) and $C=5$, then w^i is randomly drawn in $\{401,\ldots,600\}$ (resp. $\{601,\ldots,800\}$).

For all instances, we chose to set B so that the following properties hold:

- $B \geq \sum_{i=1}^{K} w^{(i)}$, where item (i) is the item with the i-th smallest weight: this inequality ensures that there is at least one feasible solution;
- $B < \sum_{i=1}^{K} w^i$: this inequality ensures that the solution consisting of the K best items is not feasible (we recall that items are first ordered decreasingly with respect to their classes, and increasingly ordered with respect to their weights within each class).

By setting $B = 0.5\sum_{i=1}^{K} w^{(i)} + 0.5\sum_{i=1}^{K} w^i$ in the tests, both properties hold (unless $\sum_{i=1}^{K} w^{(i)} = \sum_{i=1}^{K} w^i$, in which case the only non-dominated solution consists in selecting the K first items).

Table 2 shows the average computation times in seconds for both methods (DP: dynamic programming, BB: branch and bound), and the average number of non-dominated profit vectors (in other words, the size of a complete minimal set of preferred solutions) over 30 random instances for each type and size.

Symbol "-" means that no instance could be solved due to memory constraints, i.e. more than 3GB RAM were required. All generated instances have $n = 1000$ items. Notation "Un-x-y" (resp. "Co-x-y") means Uncorrelated (resp. Correlated) instances with $C = x$ and $K = y$. Since there is very little variance in the computation times for a given type and size, only the average computation times are reported.

Table 2. Average computation times of both methods, and average number of non-dominated profit vectors (ND), for uncorrelated and correlated instances of size $n = 1000$

Type	DP (sec.)	BB(sec.)	ND	Type	DP(sec.)	BB(sec.)	ND
Un-3-100	3.9	0.005	3	Co-3-100	3.9	0.004	44
Un-3-200	32.1	0.06	4	Co-3-200	32.0	0.06	75
Un-3-500	506	0.45	38	Co-3-500	505	0.5	108
Un-4-100	132	0.007	5	Co-4-100	132	0.08	1101
Un-4-150	656	0.03	12	Co-4-150	654	0.4	2166
Un-4-200	-	0.07	16	Co-4-200	-	0.8	3346
Un-5-50	117	0.003	2	Co-5-50	121	0.5	3657
Un-5-80	1114	0.004	7	Co-5-80	1263	7.0	13526
Un-5-100	-	0.018	15	Co-5-100	-	23.2	24800

First note that, for all instances, the branch and bound approach is faster than the dynamic programming one. As expected, more classes make the problem harder, and the same goes for size K of the committee. The number of non-dominated profit vectors is small for uncorrelated instances, because there are low weighted items in good classes. This number is much larger for correlated instances, because this property does not hold anymore. Comparing the results obtained for uncorrelated and correlated instances shows that the correlation has no impact on the computation times of the dynamic programming procedure. However, its impact is noticeable for the branch and bound method, since the number of nodes expanded in the enumeration tree grows with the number of non-dominated profit vectors, and this number is very high for correlated instances. The impact of the correlation on the number of non-dominated profit vectors is consistent with what can be observed in multiobjective combinatorial optimization. We will come back to the question of the size of the non-dominated set in the next subsection.

Since the branch and bound procedure is very fast, and does not have high memory requirements, we tested it on larger instances. We set $n = 10000$ and $K = 100$ for all these instances. Table 3 shows the results of those experiments for $C \in \{3, 4, 5, 10, 20, 50\}$. Resolution times are in seconds, and symbol "-" means that it exceeds 600 seconds. Most of the resolution time is now spent in the bounding part, more precisely for the comparison between the optimistic evaluation of a node and the non-dominated profit vectors. For uncorrelated instances with $3, 4, 5$ classes, the resolution times are nevertheless particularly small because the bounds enable to discard a huge amount of nodes, since there are few good feasible profit vectors (around 70% of selected items in these solutions belong to class 1). This is no longer true for correlated instances, which results in much greater resolution times.

Furthermore, as is well-known in multiobjective optimization, the number of objectives (here, the number C of classes) is a crucial parameter for the efficiency of the solution methods. For this reason, when $C = 10$, 20 or 50, the resolution is of course computationally more demanding, as can be observed in the table (for instance, for $C = 20$ and $K = 100$, the resolution time is on average 2.21 seconds for uncorrelated instances). The method seems nevertheless to scale well, though the variance in the resolution times is much higher.

Table 3. Average computation times of the BB method, and average number of non-dominated profit vectors (ND), for uncorrelated and correlated instances of size $n = 10000$ with $K = 100$ and $C \in \{3 \cdots 50\}$

Type	BB(sec.)			ND	Type	BB(sec.)			ND
	min.	avg.	max.			min.	avg.	max.	
Un-3-100	0.01	0.02	0.02	3	Co-3-100	0.03	0.05	0.06	50
Un-4-100	0.02	0.02	0.03	6	Co-4-100	1.27	1.31	1.37	4960
Un-5-100	0.02	0.03	0.04	10	Co-5-100	27.3	28.0	29.0	29418
Un-10-100	0.10	0.12	0.15	264	Co-10-100	-	-	-	-
Un-20-100	0.37	2.21	14.24	467	Co-20-100	-	-	-	-
Un-50-100	2.09	21.1*	101*	968*	Co-50-100	-	-	-	-

* Note that one instance largely exceeded the time limit, and the values indicated do not take this instance into account.

Table 4(A) (resp. 4(B)) is there to give an idea of the order of magnitude of K with respect to C in order to get tractable uncorrelated (resp. correlated) instances. For each C, the order of magnitude of parameter K in the table is the one beyond which the resolution becomes cumbersome.

Table 4. Average computation times of the BB method, and average number of non-dominated profit vectors (ND), for uncorrelated and correlated instances of size $n = 10000$ with $C \in \{3 \cdots 50\}$, for different values of K

Type	BB(sec.)			ND
	min.	avg.	max.	
Un-3-5000	375	394	425	368
Un-4-3000	208	237	266	7203
Un-5-2000	185	292	428	15812
Un-10-250	1.86	10.5	55.4	2646
Un-20-150	0.69	91.5	562	2603
Un-50-80	1.98	24.6	208	1052

(A)

Type	BB(sec.)			ND
	min.	avg.	max.	
Co-3-5000	415	419	424	1086
Co-4-1000	666	706	767	105976
Co-5-100	27.3	28.0	29.0	29418
Co-10-15	95.2	97.4	103	30441
Co-20-7	20.0	20.2	20.6	14800
Co-50-5	521	526	534	36471

(B)

5.2 IMDb Dataset

Let us now evaluate the operationality of the BB method on a real data set, namely the Internet Movie Database (www.imdb.com). On this web site, one can indeed find a top 250 movies as voted by the users. Assume that a film festival organizer wants to project K top movies within a given time limit. If the organizer refers to the IMDb Top 250 to make his/her choice (i.e., the preference classes are directly inferred from the Top 250), it amounts to a committee selection problem where the weights are the durations of the movies. The numerical tests carried out are the following:

- size K of the committee varies from 5 to 50;
- number C of classes varies from 10 to 250 (in this latter case, the setting is the same as in Klamler et al. [14], i.e. there is a linear order on the elements);
- the time limit follows the formula used for the budget constraint in the previous tests, so that both constraints (cardinality and weight) are taken into account in the choice.

Table 5 shows the computation times in seconds for the BB method, as well as the number ND of non-dominated committees (i.e., non-dominated subsets of movies). Symbol "-" means that the computation time exceeds 600 sec. Interestingly, one observes that the method remains operational even when the number of preference classes is high. The size of the non-dominated set of course increases, but this is not a real drawback if one sees the pairwise dominance relation as a first filter before an interactive exploration of the non-dominated set (by interactively adding constraints for instance, so as to reduce the set of potential selections).

Table 5. Computation times of the BB method for the IMDb data set

	$C = 10$		$C = 25$		$C = 50$		$C = 250$	
	Time	ND	Time	ND	Time	ND	Time	ND
$K = 5$	0.01	5	0.03	9	0.15	7	2.7	11
$K = 10$	0.01	8	0.08	24	0.6	108	131.6	323
$K = 15$	0.01	12	0.6	156	11.5	469	-	-
$K = 20$	0.01	16	5.17	222	295	1310	-	-
$K = 25$	0.01	14	131.3	883	-	-	-	-
$K = 50$	3.0	749	-	-	-	-	-	-

6 Conclusion

We studied the committee selection problem with a cardinality constraint, where the items are classified into ordered classes. By reducing the problem to a multiobjective knapsack problem with a cardinality constraint, we proposed two polynomial time solution algorithms: a dynamic programming scheme and a branch and bound procedure. The theoretical complexities and numerical tests tend to prove that the latter one is better, both in time and space requirements.

Note that all the results presented here naturally extends when the preference classes are only partially ordered. The only difference is that the profit vectors are then not necessarily non-decreasing. For instance, consider three partially ordered preference classes 1, 2 and 3 with: $1 \rhd 2$ and $1 \rhd 3$ (2 and 3 are not comparable). The profit vector for an item of class 2 is then $(0, 1, 0)$.

Finally, it would be interesting to study more expressive settings for ranking sets of objects. For instance, when the order relation is directly defined on the items, Fishburn [11] proposed a setting where preferences for the inclusion (resp. exclusion) of items in (resp. from) a subset can be expressed.

Acknowledgments. We would like to thank the reviewers for their helpful comments and suggestions.

References

1. Barberà, S., Bossert, W., Pattanaik, P.K.: Ranking sets of objects. In: Barberà, S., Hammond, P.J., Seidl, C. (eds.) Handbook of Utility Theory, vol. 2, Kluwer Academic Publishers, Dordrecht (2004)
2. Bartee, E.M.: Problem solving with ordinal measurement. Management Science 17(10), 622–633 (1971)
3. Bossert, W., Pattanaik, P.K., Xu, Y.: Ranking opportunity sets: An axiomatic approach. Journal of Economic Theory 63(2), 326–345 (1994)
4. Bossong, U., Schweigert, D.: Minimal paths on ordered graphs. Technical Report 24, Report in Wirtschaftsmathematik, Universität Kaiserslautern (1996)
5. Bouveret, S., Endriss, U., Lang, J.: Fair division under ordinal preferences: Computing envy-free allocations of indivisible goods. In: European Conference on Artificial Intelligence (ECAI 2010), pp. 387–392. IOS Press, Amsterdam (2010)
6. Brams, S., Edelman, P., Fishburn, P.: Fair division of indivisible items. Theory and Decision 5(2), 147–180 (2004)
7. Brams, S., King, D.: Efficient fair division – help the worst off or avoid envy? Rationality and Society 17(4), 387–421 (2005)
8. Della Croce, F., Paschos, V.T., Tsoukias, A.: An improved general procedure for lexicographic bottleneck problems. Op. Res. Letters 24, 187–194 (1999)
9. Erlebach, T., Kellerer, H., Pferschy, U.: Approximating multi-objective knapsack problems. In: Dehne, F., Sack, J.-R., Tamassia, R. (eds.) WADS 2001. LNCS, vol. 2125, pp. 210–221. Springer, Heidelberg (2001)
10. Fishburn, P.C.: Utility Theory for Decision Making. Wiley, New York (1970)
11. Fishburn, P.C.: Signed orders and power set extensions. Journal of Economic Theory 56, 1–19 (1992)
12. Halpern, J.Y.: Defining relative likelihood in partially-ordered preferential structures. Journal of Artificial Intelligence Research 7, 1–24 (1997)
13. Hansen, P.: Bicriterion path problems. In: Fandel, G., Gal, T. (eds.) Multicriteria Decision Making (1980)
14. Klamler, C., Pferschy, U., Ruzika, S.: Committee selection with a weight constraint based on lexicographic rankings of individuals. In: Rossi, F., Tsoukias, A. (eds.) ADT 2009. LNCS, vol. 5783, pp. 50–61. Springer, Heidelberg (2009)
15. Klamroth, K., Wiecek, M.M.: Dynamic programming approaches to the multiple criteria knapsack problem. Naval Research Logistics 47, 57–76 (2000)
16. Schweigert, D.: Ordered graphs and minimal spanning trees. Foundations of Computing and Decision Sciences 24(4), 219–229 (1999)

A Natural Language Argumentation Interface for Explanation Generation in Markov Decision Processes

Thomas Dodson, Nicholas Mattei, and Judy Goldsmith

University of Kentucky
Department of Computer Science
Lexington, KY 40506, USA
tcdodson@gmail.com, nick.mattei@uky.edu, goldsmit@cs.uky.edu

Abstract. A Markov Decision Process (MDP) policy presents, for each state, an action, which preferably maximizes the expected reward accrual over time. In this paper, we present a novel system that generates, in real time, natural language explanations of the optimal action, recommended by an MDP while the user interacts with the MDP policy. We rely on natural language explanations in order to build trust between the user and the explanation system, leveraging existing research in psychology in order to generate salient explanations for the end user. Our explanation system is designed for portability between domains and uses a combination of domain specific and domain independent techniques. The system automatically extracts implicit knowledge from an MDP model and accompanying policy. This policy-based explanation system can be ported between applications without additional effort by knowledge engineers or model builders. Our system separates domain-specific data from the explanation logic, allowing for a robust system capable of incremental upgrades. Domain-specific explanations are generated through case-based explanation techniques specific to the domain and a knowledge base of concept mappings for our natural language model.

1 Introduction

A Markov decision process (MDP) is a mathematical formalism which allows for long range planning in probabilistic environments [2, 15]. The work reported here uses fully observable, factored MDPs[3]. The fundamental concepts use by our system are generalizable to other MDP formalisms; we choose the factored MDP representation as it will allow us to expand our system to scenarios where we recommend a set of actions per time step. A policy for an MDP is a mapping of states to actions that defines a tree of possible futures, each with a probability and a utility. Unfortunately, this branching set of possible futures is a large object with many potential branches that is difficult to understand even for sophisticated users.

The complex nature of possible futures and their probabilities prevents many end users from trusting, understanding, and implementing the plans generated from MDP policies [9]. Recommendations and plans generated by computers are not always trusted or implemented by end users of decision support systems. Distrust and misunderstanding are two of the most often user cited reasons for not following a recommended plan

R.I. Brafman, F. Roberts, and A. Tsoukiàs (Eds.): ADT 2011, LNAI 6992, pp. 42–55, 2011.

or action [13]. For a user unfamiliar with stochastic planning, the most troublesome part of existing explanation systems is the explicit use of probabilities, as humans are demonstrably bad at reasoning with probabilities [18]. Additionally, it is our intuition that the concept of a preordained probability of success or failure at a given endeavor discomforts the average user.

Following the classifications of logical arguments and explanations given by Moore and Parker, our system generates arguments [11]. While we, as system designers, are convinced of the optimality of the optimal action the user may not be so convinced. In an explanation, two parties agree about the truth of a statement and the discussion is centered around *why* the statement is true. However, our system design is attempting to convince the user of the "goodness" of the recommended action; this is an argument.

In this paper we present an explanation system for MDP policies. Our system produces natural language explanations, generated from domain specific and domain independent information, to convince end users to implement the recommended actions. Our system generates arguments that are designed to convince the user of the "goodness" of the recommended action. While the logic of our arguments is generated in a domain independent way, there are domain specific data sources included. These are decoupled from the explanation interface, to allow a high degree of customization. This allows our base system to be deployed on different domains without additional information from the model designers. If an implementation calls for it, our system is flexible enough to incorporate domain specific language and cases to augment its generated arguments. We implement this novel, argument based approach with natural language text in order to closely connect with the user. Building this trust is essential in convincing the user to implement the policy set out by the MDP [13]. Thus, we avoid exposing the user to the specifics of stochastic planning, though we cannot entirely avoid language addressing the inherent probabilistic nature of our planning system.

Our system has been developed as a piece of a larger program working with advising college students about what courses to take and when to take them. It was tested on a subset of a model developed to predict student grades based on anonymized student records, as well as capture student preferences, and institutional constraints at the University of Kentucky [7]. Our system presents, as a paragraph, an argument as to why a student should take a specified set of courses in the next semester. The underlying policy is based on the student's preferences and abilities. This domain is interesting because it involves users who need to reason in discrete time steps about their long term benefits. Beginning students[1] at a university will have limited knowledge about utility theory and represent a good focus population for studying the effectiveness of different explanations.

Model construction, verification and validation is an extremely rich subject that we do not treat in this paper. While the quality of explanations is dependent on the quality and accuracy of a given model we will not discuss modeling accuracy or fidelity. The purpose of this work is to generate arguments in a domain-independent way, incorporating domain-specific information only to generate the explanation language. The

[1] Students may begin their college careers as Computer Science majors or switch into the major later. We consider students to *begin* with the introductory programming courses, or with the first CS course they take at the University of Kentucky.

correctness of the model is therefore irrelevant in the context of validating a method to generate explanations. Through user testing and refinement it is possible to use our work to assist in the construction, verification, and validation of models meant to be implemented with end users.

In the next section we will provide background on MDPs and a brief overview of current explanation systems. In Section 3 we define the model we use as an example domain. Section 4 provides an overview of the system design as well as specific details about the system's three main components: the model based explainer, the case based explainer, and the natural language generator. Section 5 provides examples of the output of our system and an overview of the user study we will use to verify and validate our approach. Section 6 provides some conclusions about the system development so far and our main target areas for future study.

2 Background and Related Work

Markov Decision Processes. A MDP is a formal model for planning, when actions are modeled as having probabilistic outcomes. We focus here on factored MDPs [3]. MDPs are used in many areas, including robotics, economics and manufacturing.

Definition 1. *An MDP is a tuple, $\langle S,A,T,R\rangle$, where S is a set of states and A is a set of actions, and $T(s'|s,a)$ is the probability that state s' is reached if a is taken in state s, and $R(s)$ is the reward, or utility, of being in state s. If states in S are represented by variable (attribute) vector, we say that the MDP is* factored.

A policy for an MDP is a mapping $\pi : S \rightarrow A$. The best policy for an MDP is one that maximizes the expected value (Definition 2) [15] within a specified finite or infinite time horizon, or with a guarantee of (unspecified) finiteness. In the case of academic advising, since credits become invalid at the University of Kentucky after 10 years, we assume a fixed, finite horizon [2]. Policies are computed with respect to the expected total discounted reward, where the discount rate γ is such that $0 \leq \gamma < 1$. The optimal policy with respect to discount γ is one that maximizes the total discounted expected value of the start state (see Definition 2) [2, 15].

Definition 2. *The expected value of state s with respect to policy π and discount γ is*

$$V^{\pi}(s) = R(s) + \gamma \sum_{s' \in S} T(s' \mid \pi(s), s) * V^{\pi}(s'). \tag{1}$$

The optimal value function V^* is the value function of any optimal policy π^* [2, 15]. We use the optimal policy, and other domain and model information, to generate natural language explanations for users with no knowledge of probability or utility theory.

Explanation Systems. Prior work on natural language explanation of MDP policies is sparse, and has focused primarily on what could be called "policy-based explanation," whereby the explanation text is generated solely from the policy. The nature of such systems limits the usefulness of these explanations for users who are unfamiliar with stochastic planning, as the information presented is probabilistic in nature. However, these algorithms have the advantage of being entirely domain-independent. A good example of such a system is Khan et al.'s minimal sufficient explanations [9], which chooses explanatory variables based on the occupation frequency of desired future states. Note that, while the algorithms used in policy-based explanation systems are domain-independent, the explanations generated by such systems often rely on the implicit domain-specific information encoded into the model in the form of action and variable names. Other work has focused on finding the variable which is most influential to determining the optimal action at the current state [5], while using an extensive knowledge-base to translate these results into natural language explanations.

Case-based and model-based explanation systems rely, to different extents, on domain specific information. To find literature on such systems, it is necessary to look beyond stochastic planning. Case-based explanation, which uses a database of prior decisions and their factors, called a case base, is more knowledge-light, requiring only the cases themselves and a model detailing how the factors of a case can be generalized to arbitrary cases. Care must be taken in constructing a case base in order to include sufficient cases to cover all possible inputs. Nugent et al.'s KLEF [14] is an example of a case-based explanation system. A model-based explanation system, however, relies on domain-specific information, in the form of an explicit explanation model.

An explanation interface provides explanations of the reasoning that led to the recommendation. Sinha and Swearingen [17] found that, to satisfy most users, recommendation software employing collaborative filtering must be *transparent*, i.e., must provide not only good recommendations, but also the logic behind a particular recommendation. Since stochastic planning methods are generally not well understood by our intended users, we do not restrict our explanations to cover, for example, some minimum portion of the total reward [9], and instead choose explanation primitives that, while still factual, will be most convincing to the user.

3 Model

For this paper we focus on an academic advising domain. We use a restricted domain for testing which focuses on completing courses to achieve a computer science minor focus at the University of Kentucky. Our research group is also developing a system to automatically generate complete academic advising domains that capture all classes in a university [7]. The long term goal of this ongoing research project is to develop an end-to-end system to aid academic advisors that build probabilistic grade predictors, model student preferences, plan, and explain the offered recommendations.

The variables in our factored domain are the required courses for a minor focus in computer science: Intro Computer Programming (ICP), Program Design and Problem Solving (PDPS), Software Engineering (SE), Discrete Mathematics (DM), and Algorithm Design and Analysis (ALGO). We include Calculus II (CALC2) as a predictor course for DM and ALGO due to their strong mathematical components. Each class variable can have values: (G)ood, (P)ass, (F)ail, and (N)ot Taken. An additional variable is high school grade point average, HSGPA; this can have values: (G)ood, (P)ass,

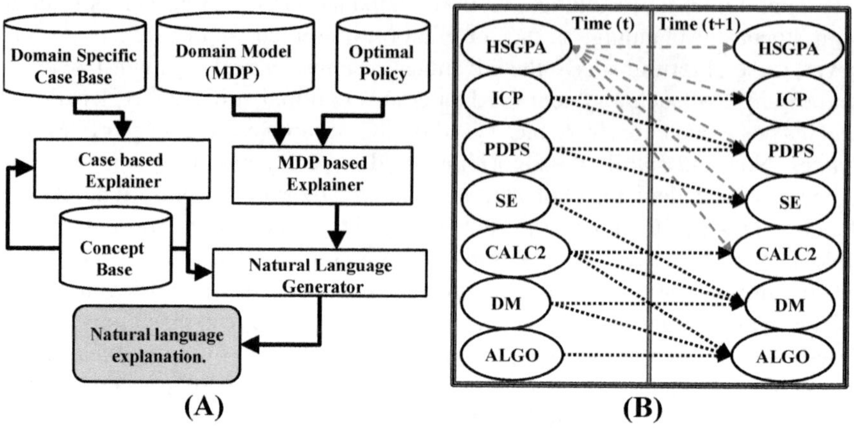

Fig. 1. System organization and data flow (A) and the dynamic decision network (temporal dependency structure) for the academic advising model (B)

(L)ow. The model was hand coded with transition probabilities derived from historic course data at the University of Kentucky.

Each action in our domain is of the form, "Take Course X," and only affects variable X. Figure 1-B shows the temporal dependencies between classes, and implicitly encodes the set of prerequisites due to the near certain probability of failure if prerequisite courses are not taken first. Complex conditional dependences exist between courses due to the possibility of failing a course. CALC2 is not required and we do not place reward on its completion. Taking it correlates with success in DM and ALGO; we want to ensure our model can explain situations where unrewarded variables are important. Most courses in the model have HSGPA, the previous class, and the current class as the priors (except ICP and CALC2 which only have HSGPA as a prior).[2]

The reward function is additive and places a value of 4.0 and 2.0 on Good and Passing grades respectively. Failure is penalized with a 0.0. A discount factor of 0.9 is used to weight early success more than later success. While our current utility function only focuses on earning the highest grades possible as quickly as possible we stress that other utility functions could be used and, in fact, are being developed as part of our larger academic advising research project.

The model was encoded using a variant of the SPUDD format [8] and the optimal policy was found using a local SPUDD implementation developed in our lab [8, 10]. We applied a horizon of 10 steps and a tolerance of 0.01. The model has about 2,400 states and the optimal value function ADD has over 10,000 leaf nodes and 15,000 edges.

4 System Overview

Our explanation system integrates a policy-based approach with case-based and model-based algorithms. However, the model-based system is constructed so the algorithm

[2] HSGPA is a strong predictor of early college success (and college graduation) and GPA's prediction power has been well studied [4].

itself is not domain-specific. Rather, the explanation model is constructed from the MDP and resulting policy and relies on domain-specific inputs and a domain-specific language, in the natural language generation module. Thus, we separate the model dependent factors from the model independent methods. This gives our methods high portability between domains.

Figure 1-A illustrates the data flow through our system. All domain specific information has been removed from the individual modules. We think of each of the modules as generating points of our argument while the natural language generator assimilates all these points into a well structured argument to the user. The assimilated argument is stronger than any of the individual points. However, we can remove modules that are not necessary for specific domains, e.g., when a case base cannot be procured. This allows our system to be flexible with respect to a single model and across multiple domains. In addition, system deployment can happen early in a development cycle while other "points" of the argument are brought online. The novel combination of a case-based explainer, which makes arguments from empirical past data, with a model-based explainer, which makes arguments from future predicted data, allows our system to generate better arguments than either piece alone.

A standard use case for our system would proceed as follows: students would access the interface either online or in an advising office. The system would elicit user preferences and course histories (these could also be gleaned from student transcripts). Once this data has been provided to the system, a natural language explanation would explain what courses to take in the coming semester. While our current model recommends one course at a time we will expand the system to include multiple actions per time step.

Our system differs from existing but similar systems such as the one designed by Elizalde et al. [5] in several important ways. First, while an extensive knowledge base will improve the effectiveness of explanations, the knowledge base required by our system to generate basic explanations is minimal, and limited to variables which can be determined from the model itself. Second, our model-based module decomposes recommendations from the MDP in a way that is more psychologically grounded in many domains, focusing on user actions instead of variables [6].

We designed with a "most convincing" heuristic; we attempt to select the factual statements and word framings that will be most influential to our target user base. This is in contrast to existing other similar systems which focus on a "most coverage" heuristic [9]. A most coverage heuristic focuses on explaining some minimal level of utility that would be accrued by the optimal policy. While this method is both mathematically grounded and convincing to individuals who understand probabilistic planning, our intuition is that it is not as convincing to the average individual.

4.1 Model Based Explanation

The model-based module extracts information from the MDP model and a policy of recommended actions on that model. This module generates explanations based on what comes next — specifically, information about why, in terms of next actions, the recommended action is best. We compare actions in terms of a set of values, called *action factored differential values (AFDVs)* for each possible action in the current state. AFDVs allow us to explain the optimal action in terms of how much better the set of actions at the next state are. E.g., we can model that taking ICP before PDPS is better because taking ICP first improves the expected value of taking PDPS in the next step. We can also highlight how the current action can affect multiple future actions and rewards. This allows our method to explain complex conditional policies without explicit knowledge of the particular conditional. Through the computation of the AFDVs we are able to extract how the current best action improves the expected assignment of one or more variables under future actions.

This method of explanation allows for a salient explanation that focuses on how the current best action will improve actions and immediate rewards in the next state (the next decision point). Many studies have shown empirically that humans use a hyperbolic discounting function and are incredibly risk adverse when reasoning about long term plans under uncertain conditions [6, 20]. This discount function places much more value on rewards realized in the short term. In contrast to human reasoning, an MDP uses an exponential discount function when computing optimal policies. The combined effects of human inability to think rationally in probabilistic terms and hyperbolic cognitive discounting means there is a fundamental disconnect between the human user and the rational policy [6, 18]. The disconnect between the two reasoning methods must be reconciled in order to communicate MDP policies to human users in terms that they will more readily understand and trust. This translation is achieved through explaining the long term plan in terms of short term gains with AFDV sets.

To generate a usable set of AFDVs from some state s, we define a method for measuring the value of taking an arbitrary two action sequence and then continuing to follow the given policy, π. Intuitively, a set of AFDVs is a set of two-step look ahead utilities for all the different possible combinations of actions and results. This is accomplished by modifying the general expression for V^{π} to accommodate deviation from the policy in the current state and the set of next states:

$$V_2^{\pi}(s,a_1,a_2) - R(s) = \gamma \sum_{s' \in S} T(s'|s,a_1) \cdot [R(s') + \gamma \sum_{s'' \in S} T(s''|s',a_2) \cdot V^{\pi}(s'')]. \quad (2)$$

Using V_2^{π}, we can then compute a single AFDV object for the action to be explained, $\pi(s)$, by computing the value of the two step sequence $\{\pi(s),a\}$ and the value of another two step sequence $\{a_i,a\}$ and taking the difference,

$$\Delta^{\pi}(s,\pi,a_i,a) = V_2^{\pi}(s,\pi(s),a) - V_2^{\pi}(s,a_i,a). \quad (3)$$

To compute a full set of AFDVs for the explanation action, $\pi(s)$, this computation is done for all $a_i \in A \setminus \pi(s)$ and for all $a \in A$.

In order to choose variables for explanation, we compute, for each i, $\Delta^{\pi}(s,\pi,a_i,a)$, to find out how many actions' utilities will increase after having taken the recommended action. This set of counts gives the number of actions in the current state which cause a greater increase in utility of the action a than the recommended action. We define

$$x_s^{\pi}(a) = |\{i : \Delta^{\pi}(s,\pi,a_i,a) < 0\}|. \quad (4)$$

Note that we may have for all $a \in A : x_s^{\pi}(a) > 0$, since only the sum of the AFDV set over a_i for the optimal action is guaranteed to be greater than or equal to the sum for any other action. We choose the subset of A for which $x_s^\pi(a)$ is minimal as our explanation variables, and explain $\pi(s)$ in terms of its positive effects on those actions. We can also decompose the actions into corresponding variable assignments and explain how those variables change, leading to higher reward. By focusing on actions we reduce the overall size of the explanation in order to avoid overwhelming the user, while still allowing the most salient variables of the recommended action to be preserved. If more variables are desired, another subset of A can be chosen for which $x_s^\pi(a)$ is greater than the minimum, but less than any other value. While the current method of choosing explanation variables relies on knowledge of the *optimal* policy, the AFDV objects are meaningful for any policy. However, our particular method for choosing the subset of AFDVs for explanation relies on the optimality of the action $\pi(s)$, and would have to be adapted for use with a heuristic policy.

For example, the explanation primitive for a set of future actions with $\pi(s) = act\_PDPS$, $x_s^\pi(act\_SE) = x_s^\pi(act\_DM) = 0$, $x_s^\pi(act\_ALGO) = 1$, and $x_s^\pi(a) = 2$ for all other a is:

> The recommended action is *act_PDPS*, generated by examining long-term future reward. It is the optimal action with regards to your current state and the actions available to you. Our model indicates that this action will best prepare you for *act_SE* and *act_DM* in the future. Additionally, it will prepare you for *act_ALGO*.

It is possible to construct pathological domains where our domain independent explainer fails to select a best action. In these rare cases, the explainer will default to stating that the action prescribed by the given policy is the best because it leads to the greatest expected reward; this prevents contradictions between the explanation and policy. The AFDV method will break down if domains are constructed such that the expected reward is 0 within the horizon (2 time steps). This can happen when there are balanced positive and negative rewards. For this reason, we currently restrict our domain independence claims to those domains with only non-negative rewards.

4.2 Case-Based Explanation

Case-based explanation (CBE) uses past performance in the same domain in order to explain conclusions at the present state. It is advantageous because it uses real evidence, which enhances the transparency of the explanation, and analogy, a natural form of explanation in many domains [14]. This argument from past data combined with our model-based argument from predicted future outcomes creates a strong complete argument for the action recommended by the optimal policy. Our case base consists of 2693 distinct grade assignments in 6 distinct courses taken by 955 unique students. This anonymized information was provided by the University of Kentucky, about all courses taken by students who began their academic tenure between 2001 and 2004.

In a typical CBE system, such as KLEF [14], *a fortiori* argumentation is used in the presentation of individual cases. This presents evidence of a strong claim in order to support a weaker claim. In terms of academic achievement, one could argue that if there is a case of a student receiving a “Fair” in PDPS and a “Good” in SE, then a student who has received a “Good” in PDPS should expect to do at least as well.

In our system, a single case takes the form of: $scenario1 \rightarrow action \rightarrow scenario2$, where a scenario is a partial assignment of state variables, and $scenario2$ occurs immediately after action, which occurs at any time after $scenario1$. In particular, we treat a single state variable assignment, followed by an action, followed by an assignment to single state variable, usually differing from the first, as a single case. For example, a student having received an A in ICP and a B in PDPS in a later semester comprises a single case with $scenario1 = \{var\_ICP = A\} \rightarrow action = take\_PDPS \rightarrow scenario2 = \{var\_PDPS = B\}$. If the same student had also taken CALC2 after having taken ICP, that would be considered a distinct case.

In general, the number of state variables used to specify a case depends on the method in which the case base is used. Two such methods of using a case base are possible: case aggregation and case matching [1]. When using case aggregation, which is better suited to smaller scenarios, the system combines all matching cases into relevant statistics in order to generate arguments. For example, case aggregation in our system would report statistics on groups of students who have taken similar courses to the current student and explain the system recommendation using the success or failure of these groups of students. When using case matching, a small number of cases, whose scenarios match the current state closely, would be selected to generate arguments [14]. Case matching methods are more suited to larger scenarios, and ideally use full state assignments [1]. For example, case matching in our system would show the user one or two students who have identical or nearly identical transcripts and explain the system recommendation using the selected students' transcripts.

Our system uses a case aggregation method, as our database does not have the required depth of coverage of our state-space. There are some states which can be reached by our MDP which have few or no cases. With a larger case base, greater specificity in argumentation is possible by considering an individual case to be the entirety of a single student's academic career. However, presenting individual cases still requires that the case base be carefully pruned to generate relevant explanations. Our system instead presents explanations based on dynamically generated statistics over all relevant cases (i.e., assignments of the variables affected by the recommended action). We select the relevant cases and compute the likelihood of a more rewarding variable assignment under a given action. This method allows more freedom to chose the action for which we present aggregated statistics; the system can pick the most convincing statistics from the set of all previous user actions instead of attempting to match individual cases.

Our method accomplishes this selection in a domain-independent way using the ordered variable assignments stored in the concept base. We use a separate configuration file, called a concept base, to store any domain specific information. We separate this data from the explanation system in order to maintain domain independence. In our system, there is a single required component of the concept base which must be defined by the system implementer; an ordering in terms of reward value over the assignments for each variable, with an extra marker for a valueless assignment that allows us to easily generate meaningful and compelling case-based explanations. The mapping could also be computed from the model on start-up, but explicitly enumerating the ordering in the concept base allows the system designer to tweak the case-based explanations in response to user preferences by reordering the values and repositioning the zero-value marker.

For a given state, s, for each variable v_i affected by $\pi(s)$, we consider the naïve distribution, $\phi(v_i)$, over the values of v_i from cases in the database. We compute the conditional distribution, $\phi(v_i|s)$, over the values of v_i given the values to all other variables in s. Then, for each conditional distribution, we examine the probability of a rewarding assignment. We then sort the distributions in order from most rewarding to least, by comparing each one to the probability of receiving the assignment from any of the naïve distributions. Conditional distributions which have increased probability of rewarding assignments over the naïve distributions are then chosen to be used for explanation.

For a student in a state such that $var\_ICP = Good$, $var\_CALC2 = Good$, and $\pi(s_e) = act\_PDPS$: since act_PDPS influences only var_PDPS, three grade distributions will be generated over its values: one distribution for all pairs with $var\_ICP = Good$, one with $var\_CALC2 = Good$, and one over all cases which have some assignment for var_PDPS. If, in the case base, 200 students had $var\_ICP = Good$ and $var\_PDPS \neq NotTaken$ with 130 "Good" assignments, 40 "Fair", and 30 "Poor", giving a [0.65, 0.20, 0.15] distribution; 150 students had $var\_CALC2 = Good$ and $var\_PDPS \neq NotTaken$ with 100 "Good", 30 "Fair", and 20 "Poor", giving a [0.67, 0.20, 0.13] distribution; while 650 students had $var\_PDPS \neq NotTaken$ with 300 "Good", 250 "Fair", and 100 "Poor", giving a [0.47, 0.38, 0.15] distribution, then the distributions indicate that such assignments increase the probability of receiving $var\_PDPS = Good$, and the generated explanation primitive is:

> Our database indicates that with either $var\_ICP = Good$ or $var\_CALC2 = Good$, you are more likely to receive $var\_PDPS = Good$ in the future.

4.3 Natural Language Generator

In explanations generated by our system, particular emphasis is placed on displaying probabilities in terms that are more comfortable to the target user base, undergraduate students. A verbal scale has some inherent problems. In medical decision making, Witterman et al. found that experienced doctors were more confident using a verbal, rather than numeric, scale [21]. Unfortunately, Renooij [16] reports large variability of the numerical values assigned to verbal expressions between subjects. However, Renooij found that there was a high level of inter-subject consistency and intra-subject consistency over time, in the ordering of such verbal expressions. Additionally, numerical interpretations of ordered lists of verbal expressions were less variable than interpretations of randomly ordered lists [16]. Thus, our explanations replace numerical probabilities with a system of intuitively ordered adverb phrases: *very likely* ($p > 0.8$), *likely* ($p > 0.5$), *unlikely* ($p < 0.5$), and *very unlikely* ($p < 0.2$). Since words at the extremes of the scale are less likely to be misinterpreted, *nearly certain* ($p > 0.95$) and *nearly impossible* ($p < 0.05$) could also be added to the scale.

Though these cutoffs work well for expressing the probabilities of state changes predicated on some action in an MDP model, they are not well suited for expressing the probability of a particular variable assignment with some underlying distribution. In this case, our system simply uses *less likely* and *more likely* for effects which cause the probability of the particular value to be less than or greater than the probability in the naïve distribution.

While MDP-based explanations can be generated in a domain-independent way, producing domain-independent natural language explanations is more problematic. The only domain semantics available from the MDP are the names of the actions, variables, and values. These labels, however, tend to be abbreviated or otherwise distorted to conform to technical limitations. Increasing the connection between the language and domain increases the user trust and relation to the system by communicating in language specific to the user [13, 17]. Our system uses a relatively simple concept base which provides mappings from variable names and assignments to noun phrases, and action names to verb phrases. This is an optional system component; the domain expert should be able to produce this semantic mapping when constructing the MDP model.

All of these mappings are stored in the concept base as optional components. The template arguments that are populated by the explanation primitives are also stored in the concept base. Each explanation module only computes the relations between variables. It is up to the interface designer to establish the mappings and exact wordings in the concept base. We allow for multiple templates and customizable text, based on state or variable assignment, to be stored in the concept base. This flexible component allows for as much or as little domain tailoring as is required by the application.

5 Discussion and Study Proposal

Our system successfully generates natural language explanations in real time using domain-independent methods, while incorporating domain specific language for the final explanation. The concept base allows designers to insert custom language as a preamble to any or all of the recommendations. This allows the user interface designer flexibility as to how much domain, modeling, and computational information to reveal to the end user.

The runtime complexity of our system, to generate an explanation for a given state, is $\mathcal{O}(n^2)$ where n is the number of actions in the MDP model. Almost all the computational burden is experienced when computing the AFDVs. These could, for very large domains, be precomputed and stored in a database if necessary. This complexity is similar to the computational requirements imposed by other MDP explanation systems [9] and is easily within the abilities of most modern systems for domains with several thousand states.

Our concept base includes text stating that recommendations depend on grades (outcomes) the student has received previously, and on the user's preferences. In many applications we expect that users do not want to know how every decision in the system is made; we are building convincing arguments for a general population, not computer scientists. While technically inclined people may want more information regarding the model construction and planning, it is our feeling that most users want to understand what they should do now. Thus, our example explanation does not explain or exhibit the entire policy. The important concept for our end users is not the mathematical structure of a policy, but that future advice will depend on current outcomes. After language substitution, the generated explanations look like:

> The recommended action is taking Introduction to Program Design and Problem Solving, generated by examining possible future courses. It is the optimal course with regards to your current grades and the courses available to you. Our model indicates that this action will best prepare you for taking Introduction to Software Engineering and taking Discrete Mathematics in the future. Additionally, it will prepare you for taking

Algorithm Design and Analysis. Our database indicates that with either a grade of A or B in Introductory Computer Programming or a grade of A or B in Calculus II, you are more likely to receive a grade of A or B in Introduction to Program Design and Problem Solving, the recommended course.

This form of explanation offers the advantage of using multiple approaches. The first statement explains the process of generating an MPD policy, enhancing the transparency of the recommendation in order to gain the trust of the user [17]. It makes clear that the planning software is considering the long-term future, which may inspire confidence in the tool. The second statement relies solely on the optimal policy and MDP model. It offers data about expected future performance in terms of the improvement in value of possible future actions, the AFDVs. The AFDVs are computed using an optimal policy. That means the policy maximizes expected, long term reward. This part of the explanation focuses on the near future to explain actions which may only be preferable because of far future consequences. The shift in focuses leverages the users inherent bias towards hyperbolic discounting of future rewards [6]. The last statement focuses on the student's past performance in order to predict performance at the current time step and explains that performance in terms of variable assignments. This paragraph makes an analogy between the user's performance and the aggregated performance of past students. Argument from analogy is very relevant to our domain — academic advisors often suggest, for example, that advisees talk to students who have taken the course from a particular professor. Additionally, the case-based explanation module can be adapted to take into account user preferences, and therefore make more precise analogies.

User Study. We have recently received institutional approval for a large, multi-staged user study. We informally piloted the system with computer science students at our university, but this informal test fails to address the real issues surrounding user interfaces. Our study will use students from disciplines including psychology, computer science, and electrical engineering, and advisors from these disciplines. We will compare the advice generated by our system and its "most convincing" approach to other systems which use a "most coverage" (with respect to rewards) approach. We will survey both students and advisors to find what, if any, difference exists between these two approaches. We will also test differences in framing advice in positive and negative lights. There is extensive literature about the effects of goal framing on choice and we hope to leverage this idea to make our recommendations more convincing [19].

By approaching a user study from both the experts' and users' viewpoints we will learn about what makes good advice in this domain and what makes convincing arguments in many more domains. A full treatment of this study, including pilot study, methodology, instrument development, and data analysis will fill another complete paper. We did not want to present a token user study. Quality evaluation methods must become the standard for, and not the exception to, systems that interact with non-expert users such as the one developed here.

6 Conclusion and Future Work

In this work we have presented a system and design which generates natural language explanations for actions generated by MDPs. This system uses a novel mix of case-based and model-based techniques to generate highly salient explanations. The system design abstracts the domain dependent knowledge from the explanation system, allowing it to be ported to other domains with minimal work by the domain expert. The generated explanations are grounded both psychologically and mathematically for maximum impact, clarity, and correctness. The system operates in real time and is scalable based on the amount of domain specific information available.

Automatic planning and scheduling tools generate recommendations that are often not followed by end users. As computer recommendations integrate deeper into everyday life it becomes imperative that we, as computer scientists, understand why and how users implement recommendations generated by our systems. The framework here starts to bridge the gap between mathematical fundamentals and user expectations.

Our current model recommends one course at a time. We will be expanding the system to include multiple actions per time step. This requires a planner that can handle factored actions, and requires that we adjust the explanation interface. We expect that explanations will consist of three parts, not necessarily all present in each response. The first will answer the question, "Why this particular course/atomic action?" The second will answer, "Why these two/few courses/atomic actions together?" And the third will look at the entire set. Answers to the first type of query will be very similar to what is described here, but will take into account whether the effects are on simultaneous or future courses. Answers to the second type will build directly on the information generated to answer the first type. We expect that answers to "Why this set of courses" will depend on the constraints given on sets of courses/atomic actions, such as "You are only allowed to take 21 credits per semester, and your transcript indicates that you/people with records like yours do best with about 15 per semester."

Our model based module extracts information from the MDP model and a policy of recommended actions on that model. Finding optimal policies for factored MDPs is PSPACE-hard [12]. We assumed, in the development of this system, that the optimal policy is available. Given a heuristic policy, our system will generate consistent explanations, but they will not necessarily be as convincing. We would like to extend our work and improve the argument interface when only heuristic policies are available.

Acknowledgements. This work is partially supported by NSF EAGER grant CCF-1049360. We would like to thank the members of the UK-AILab, especially Robert Crawford, Joshua Guerin, Daniel Michler, and Matthew Spradling for their support and helpful discussions. We are also grateful to the anonymous reviewers who have made many helpful recommendations for the improvement of this paper.

References

1. Aamodt, A., Plaza, E.: Case-based reasoning: Foundational issues, methodological variations, and system approaches. AI communications 7(1), 39–59 (1994)
2. Bellman, R.: Dynamic Programming. Princeton University Press, Princeton (1957)
3. Boutilier, C., Dean, T., Hanks, S.: Decision-theoretic planning: Structural assumptions and computational leverage. Journal of Artificial Intellgence Research 11, 1–94 (1999)
4. Camara, W.J., Echternacht, G.: The SAT I and high school grades: utility in predicting success in college. RN-10, College Entrance Examination Board, New York (2000)

5. Elizalde, F., Sucar, E., Noguez, J., Reyes, A.: Generating explanations based on markov decision processes. In: Aguirre, A.H., Borja, R.M., Garciá, C.A.R. (eds.) MICAI 2009. LNCS, vol. 5845, pp. 51–62. Springer, Heidelberg (2009)
6. Frederick, S., Loewenstein, G., O'Donoghue, T.: Time discounting and time preference: A critical review. Journal of Economic Literature 40, 351–401 (2002)
7. Guerin, J.T., Crawford, R., Goldsmith, J.: Constructing dynamic bayes nets using recommendation techniques from collaborative filtering. Tech report, University of Kentucky (2010)
8. Hoey, J., St-Aubin, R., Hu, A., Boutilier, C.: SPUDD: Stochastic planning using decision diagrams. In: Proc. UAI, pp. 279–288 (1999)
9. Khan, O., Poupart, P., Black, J.: Minimal sufficient explanations for factored Markov decision processes. In: Proc. ICAPS (2009)
10. Mathias, K., Williams, D., Cornett, A., Dekhtyar, A., Goldsmith, J.: Factored mdp elicitation and plan display. In: Proc. ISDN. AAAI, Menlo Park (2006)
11. Moore, B., Parker, R.: Critical Thinking. McGraw-Hill, New York (2008)
12. Mundhenk, M., Lusena, C., Goldsmith, J., Allender, E.: The complexity of finite-horizon Markov decision process problems. JACM 47(4), 681–720 (2000)
13. Murray, K., Häubl, G.: Interactive consumer decision aids. In: Wierenga, B. (ed.) Handbook of Marketing Decision Models, pp. 55–77. Springer, Heidelberg (2008)
14. Nugent, C., Doyle, D., Cunningham, P.: Gaining insight through case-based explanation. JIIS 32, 267–295 (2009)
15. Puterman, M.: Markov Decision Processes. Wiley, Chichester (1994)
16. Renooij, S.: Qualitative Approaches to Quantifying Probabilistic Networks. Ph.D. thesis, Institute for Information and Computing Sciences, Utrecht University, The Netherlands (2001)
17. Sinha, R., Swearingen, K.: The role of transparency in recommender systems. In: CHI 2002 Conference Companion, pp. 830–831 (2002)
18. Tversky, A., Kahneman, D.: Judgement under uncertainty: Heuristics and biases. Science 185, 1124–1131 (1974)
19. Tversky, A., Kahneman, D.: Rational choice and the framing of decisions. The Journal of Business 59(4), 251–278 (1986)
20. Tversky, A., Kahneman, D.: Advances in prospect theory: Cumulative representation of uncertainty. Journal of Risk and uncertainty 5(4), 297–323 (1992)
21. Witteman, C., Renooij, S., Koele, P.: Medicine in words and numbers: A cross-sectional survey comparing probability assessment scales. BMC Med. Informatics and Decision Making 7(13) (2007)

A Bi-objective Optimization Model to Eliciting Decision Maker's Preferences for the PROMETHEE II Method

Stefan Eppe, Yves De Smet, and Thomas Stützle

Computer & Decision Engineering (CoDE) Department
Université Libre de Bruxelles, Belgium (ULB)
{stefan.eppe,yves.de.smet,stuetzle}@ulb.ac.be

Abstract. Eliciting the preferences of a decision maker is a crucial step when applying multi-criteria decision aid methods on real applications. Yet it remains an open research question, especially in the context of the PROMETHEE methods. In this paper, we propose a bi-objective optimization model to tackle the preference elicitation problem. Its main advantage over the widely spread linear programming methods (traditionally proposed to address this question) is the simultaneous optimization of (1) the number of inconsistencies and (2) the robustness of the parameter values. We experimentally study our method for inferring the PROMETHEE II preference parameters using the NSGA-II evolutionary multi-objective optimization algorithm. Results obtained on artificial datasets suggest that our method offers promising new perspectives in that field of research.

1 Introduction

To solve a multi-criteria decision aid problem, the preferences of a decision maker (DM) have to be formally represented by means of a model and its *preference parameters* (PP) [13]. Due to the often encountered difficulty for decision makers to provide values for these parameters, methods for inferring PP's have been developed over the years [1,3,9,10].

In this paper, we follow the aggregation/disaggregation approach [11] for preference elicitation: given a set A of actions, the DM is asked to provide holistic information about his preferences. She states her overall preference of one action over another rather than giving information at the preference parameter level, since the former seems to be a cognitively easier task. The inference of the decision maker's (DM) preferences is a crucial step of multi-criteria decision aid, having great practical implications on the use of a particular MCDA method. In this paper, we work with the PROMETHEE outranking method. To the best of our knowledge, only few works on preference elicitation exist for that method. Frikha *et al.* [8] propose a method for determining the criteria's relative weights. They consider two sets of partial information provided by the DM: (i) ordinal preference between two actions, and (ii) a ranking of the relative weights. These

R.I. Brafman, F. Roberts, and A. Tsoukiàs (Eds.): ADT 2011, LNAI 6992, pp. 56–66, 2011.

are formalized as constraints of a first linear program (LP) that may admit multiple solutions. Then, for each criterion independently, an interval of weights that satisfies the first set of constraints is determined. Finally, a second LP is applied on the set of weight intervals to reduce the number of violations of the weights' partial pre-order constraint. Sun and Han [14] propose a similar approach that also limits itself to determine the weights of the PROMETHEE preference parameters. These, too, are determined by resolving an LP. Finally, Özerol and Karasakal [12] present three interactive ways of eliciting the parameters of the PROMETHEE preference model for PROMETHEE I and II.

Although most methods for inferring a DM's preferences found in the MCDA literature are based on the resolution of linear programs [1,10], some recent works also explore the use of meta-heuristics to tackle that problem [4]. In particular, [6] uses the NSGA-II evolutionary multi-objective optimization (EMO) algorithm to elicit ELECTRE III preference parameters in the context of sorting problems.

The goal of this work is to contribute to exploring the possible use of multi-objective optimization heuristics to elicit a decision maker's preferences for the PROMETHEE II outranking method. In addition to minimizing the constraint violations induced by a set of preference parameters (PP), we consider robustness of the elicited PP's as a second objective. The experimental setup is described in detail in Sec. 2.

Before going further in the description of our experimental setup, let us define the notation used in the following. We consider a set $A = \{a_1, \ldots, a_n\}$ of $n = |A|$ potential actions to be evaluated over a set of m conflicting criteria. Each action is evaluated on a given criterion by means of an evaluation function $f_h : A \to \mathbb{R} : a \to f_h(a)$. Let $F(a) = \{f_1(a), \ldots, f_m(a)\}$ be the evaluation vector associated to action $a \in A$.

Let Ω be the set of all possible PP sets and let $\omega \in \Omega$ be one particular PP set. Asking a DM to provide (partial) information about his preferences is equivalent to setting constraints on Ω[1], each DM's statement resulting in a constraint. We denote $C = \{c_1, \ldots, c_k\}$ the set of k constraints.

In this paper, we focus on the PROMETHEE II outranking method [2], which provides the DM with a complete ranking over the set A of potential actions. The method defines the net flow $\Phi(a)$ associated to action $a \in A$ as follows:

$$\Phi(a) = \frac{1}{n-1} \sum_{b \in A \setminus a} \sum_{h=1}^{m} w_h \left(P_h(a,b) - P_h(b,a) \right),$$

where w_h and $P_h(a,b)$ are respectively the relative weight and the preference function (Fig. 1) for criteria $h \in \{1, \ldots, m\}$. For any pair of actions $(a,b) \in A \times A$, we have one of the following relations: (i) the rank of action a is better than

[1] The constraint can be direct or indirect, depending on the type of information provided. Direct constraints will have an explicit effect on the preference model's possible parameter values (e.g., *the relative weight of the first criterion is greater than* $\frac{1}{2}$), while indirect constraints will have an impact on the domain Ω (e.g., *the first action is better than the fifth one*).

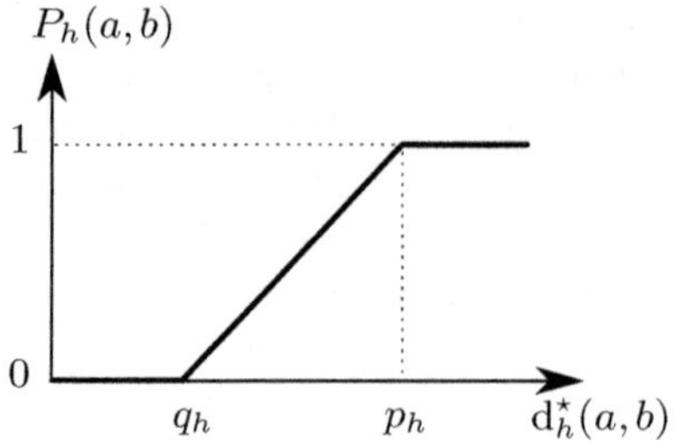

Fig. 1. Shape of a PROMETHEE preference function type V, requiring the user to define, for each objective h, a indifference threshold q_h, and a preference threshold p_h. We have chosen to slightly modify the original definition of the preference function, replacing the difference $d_h(a,b) = f_h(a) - f_h(b)$, by a relative difference, defined as follows: $d^\star_h(a,b) = \frac{f_h(a)-f_h(b)}{\frac{1}{2}(f_h(a)+f_h(b))}$, i.e., we divide the difference by the mean value of both evaluations. For $d^\star_h(a,b) \in [0, q_h]$, both solutions a and b are considered indifferently; for a relative difference greater than p_h, a strict preference (with value 1) of a over b is stated. Between the two thresholds, the preference evolves linearly with increasing evaluation difference.

the rank of action b, iff $\Phi(a) > \Phi(b)$; (ii) the rank of action b is better than the rank of action a, iff $\Phi(a) < \Phi(b)$; (iii) action a has the same rank as action b, iff $\Phi(a) = \Phi(b)$. Although six different types of preference functions are proposed [2], we will limit ourselves to the use of a relative version of the "*V-shape*" preference function $P : A \times A \rightarrow \mathbb{R}_{[0,1]}$ (Fig. 1). For the sake of ease, we will sometimes write the PROMETHEE II specific parameters explicitly: $\omega = \{w_1, q_1, p_1, \ldots, w_m, q_m, p_m\}$, where w_h, q_h, and p_h are respectively the relative weight, the indifference threshold, and the preference threshold associated to criterion $h \in \{1, \ldots, m\}$. The preference parameters have to satisfy the following constraints: $w_h \geq 0, \forall h \in \{1, \ldots, m\}$, $\sum_{h=1}^{m} w_h = 1$, and $0 \leq q_h \leq p_h, \forall\ h \in \{1, \ldots, m\}$.

2 Experimental Setup

The work flow of our experimental study is schematically represented in Fig. 2: (1) For a given set of actions A, a reference preference parameter set ω_{ref} is chosen. (2) Based on A and ω_{ref}, in turn a set $C = \{c_1, \ldots, c_k\}$ of constraints is generated. A fraction p_{cv} of the constraints will be incompatible with ω_{ref}, in order to simulate inconsistencies in the information provided by the DM. (3) By means of an evolutionary multi-objective algorithm, the constraints are then used to optimize a population of parameter sets on two objectives: constraint violation and robustness. (4) The obtained population of parameter sets is clustered. (5) The clusters are analysed and compared with ω_{ref}. In the following paragraphs, we explain in more detail the different components of the proposed approach.

We consider non-dominated **action sets** of constant size (100 actions), ranging from 2 to 5 objectives. The use of non-dominated actions seems intuitively

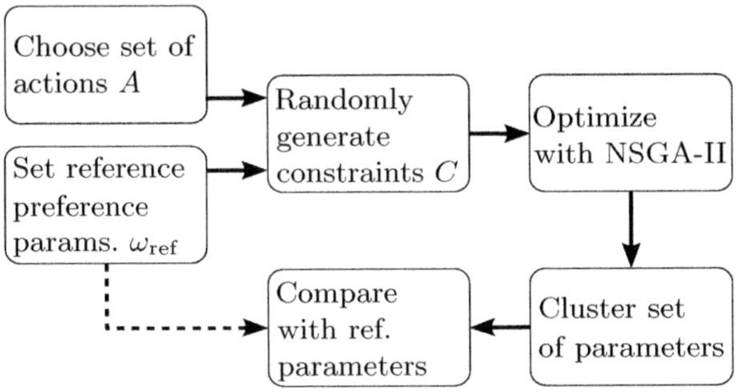

Fig. 2. The work flow of our experimental study

meaningful, but the impact of that choice on the elicitation process should be further investigated, since it does not necessarily correspond to real-life conditions. For our convenience, we have used approximations of the Pareto optimal frontier of multi-objective TSP instances that we already had.[2] Nevertheless, the results presented in the following are in no way related to the TSP.

The **reference preference parameters** ω_{ref} are chosen manually for this approach in order to be representative and allow us to draw some conclusions. We perform the optimization process exclusively on the weight parameters $\{w_1, \ldots, w_m\}$. Unless otherwise specified, we use the following values for the relative thresholds: $q_h = 0.02$ and $p_h = 0.10$, $\forall h \in \{1, \ldots, m\}$. This means that the indifference threshold for all criteria is considered as 2% of the relative difference of two action's evaluations (Fig. 1). The preference threshold is similarly set to 10% of the relative distance.

We will consider **constraints** of the following form: $\Phi(a) - \Phi(b) > \Delta$, where $(a, b) \in A \times A$ and $\Delta \geq 0$.

Constraints on the threshold parameters q_h and p_h, $\forall h \in \{1 \ldots m\}$ have not been considered in this work. We could address this issue in a future paper (e.g. stating that the indifference threshold of the third criterion has to be higher than a given value: $q_3 > 0.2$).

We have chosen to **randomly generate** a given number of **constraints** that will be consistent (i.e., compatible) with the reference preference parameters ω_{ref}. More specifically, given ω_{ref} and the action set A, the net flow $\Phi_{\omega_{\text{ref}}}(a)$ of each action $a \in A$ is computed. Two distinct actions a and b are randomly chosen and a constraint is generated on their basis, that is compatible with their respective net flow values $\Phi_{\omega_{\text{ref}}}(a)$ and $\Phi_{\omega_{\text{ref}}}(b)$. For instance, if $\Phi_{\omega_{\text{ref}}}(a) > \Phi_{\omega_{\text{ref}}}(b)$, the corresponding compatible constraint will be given by $\Phi(a) > \Phi(b)$. A fraction of incompatible constraints will also be generated, with a probability that is defined by the parameter p_{cv}. For these, the previous inequality becomes $\Phi(a) < \Phi(b)$ (for $\Phi_{\omega_{\text{ref}}}(a) > \Phi_{\omega_{\text{ref}}}(b)$).

[2] We have taken solution sets of multi-objective TSP instances from [5], available on-line at `http://iridia.ulb.ac.be/supp/IridiaSupp2011-006`

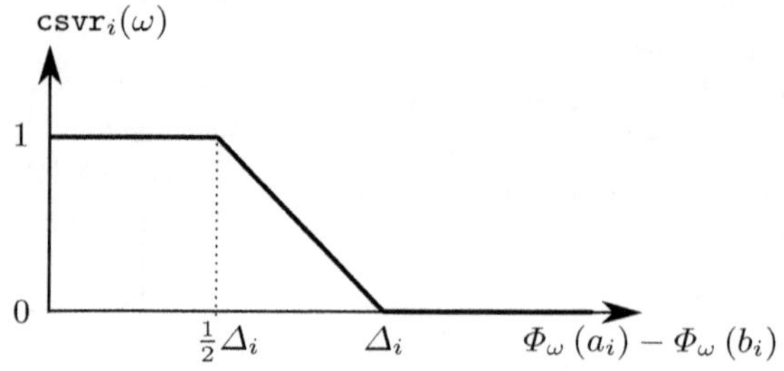

Fig. 3. Shape of the constraint violation rate function $\texttt{csvr}_i(\omega)$ associated with a given constraint $c_i \in C$ and a set of preference parameters ω. The constraint c_i expresses the inequality $\Phi_\omega(a_i) - \Phi_\omega(b_i) \geq \Delta_i$, linking together actions a_i and $b_i \in A$.

As already mentioned, we take a bi-objective optimization point of view on the preference elicitation problem. We hereafter define the objectives that we will consider for the optimization process.

Global constraint violation rate (csvr). Each constraint $c_i \in C$, with $i \in \{1 \ldots k\}$, expresses an inequality relation between a pair of actions $(a_i, b_i) \in A \times A$ by means of a minimal difference parameter Δ_i. We define the *violation rate* of the i-th constraint as follows (Fig. 3):

$$\texttt{csvr}_i(\omega) = \zeta\left(\frac{\Delta_i - (\Phi_\omega(a_i) - \Phi_\omega(b_i))}{\frac{1}{2}\Delta_i}\right),$$

where $\zeta(x) = \min(1, \max(0, x))$ is a help function that restrains the values of its argument x to the interval $[0, 1]$.
Finally, the set of measures is aggregated on all constraints to compute a global constraint representing the average violation rate:

$$\texttt{csvr}(\omega) = \frac{1}{k}\sum_{i=1}^{k}\texttt{csvr}_i(\omega)$$

Example. Let us consider the first constraint given by $\Phi(a_1) - \Phi(a_4) \geq 0.3$. We thus have $\Delta_1 = 0.3$. Let the pair of actual net flows (that directly depend on the associated preference parameter set ω_1) be as follows:
$\Phi_{\omega_1}(a_1) = 0.2$ and $\Phi_{\omega_1}(a_4) = -0.1$. Considering only one constraint for the sake of simplicity, the global constraint violation rate becomes

$$\texttt{csvr}(\omega) = \texttt{csvr}_1(\omega) = \zeta\left(\frac{0.3-(0.2-(-0.1))}{\frac{1}{2}0.3}\right) = 0.$$

For the given parameter set ω, the constraint is thus fully satisfied.

Promethee II sampled sensitivity (p2ss). Given a preference parameter set ω, we compute its p2ss value by sampling a given number N_{p2ss} of parameter sets $\omega^{\star s} = \{\omega_1^{\star s}, \ldots, \omega_m^{\star s}\}$, $\forall s \in \{1, \ldots, N_{\texttt{p2ss}}\}$ "around" ω. Practically, we take $N_{\texttt{p2ss}} = 10$ and we generate each parameter $\omega_j^{\star s}$, with $j \in \{1, \ldots, m\}$, of

Table 1. Parameter values used for the NSGA-II algorithm

Parameter		Value(s)
Population size	n_{pop}	50
Termination condition	t_{max}	120 sec
Probability of cross-over	p_{xover}	0.8
Probability of mutation	p_{mut}	0.2

the sample by randomly evaluating a normally distributed stochastic variable that is centred on the value ω_j and has a relative standard deviation of 10%: $\omega_j^{\star s} \sim \mathcal{N}(\omega_j, (\frac{\omega_j}{10})^2)$.

We define the sensitivity as the square root of the average square distance to the reference constraint violation $\texttt{csvr}(\omega)$:

$$\texttt{p2ss}(\omega) = \sqrt{\frac{1}{N_{\texttt{p2ss}}} \sum_{s=1}^{N_{\texttt{p2ss}}} \left(\texttt{csvr}(\omega^{\star s}) - \texttt{csvr}(\omega)\right)^2}$$

As some first results have shown that the resulting set of preference parameters presented a clustered structure of sub-sets, we have decided to apply a **clustering procedure** (with regard to the weight parameters) on the set of results. Practically, we use the `pamk` function of `R`'s 'fpc' package, performing a partitioning around medoids clustering with the number of clusters estimated by optimum average silhouette width.

Finally, we compare the obtained results, i.e., a set of preference parameter sets, with the reference parameter set ω_{ref}. The quality of each solution is quantified by means of the following fitness measure:

Correlation with the reference ranking τ_K. We use Kendall's τ to measure the distance between a ranking induced by a parameter set ω_i and the reference parameter set ω_{ref}.

3 Results

The aim of the tests that are described below is to provide some global insight into the behaviour of the proposed approach. Further investigations should be carried out in order to gain better knowledge, both on a larger set of randomly generated instances and on real case studies. In the following, main parameters of the experimental setup are systematically tested. We assume that the parameters of the tests, i.e., instance size, number of objectives, number of constraints, etc., are independent from each other, so that we can study the impact each of them has on the results of the proposed model. The values used for the parameters of the experiments are given in Table 2. In the following, we only present the most noticeable results.

Figure 4 shows the effect of changing the **proportion of incompatible constraints** with respect to the total number of constraints. As expected, higher

Table 2. This table provides the parameter values used for the experiments. For each parameter, the value in bold represents its default value, i.e., the value that is taken in the experiments, if no other is explicitly mentioned.

Parameter		Value(s)
Size of the action set	n	**100**
Number of criteria of the action set	m	**2**, 3, 4, 5
Number of constraints	k	2, 10, **20**, 30, 40, 50
Constraint violation rate	p_{cv}	0, 0.05, 0.10, **0.20**, 0.30
Scalar weight parameter	w	0.10, 0.20, 0.30, 0.40, 0.50

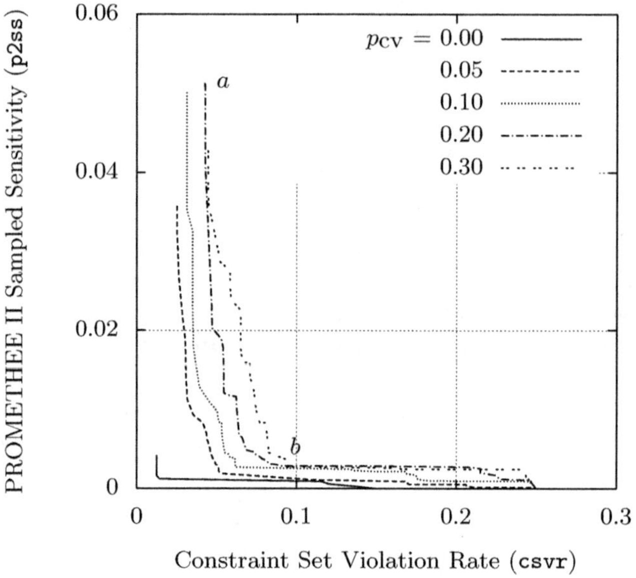

Fig. 4. This plot represents the approximated Pareto frontiers in the objective space, for 20 constraints and several values of the constraint violation rate p_{cv}, i.e., the proportion of inconsistent constraints with respect to the total number of constraints. As expected, increasing the value of p_{cv} has the effect of deteriorating the quality of the solution set both in terms of constraint violation rate and PROMETHEE II sampled sensitivity.

values of the constraint incompatibility ratios induce worse results on both objectives (`csvr` and `p2ss`). Thus, the more consistent the information provided by the decision maker, the higher the possibility for the algorithm to reach stable sets of parameters that do respect the constraints.[3] The second and more noteworthy

[3] We investigate the impact of inconsistencies in partial preferential information provided by the DM. We would like to stress that the way we randomly generate inconsistent constraints (with respect to the reference preference parameters ω_{ref}) induces a specific type of inconsistencies. Other types should be studied in more depth in a future work.

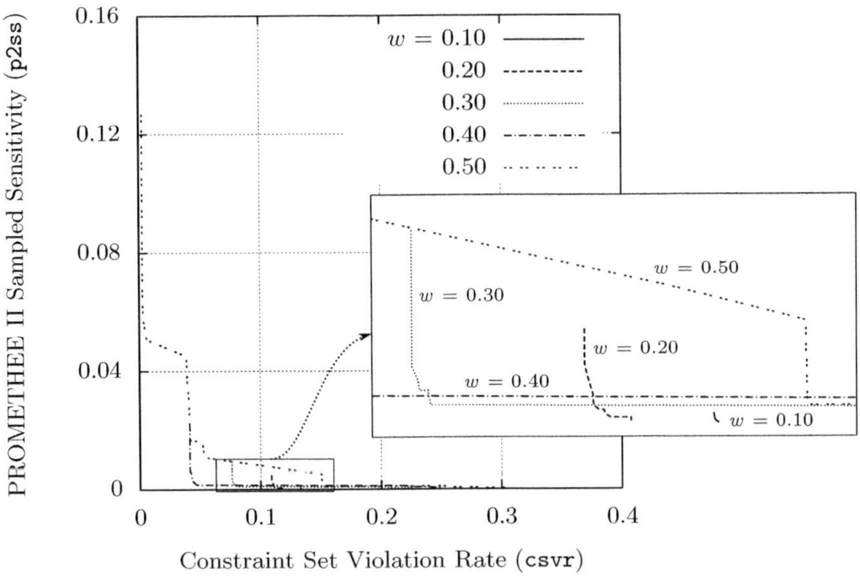

Fig. 5. Approximations of the Pareto optimal frontier are shown for different values of the reference weight parameters $w = \{0.1, \ldots, 0.5\}$ for an action set with two criteria. The weights of the reference preference model ω_{ref} are given by $w_1 = w$ and $w_2 = 1 - w$.

observation that can be made on that plot is related to the advantage of using a multi-objective optimization approach for the elicitation problem. Indeed, as can be seen, optimizing only the constraint violation rate (`csvr`) would have led to solutions with comparatively poor performances with regard to sensitivity (area marked with an *a* on the plot). This would imply that small changes to `csvr`-well-performing preference parameters might induce important alteration of the constraint violation rate. However, due to the steepness of the approximated Pareto frontier for low values of `csvr` the DM is able to select much more robust solutions at a relatively small cost on the `csvr` objective (area *b*).

For action sets that are evaluated on two criteria[4], we also observe the effects of **varying the value of the weight preference parameter** w, where $w_1 = w$ and $w_2 = 1 - w$. As shown in Fig. 5, the underlying weight parameter w has an impact on the quality of the resulting Pareto set of approximations. It suggests that the achievable quality for each objective (i.e., `csvr` and `p2ss`) is related to the "distance" from an equally weighted set of criteria ($w = 0.5$): lowering the values of w makes it harder for the algorithm to optimize on the constraint violation objective `csvr`. On the other hand, having a underlying preference model with a low value of w seems to decrease the sampled sensitivity `p2ss`, making the model more robust to changes on parameter values. It should be noted that for $w = 0.5$ there appears to be an exception in the central area of the Pareto frontier. This effect has not been studied yet.

[4] Similar results have been observed for higher number of criteria.

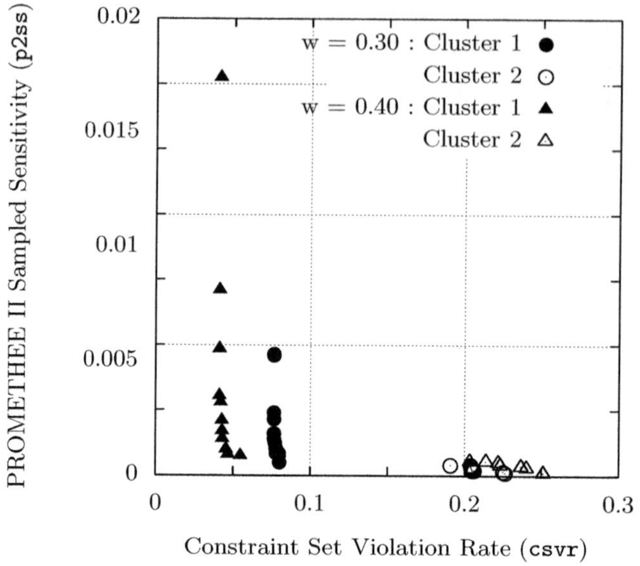

Fig. 6. Results of the clustering applied on two different reference parameter sets (for a action set with two criteria), that are characterized by respective weight parameters $w = 0.30$ and 0.40. For each set, 2 clusters have been automatically identified. The proximity of the centroid parameter set of each cluster to the reference parameter set is measured by means of Kendall's τ (Fig. 7) to compare the clusters for each weight parameter. The filled symbol (cluster 1) corresponds to the better cluster, i.e., the one that best fits the reference weights.

In this experimental study, we will compare the obtained results with the reference parameters ω_{ref}. To that purpose, we partition the set of obtained preference parameters based on their weights into a reduced number of **clusters**. The clustering is thus performed in the solution space (on the weights) and represented in the objective space (`csvr` - `p2ss`). Figure 6 shows the partition of the resulting set for a specific instance, for two different weights of the reference preference parameters: (1) $\omega_{\text{ref}} = 0.30$ and (2) $\omega_{\text{ref}} = 0.40$. Both cases suggest that there is a strong relationship between ω_{ref} and the objective values (`csvr` and `p2ss`). Indeed, in each case, two separated clusters are detected: cluster 1, with elements that are characterized by relatively small `csvr` values and a relatively large dispersion of `p2ss` values; cluster 2, with elements that have relatively small `p2ss` values and a relatively higher dispersion of `csvr` values. In both cases, too, the centroid associated to cluster 1 has a weight vector that is closer, based on an Euclidean distance, the the weight vector of ω_{ref} than the centroid of cluster 2.

Although this has to be verified through more extensive tests, this result could suggest a reasonable criterion for deciding which cluster to choose from the set of clusters, and therefore provide the DM with a sensible set of parameters that is associated to that cluster.

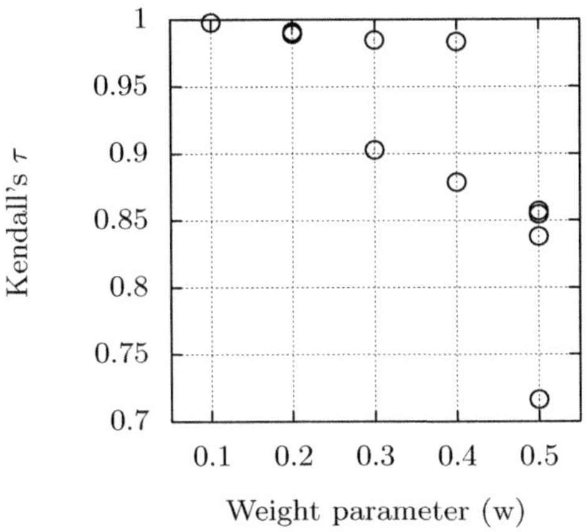

Fig. 7. Kendall's τ represented for different reference parameter weights $w \in \{0.1, \ldots, 0.5\}$. For each weight, the mean values of all clusters are shown. For $w = 0.30$, for instance, the upper circle represents the first (best) cluster, and the lower one represents the other cluster of one same solution set.

Finally, in order to assess the quality of the result with respect to the reference parameter set, we plot (Fig. 7) the values of Kendall's τ for each cluster that has been determined for a range of reference weight parameters $w \in \{0.1, 0.2, 0.3, 0.4, 0.5\}$. For each weight w, we plot Kendall's τ for each cluster's medoid (compared to the reference parameter set ω_{ref}). We first observe that we have between 2 and 6 clusters depending on the considered weight. Although the results worsen (slightly, except for $w = 0.5$), the best values that correspond to the previously identified "best" clusters remain very high: The ranking induced by the reference parameter set are reproduced to a large extent. These results are encouraging further investigations, because they tend to show that our approach converges to good results (which should still be quantitatively measured by comparing with other existing methods).

4 Conclusion

Eliciting DM's preferences is a crucial step of multi-criteria decision aid that is commonly tackled in the MCDA community by solving linear problems. As some other recent papers, we explore an alternative bi-objective optimization based approach to solve it. Its main distinctive feature is to explicitly integrate the sensitivity of the solution as an objective to be optimized. Although this aspect has not been explored yet, our approach should also be able, without any change, to integrate constraints that are more complicated than linear ones.

Finally, and although we have focused on the PROMETHEE II outranking method in this paper, we believe that the approach could potentially be extended to a wider range of MCDA methodologies.

Future directions for this work should include a more in-depth analysis of our approach, as well as an extension to real, interactive elicitation procedures. A further goal could also be to determine additional objectives that would allow eliciting the threshold values of the PROMETHEE preference model. Finally, investigating other ways of expressing robustness would probably yield interesting new paths for the future.

Acknowledgments. Stefan Eppe acknowledges support from the META-X Arc project, funded by the Scientific Research Directorate of the French Community of Belgium.

References

1. Bous, G., Fortemps, P., Glineur, F., Pirlot, M.: ACUTA: A novel method for eliciting additive value functions on the basis of holistic preference statements. European J. Oper. Res. 206(2), 435–444 (2010)
2. Brans, J.P., Mareschal, B.: PROMETHEE methods. In: [7], ch. 5, pp. 163–195
3. Dias, L., Mousseau, V., Figueira, J.R., Clímaco, J.: An aggregation/disaggregation approach to obtain robust conclusions with ELECTRE TRI. European J. Oper. Res. 138(2), 332–348 (2002)
4. Doumpos, M., Zopounidis, C.: Preference disaggregation and statistical learning for multicriteria decision support: A review. European J. Oper. Res. 209(3), 203–214 (2011)
5. Eppe, S., López-Ibáñez, M., Stützle, T., De Smet, Y.: An experimental study of preference model integration into multi-objective optimization heuristics. In: Proceedings of the 2011 Congress on Evolutionary Computation (CEC 2011), IEEE Press, Piscataway (2011)
6. Fernandez, E., Navarro, J., Bernal, S.: Multicriteria sorting using a valued indifference relation under a preference disaggregation paradigm. European J. Oper. Res. 198(2), 602–609 (2009)
7. Figueira, J.R., Greco, S., Ehrgott, M. (eds.): Multiple Criteria Decision Analysis, State of the Art Surveys. Springer, Heidelberg (2005)
8. Frikha, H., Chabchoub, H., Martel, J.M.: Inferring criteria's relative importance coefficients in PROMETHEE II. IJOR Int. J. Oper. Res. 7(2), 257–275 (2010)
9. Greco, S., Kadzinski, M., Mousseau, V., Słowiński, R.: ELECTRE^{GKMS}: Robust ordinal regression for outranking methods. European J. Oper. Res. 214(1), 118–135 (2011)
10. Mousseau, V.: Elicitation des prfrences pour l'aide multicritre la dcision. Ph.D. thesis, Université Paris-Dauphine, Paris, France (2003)
11. Mousseau, V., Słowiński, R.: Inferring an ELECTRE TRI model from assignment examples. J. Global Optim. 12(2), 157–174 (1998)
12. Özerol, G., Karasakal, E.: Interactive outranking approaches for multicriteria decision-making problems with imprecise information. JORS 59, 1253–1268 (2007)
13. Öztürk, M., Tsoukiàs, A., Vincke, P.: Preference modelling. In: [7], ch. 2, pp. 27–72
14. Sun, Z., Han, M.: Multi-criteria decision making based on PROMETHEE method. In: Proceedings of the 2010 International Conference on Computing, Control and Industrial Engineering, pp. 416–418. IEEE Computer Society Press, Los Alamitos (2010)

Strategy-Proof Mechanisms for Facility Location Games with Many Facilities*

Bruno Escoffier[1], Laurent Gourvès[1], Nguyen Kim Thang[1], Fanny Pascual[2], and Olivier Spanjaard[2]

[1] Université Paris-Dauphine, LAMSADE-CNRS, UMR 7243, F-75775 Paris, France
[2] UPMC, LIP6-CNRS, UMR 7606, F-75005 Paris, France
{bruno.escoffier,laurent.gourves,kim-thang.nguyen}@lamsade.dauphine.fr,
{fanny.pascual,olivier.spanjaard}@lip6.fr

Abstract. This paper is devoted to the location of public facilities in a metric space. Selfish agents are located in this metric space, and their aim is to minimize their *own* cost, which is the distance from their location to the nearest facility. A central authority has to locate the facilities in the space, but she is ignorant of the true locations of the agents. The agents will therefore report their locations, but they may lie if they have an incentive to do it. We consider two social costs in this paper: the sum of the distances of the agents to their nearest facility, or the maximal distance of an agent to her nearest facility. We are interested in designing *strategy-proof* mechanisms that have a small approximation ratio for the considered social cost. A mechanism is strategy-proof if no agent has an incentive to report false information. In this paper, we design strategy-proof mechanisms to locate $n-1$ facilities for n agents. We study this problem in the general metric and in the tree metric spaces. We provide lower and upper bounds on the approximation ratio of deterministic and randomized strategy-proof mechanisms.

Keywords: Facility location games, Strategy-proof mechanisms, Approximation guarantee.

1 Introduction

We study *Facility Location Games* that model the following problem in economics. Consider installation of public service facilities such as hospitals or libraries within the region of a city, represented by a metric space. The authority announces that some locations will be chosen within the region and runs a survey over the population; each inhabitant may declare the spot in the region that she prefers some facility to be opened at. Every inhabitant wishes to minimize her individual distance to the closest facility, possibly by misreporting her preference to the authorities. The goals of the authority are twofold: avoiding such

* This work is supported by French National Agency (ANR), project COCA ANR-09-JCJC-0066-01.

R.I. Brafman, F. Roberts, and A. Tsoukiàs (Eds.): ADT 2011, LNAI 6992, pp. 67–81, 2011.

misreports and minimizing some social objectives. The authority needs to design a *mechanism*, that maps the reported preferences of inhabitants to a set of locations where the facilities will be opened at, to fulfill the purposes. The mechanism must be *strategy-proof*, i.e., it ensures that no inhabitant can benefit by misreporting her preference. At the same time, the mechanism should guarantee a reasonable approximation to the optimal social cost. The model has many applications in telecommunication networks where locations may be easily manipulated by reporting false IP addresses, false routers, etc.

1.1 Facility Location Games

We consider a metric space (Ω, d), where $d : \Omega \times \Omega \rightarrow \mathbb{R}$ is the metric function. Some usual metrics are the line, circle and tree metrics where the underlying spaces are an infinite line, a circle and an infinite tree, respectively. The distance between two positions in such metrics is the length of the shortest path connecting those positions. Let n be the number of agents, each agent i has a location $x_i \in \Omega$. A *location profile (or strategy profile)* is a vector $\mathbf{x} = (x_1, \ldots, x_n) \in \Omega^n$. Let k be the number of facilities that will be opened. A *deterministic* mechanism is a mapping f from the set of location profiles Ω^n to k locations in Ω. Given a reported location profile $\mathbf{x}$ the mechanism's output is $f(\mathbf{x}) \in \Omega^k$ and the individual cost of agent i under mechanism f and profile $\mathbf{x}$ is the distance from its location to the closest facility, denoted by $c_i(f, \mathbf{x})$:

$$c_i(f, \mathbf{x}) := d(f(\mathbf{x}), x_i) := \min\{d(F, x_i) : F \in f(\mathbf{x})\}$$

A *randomized* mechanism is a function f from the set of location profiles to $\Delta(\Omega^k)$ where $\Delta(\Omega^k)$ is the set of probability distributions over Ω^k. The cost of agent i is now the expected distance from its location to the closest facility over such distribution:

$$c_i(f, \mathbf{x}) := \mathbb{E}\left[d(f(\mathbf{x}), x_i)\right] := \mathbb{E}\left[\min\{d(F, x_i) : F \in f(\mathbf{x})\}\right]$$

We are interested in two standard *social objectives*: (i) the *utilitarian objective* defined as the total individual costs (total individual expected cost for a randomized mechanism), i.e., $C(f, \mathbf{x}) = \sum_{i=1}^{n} c_i(f, \mathbf{x})$; and (ii) the *egalitarian objective* defined as the maximal individual cost (expected maximal individual cost for a randomized mechanism), i.e., $C(f, \mathbf{x}) = \mathbb{E}\left[\max_{1 \leq i \leq n} d(f(\mathbf{x}), x_i)\right]$. This is thus simply $\max_{1 \leq i \leq n} c_i(f, \mathbf{x})$ for deterministic mechanisms.

We say that a mechanism f is r*-approximate* with respect to profile $\mathbf{x}$ if

$$C(f, \mathbf{x}) \leq r \cdot OPT(\mathbf{x})$$

where $OPT(\mathbf{x})$ is the social cost of an optimal facility placement (for the egalitarian or utilitarian social cost). Note that since for a randomized mechanism the social cost is the expectation of the social cost on each chosen set of locations, there always exists an optimal deterministic placement.

We will be concerned with *strategy-proof* (SP) mechanisms, which render truthful revelation of locations a dominant strategy for the agents.

Definition 1. (Strategyproofness) *Let* $\mathbf{x} = (x_1, \ldots, x_n)$ *denote the location profile of* n *agents over the metric space* (Ω, d)*. A mechanism* f *is strategy-proof (SP) if for every agent* $1 \leq i \leq n$ *and for every location* $x'_i \in \Omega$, $c_i(f, (x'_i, x_{-i})) \geq c_i(f, \mathbf{x})$ *where* x_{-i} *denotes the locations of the agents other than* i *in* $\mathbf{x}$.

1.2 Previous Work

The facility locations game where only *one* facility will be opened is widely-studied in economics. On this topic, Moulin [6] characterized all strategy-proof mechanisms in the line metric space. Subsequently, Schummer and Vohra [10] gave a characterization of strategy-proof mechanisms for the circle metric space. More recently, Procaccia and Tennenholtz [9] initiated the study of approximating an optimum social cost under the constraint of strategy-proofness. They studied deterministic and randomized mechanisms on the line metric space with respect to the utilitarian and egalitarian objectives. Several (tight) approximation bounds for strategy-proof mechanisms were derived in their paper. For general metric space, Alon et al. [1] and Nguyen Kim [7] proved randomized tight bounds for egalitarian and utilitarian objectives, respectively.

Concerning the case where *two* facilities are opened, Procaccia and Tennenholtz [9] derived some strategy-proof mechanisms with guaranteed bounds in the line metric space for both objectives. Subsequently, Lu et al. [5] proved tight lower bounds of strategy-proof mechanisms in the line metric space with respect to the utilitarian objective. Moreover, they also gave a randomized strategy-proof mechanism, called Proportional Mechanism, that is 4-approximate for general metric spaces. It is still unknown whether there exists a deterministic strategy-proof mechanism with bounded approximation ratio in a general metric space.

Due to the absence of any positive result on the approximability of multiple facility location games for *more than two* facilities, Fotakis and Tzamos [3] considered a variant of the game where an authority can impose on some agents the facilities where they will be served. With this restriction, they proved that the Proportional Mechanism is strategy-proof and has an approximation ratio linear on the number of facilities.

1.3 Contribution

Prior to our work, only extreme cases of the game where the authority opens one or two facilities have been considered. No result, positive or negative, has been known for the game with three or more facilities. Toward the general number of facilities, we need to understand and solve the extreme cases of the problem. We consider here the extreme case where many facilities will be opened.

This type of situation occurs when every agent would like to have its own personal facility. The problem becomes interesting when it lacks at least one facility to satisfy everyone, i.e. $k = n - 1$. For instance, consider a blood collection agency that wishes to install 19 removable collection centers in the city of Paris, which consists of 20 districts. The agency asks every district council for the most

Table 1. Summary of our results. In a cell, UB and LB mean the upper and lower bounds on the approximation ratio of strategy-proof mechanisms. Abbreviation *det* (resp. *rand*) refers to deterministic (resp. randomized) strategy-proof mechanisms.

Objective	Tree metric space	General metric space
Utilitarian	UB: $n/2$ (rand) LB: 3/2 (det), 1.055 (rand)	UB: $n/2$ (rand) LB: 3 (det), 1.055 (rand)
Egalitarian	UB: 3/2 (rand) LB: 3/2 (rand) [9]	UB: n (rand) LB: 2 (det)

frequented spot in the district, and will place the facilities so as to serve them at best (minimize the sum of the distances from these spots to the nearest centers).

Another example, more related to computer science, is the service of k servers for online requests in the metric of n points. This issue, which is the k-servers problem [4], has been extensively studied and plays an important role in Online Algorithms. The special case of k servers for the metric of $(k+1)$ points is widely studied [2]. Similar problematics have also been adressed in Algorithmic Game Theory for the replication of data in a network, from the viewpoint of Price of Anarchy and Stability [8]. These issues are also interesting from the viewpoint of strategy-proofness. Assume that each server replicates some data to optimize the requests of the clients, but the positions of the clients in the network are private. The efficiency of the request answer depends on the distance from the client to the nearest server. The clients are thus asked for their positions, and one wishes to minimize the sum of the distances from the clients to the nearest servers.

In this paper, we study strategy-proof mechanisms for the game with n agents and $n-1$ facilities in a general metric space and in a tree metric space. Our main results are the following ones. For general metric spaces, we give a randomized strategy-proof mechanism, called *Inversely Proportional Mechanism*, that is an $n/2$-approximation for the utilitarian objective and an n-approximation for the egalitarian one. For tree metric spaces, we present another randomized strategy-proof mechanism that particularly exploit the property of the metric. This mechanism is also an $n/2$-approximation under the utilitarian objective but it induces a 3/2-approximation (tight bound) under the egalitarian objective.

Besides, several lower bounds on the approximation ratio of deterministic/randomized strategy-proof mechanisms are derived (see Table 1 for a summary). We proved that any randomized strategy-proof mechanism has ratio at least 1.055 even in the tree metric space. The interpretation of this result is that no mechanism, even randomized one, is both socially optimal and strategy-proof. Moreover, deterministic lower bounds of strategy-proof mechanisms are shown to be: at least 3/2 in a tree metric space, utilitarian objective; at least 3 in a general metric space, utilitarian objective; and at least 2 in a general metric space, egalitarian objective. Note that the lower bounds given for a tree metric space hold even for a line metric space.

Organization. We study the performance of randomized SP mechanisms in general metric spaces and in tree metric spaces in Section 2, and Section 3, respectively. Due to lack of space, some claims are only stated or partially proved.

2 SP Mechanisms for General Metric Spaces

2.1 Inversely Proportional Mechanism

Consider the setting of n agents whose true locations are $\mathbf{x} = (x_1, \ldots, x_n)$. For each location profile $\mathbf{y} = (y_1, \ldots, y_n)$, define $P_i(\mathbf{y})$ as the placement of $(n-1)$ facilities at the reported locations of all but agent i, i.e., $P_i(\mathbf{y}) = \{y_1, \ldots, y_{i-1}, y_{i+1}, \ldots, y_n\}$. Moreover, $d(y_i, P_i(\mathbf{y}))$ is the distance between y_i and her closest location in $P_i(\mathbf{y})$. The idea of the mechanism is to choose with a given probability a location y_i where no facility is open (and to put $n-1$ facilities precisely on the $n-1$ locations of the other agents), i.e., to choose with a given probability the placement $P_i(\mathbf{y})$. The main issue is to find suitable probabilities such that the mechanism is strategy-proof, and such that the expected cost is as small as possible.

Inversely proportional mechanism. Let $\mathbf{y}$ be a reported location profile. If there are at most $(n-1)$ distinct locations in profile $\mathbf{y}$ then open facilities at the locations in $\mathbf{y}$. Otherwise, choose placement $P_i(\mathbf{y})$ with probability

$$p_i(\mathbf{y}) = \frac{\frac{1}{d(y_i, P_i(\mathbf{y}))}}{\sum_{j=1}^{n} \frac{1}{d(y_j, P_j(\mathbf{y}))}}$$

Lemma 1. *The Inversely Proportional Mechanism is strategy-proof in a general metric space.*

Sketch of the proof. Let $\mathbf{x} = (x_1, \ldots, x_n)$ be the true location profile of the agents, and let $d_j := d(x_j, P_j(\mathbf{x}))$ for $1 \leq j \leq n$.

If there are at most $(n-1)$ distinct locations in profile $\mathbf{x}$ then the mechanism locates one facility on each position: no agent has incentive to misreports its location. In the sequel, we assume that all the agent locations in $\mathbf{x}$ are distinct. If all the agents report truthfully their locations, the cost of agent i is

$$c_i := c_i(f, \mathbf{x}) = \sum_{j=1}^{n} p_j(\mathbf{x}) \cdot d(x_i, P_j(\mathbf{x})) = p_i(\mathbf{x}) \cdot d_i = \frac{1}{\sum_{j=1}^{n} 1/d_j}$$

Thus $c_i < d_i$. Let us now suppose that i misreports its location and bids x'_i. Let $\mathbf{x}' = (x'_i, x_{-i})$ be the location profile when i reports x'_i and the other agents report truthfully their locations. Let $d'_j = d(P_j(x_j, \mathbf{x}'))$ for $j \neq i$ and $d'_i = d(P_i(x'_i, \mathbf{x}'))$. We will prove that $c'_i := c_i(f, \mathbf{x}') \geq c_i$. The new cost of agent i is:

$$c'_i = \sum_{j=1}^{n} p_j(\mathbf{x}') \cdot d(x_i, P_j(\mathbf{x}')) \geq p_i(\mathbf{x}') \cdot d_i + (1 - p_i(\mathbf{x}')) \min\{d_i, d(x_i, x'_i)\}$$

where the inequality is due to the fact that in $P_j(\mathbf{x}')$ (for $j \neq i$), agent i can choose either some facility in $\{x_1, \ldots, x_{i-1}, x_{i+1}, \ldots, x_n\}$ or the facility opened at x_i'. Define $T := \{j : d_j' \neq d_j, j \neq i\}$. Note that

$$p_i(\mathbf{x}') = \frac{1/d_i'}{\sum_{j \notin T} 1/d_j + \sum_{j \in T} 1/d_j' + 1/d_i'}$$

Let $e := d(x_i, x_i')$. Remark that i has no incentive to report its location x_i' in such a way that $e \geq d_i$ since otherwise $c_i' \geq p_i(\mathbf{x}') \cdot d_i + (1 - p_i(\mathbf{x}'))d_i = d_i > c_i$. In the sequel, consider $e < d_i$. In this case,

$$c_i' \geq p_i(\mathbf{x}') \cdot d_i + (1 - p_i(\mathbf{x}')) \cdot e$$

We also show that $e \geq |d_i' - d_i|$ by using the triangle inequality. Then, by considering two cases (whether $\sum_{j \in T} \frac{1}{d_j'}$ is larger than $\sum_{j \in T} \frac{1}{d_j}$ or not), we show that in both case $c_i' \geq c_i$ (technical details are omitted): any agent i has no incentive to misreport its location, i.e., the mechanism is strategy-proof. □

Theorem 1. *The Inversely Proportional Mechanism is strategy-proof, an $n/2$-approximation with respect to the utilitarian social cost and an n-approximation with respect to the egalitarian one. Moreover, there exists an instance in which the mechanism gives the approximation ratio at least $\frac{n}{2} - \epsilon$ for the utilitarian social cost, and $n - \epsilon$ for the egalitarian one, where $\epsilon > 0$ is arbitrarily small.*

Proof. By the previous lemma, the mechanism is strategy-proof. We consider the approximation ratio of this mechanism. Recall that $\mathbf{x} = (x_1, \ldots, x_n)$ is the true location profile of the agents. Let $P_i := P_i(\mathbf{x})$, $d_i := d(x_i, P_i)$ and $p_i = p_i(\mathbf{x})$. Let $\ell := \arg\min\{d_i : 1 \leq i \leq n\}$. For the egalitarian social cost, due to the triangle inequality at least one agent has to pay $d_\ell/2$, while the optimal solution for the utilitarian objective has cost d_ℓ (placement P_ℓ for instance).

The mechanism chooses placement P_i with probability p_i. In P_i, agent i has cost d_i and the other agents have cost 0. Hence, the social cost induced by the mechanism (in both objectives) is $\sum_j p_j(\mathbf{x})d_j = \frac{n}{\sum_j 1/d_j}$. For the utilitarian objective, the approximation ratio is $\frac{n}{d_\ell \sum_j 1/d_j} < \frac{n}{2}$ since in the sum of the denominator, there are two terms $1/d_\ell$. Similarly, it is at most $\frac{2n}{d_\ell \sum_j 1/d_j} < n$ for the egalitarian objective.

We describe an instance on a line metric space in which the bounds $n/2$ and n are tight. Let M be a large constant. Consider the instance on a real line in which $x_1 = 1, x_2 = 2, x_{i+1} = x_i + M$ for $2 \leq i \leq n$. We get $d_1 = d_2 = 1$ and $d_i = M$ for $3 \leq i \leq n$. An optimal solution chooses to put a facility in each x_i for $i \geq 2$ and to put the last one in the middle of $[x_1, x_2]$. Its social cost is 1 for the utilitarian objective and 1/2 for the egalitarian one. The cost (in both objectives) of the mechanism is

$$\frac{n}{\sum_{j=1}^{n} 1/d_j} = \frac{n}{2 + (n-2)/M} = \frac{nM}{2M + n - 2}$$

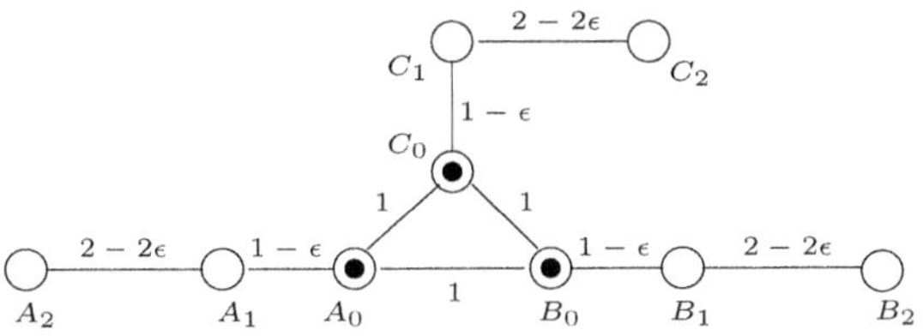

Fig. 1. Graph metric that gives a lower bound on the ratio of strategy-proof mechanisms in a general metric space (dots are the agents' locations in profile $\mathbf{x}$)

Hence, for any $\epsilon > 0$, one can choose M large enough such that the approximation ratio is larger than $\frac{n}{2} - \epsilon$ for the utilitarian objective and to $n - \epsilon$ for the egalitarian one. □

2.2 Lower Bounds on the Approximation Ratio for SP Mechanisms

Proposition 1. *Any deterministic strategy-proof mechanism has approximation ratio at least* $3 - 2\epsilon$ *for the utilitarian objective and* $2 - 2\epsilon$ *for the egalitarian objective where* $\epsilon > 0$ *is arbitrarily small.*

Proof. We consider the metric space induced by the graph in Figure 1. Note that this is a discrete space where agents and possible locations for facilities are restricted to be on vertices of the graph, i.e., $\Omega = V$. There are three agents and two facilities to be opened. Let f be a deterministic strategy-proof mechanism. Let $\mathbf{x}$ be a profile where $x_1 = A_0, x_2 = B_0, x_3 = C_0$. For any (deterministic) placement of two facilities, there is one agent with cost at least 1. By symmetry of the graph as well as profile $\mathbf{x}$, suppose that agent 1 has cost at least 1.

Consider another profile $\mathbf{y}$ where $y_1 = A_1, y_2 = B_0, y_3 = C_0$ ($\mathbf{y}$ and $\mathbf{x}$ only differ on the location of agent 1). In this profile, no facility is opened neither at A_0 nor at A_1 since otherwise agent 1 in profile $\mathbf{x}$ could report its location as being A_1 and reduce its cost from 1 to $1 - \epsilon$ or 0. We study two cases: (i) in profile $f(\mathbf{y})$, there is a facility opened at A_2; and (ii) in profile $f(\mathbf{y})$, no facility is opened at A_2.

In the former, a facility is opened at A_2, no facility is opened at A_0, A_1. For the egalitarian objective, the social cost is at least $2 - 2\epsilon$. For the utilitarian objective, the total cost of agents 2 and 3 is at least 1 and the cost of agent 1 is $2 - 2\epsilon$, that induces a social cost at least $3 - 2\epsilon$. An optimal solution has cost 1 (for both objective) by opening a facility at A_1 and a facility at B_0.

In the latter, the cost of agent 1 is at least $2 - \epsilon$ (since no facility is opened at A_0, A_1, A_2). Consider a profile $\mathbf{z}$ similar to $\mathbf{y}$ but the location of agent 1 is now at A_2. By strategy-proofness, no facility is opened at A_0, A_1, A_2 in $f(\mathbf{z})$ (since otherwise, agent 1 in profile $\mathbf{y}$ can decrease its cost by reporting its location as A_2). So, the social cost induced by mechanism f in $\mathbf{z}$ is at least $4 - 3\epsilon$ (for both objectives), while optimal is 1 (for both objectives) by placing a facility at A_2 and other at B_0.

Therefore, in any case, the approximation ratio of mechanism f is at least $3 - 2\epsilon$ for the utilitarian objective and $2 - 2\epsilon$ for the egalitarian objective. □

3 Randomized SP Mechanisms on Trees

We study in this section the infinite tree metric. This is a generalization of the (infinite) line metric, where the topology is now a tree. Infinite means that, like in the line metric, branches of the tree are infinite. As for the line metric, the locations (reported by agents or for placing facilities) might be anywhere on the tree. We first devise a randomized mechanism. To achieve this, we need to build a partition of the tree into subtrees that we call components, and to associate a status even or odd to each component. This will be very useful in particular to show that the mechanism is strategy-proof. In the last part of this section, we propose a lower bound on the approximation ratio of any strategy-proof mechanism.

3.1 Preliminary Tool: Partition into Odd and Even Components

Partition procedure. Given a tree T and a set of vertices V on this tree, we partition T into subtrees with respect to V. For the ease of description, consider also some virtual vertices, named ∞, which represent the extremities of the branches in T. We say that two vertices i and j are *neighbor* if the unique path in T connecting i and j contains no other vertex ℓ. A *component* T_t is a region of the tree delimited by a *maximal* set of pairwise neighbor vertices (see below for an illustration). The maximality is in the sense of inclusion: T_t is maximal means that there is no vertex $i \notin T_t$ such that vertex i is a neighbor of all vertices in T_t. The set $\{T_1, \dots, T_m\}$ of all components is a cover of the tree T. Note that a vertex i can appear in many sets T_t. As T is a tree, the set of all T_t's is well and uniquely defined.

For instance, in Figure 2, the components are the subtrees delimited by the following sets of vertices: $\{1,2,3\}$, $\{1,4\}$, $\{2,5\}$, $\{2,6\}$, $\{6,10\}$, $\{4,7\}$, $\{4,8,9\}$, $\{3,\infty\}$, $\{5,\infty\}$, $\{7,\infty\}$, $\{8,\infty\}$, $\{9,\infty\}$, $\{10,\infty\}$.

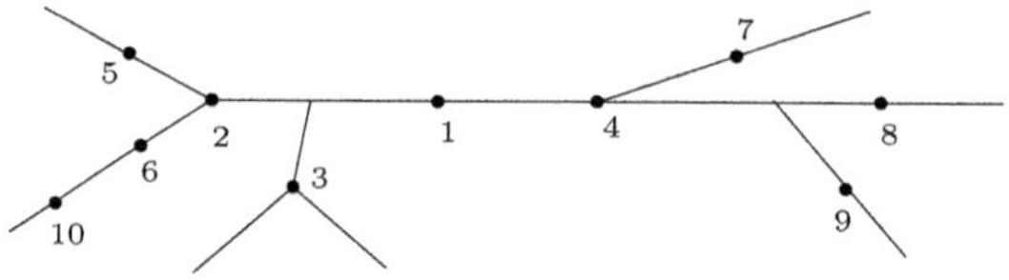

Fig. 2. An illustration of the partition procedure

Odd and even components. Root the tree at some vertex i_0, and define the depth of a vertex j as the number of vertices in the unique path from i_0 to j (i_0 has depth 1). Then each component T corresponds to the region of the tree between a vertex j (at depth p) and some of its sons (at depth $p+1$) in the tree. We say that T is odd (resp. even) if the depth p of j is odd (resp. even). This obviously depends on the chosen root.

For instance, in Figure 2 vertices of the same depth are in the same horizontal position (the tree is rooted at vertex 1). Then the components corresponding to $\{1,2,3\},\{1,4\},\{5,\infty\},\{6,10\},\ldots$ are odd while the ones corresponding to $\{2,5\},\{2,6\},\{3,\infty\},\{4,8,9\},\ldots$ are even.

Note that each vertex except the root — and the ∞-vertices — is both in (at least) one even component and in (at least) one odd component. The root is in (at least) one odd component.

3.2 A Randomized Mechanism

Given a reported profile $\mathbf{y}$ and a tree $\mathcal{T}$ as a metric space, let $2\alpha = 2\alpha(\mathbf{y})$ be the minimum distance between any two neighbor agents. Let $i^* = i^*(\mathbf{y})$ and $j^* = j^*(\mathbf{y})$ be neighbor agents such that $d(y_{i^*}, y_{j^*}) = 2\alpha$ (if there are more than one choice, break ties arbitrarily). We partition $\mathcal{T}$ into its components as described previously, considering as vertices the set of locations $\mathbf{y}$. Let T^* be the component containing y_{i^*} and y_{j^*}, and let U be the set of agents in T^*. For instance, in Figure 3, the components are $\{7,10,11,12\}$, $\{4,6,7,8\}$, $\{6,13\}$, $\{13,\infty\},\cdots$ Suppose that $i^* = 4$ and $j^* = 7$. Then T^* is the component whose set of agents is $U = \{4,6,7,8\}$.

We design a mechanism made of four deterministic placements P_1, P_2, P_3 and P_4; each P_i occurs with probability 1/4. Intuitively, the mechanism satisfies the following properties: (i) all agents have the same expected cost α, and (ii) for any component in $\mathcal{T}$, with probability 1/2, no facility is opened inside the component (but possibly at its extremities). To get this, each agent i different from i^* and j^* will have its own facility F_i open at distance α, while i^* and j^* will "share" a facility open either at y_{i^*}, or at y_{j^*}, or in the middle of the path between y_{i^*} and y_{j^*}. However, to ensure strategy-proofness, we need to carefully combine these positions.

If we remove the component T^* (while keeping its vertices) from $\mathcal{T}$, we now have a collection of subtrees $\mathcal{T}_i$ for $i \in U$, where $\mathcal{T}_i$ is rooted at y_i (the location of agent i). For each rooted-subtree $\mathcal{T}_i$, assign the status odd or even to its components according to the procedure previously defined. In Figure 3 (B) if we remove T^* we have four subtrees rooted at 4, 6, 7 and 8. Bold components are odd.

We are now able to define the four placements P_1, P_2, P_3, P_4. Nevertheless, recall that a node is in at least one odd component and at least one even component.

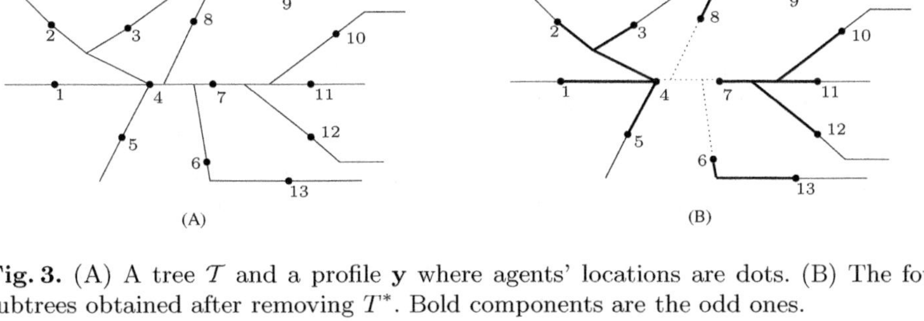

Fig. 3. (A) A tree $\mathcal{T}$ and a profile $\mathbf{y}$ where agents' locations are dots. (B) The four subtrees obtained after removing T^*. Bold components are the odd ones.

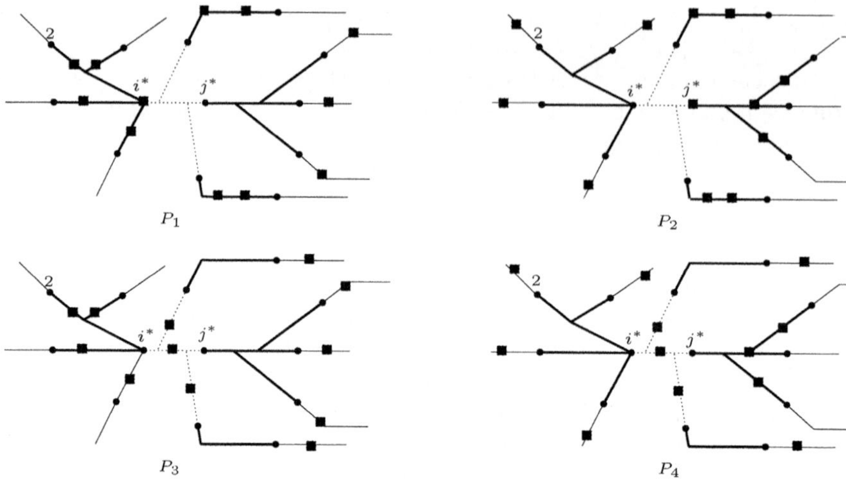

Fig. 4. Placements P_1, P_2, P_3, P_4 for the instance in Figure 3. Agents i^*, j^* are 4, 7. Facilities are represented by squares.

Each agent $i \neq i^*, j^*$ is associated with a facility F_i, while i^* and j^* share a common facility. We describe in the following the placements of these facilities. We distinguish the agents with respect to the subtree $\mathcal{T}_i$ where they are.

Table 2. Placements of facilities associated with agents

Placement	i^*	$i \in \mathcal{T}_{i^*}$	j^*	$i \in \mathcal{T}_{j^*}$	$i \in U \setminus \{i^*, j^*\}$	$i \in \mathcal{T}_\ell \setminus U$ for $\ell \in U$
P_1	at y_{i^*}	O	no facility	E	O	O
P_2	no facility	E	at y_{j^*}	O	O	O
P_3	mid. y_{i^*}, y_{j^*}	O	no facility	E	T^*	E
P_4	no facility	E	mid. y_{i^*}, y_{j^*}	O	T^*	E

In Table 2, E (resp. O) means that we open a facility F_i in an even component (resp. odd component) at distance α of y_i for agent i; T^* means that the facility F_i is opened in the component T^*, with distance α from y_i. For the location of any facility, if there are several choices, pick one arbitrarily. In placements P_3 and P_4 "mid. i^*, j^*" means that the position is the middle of the path connecting y_{i^*} and y_{j^*}. We denote by $F^*(\mathbf{y})$ the facility opened at this position. In this case, i^* and j^* share the same facility $F^*(\mathbf{y})$.

An illustration is shown in Figure 4. For instance, since y_2 is in the subtree $\mathcal{T}_4 = \mathcal{T}_{i^*}$, the facility F_2 associated with agent 2 is opened in an odd (bold) component in placements P_1 and P_3 and in an even one in placements P_2 and P_4.

Analysis. By definition, all the placements P_1, P_2, P_3, P_4 are well defined, i.e., there are at most $n-1$ opening facilities in each placement (one associated to each agent $i \neq i^*, j^*$, plus only one shared by i^* and j^*). The following lemma shows some properties of the mechanism.

Lemma 2. *Given a reported profile* $\mathbf{y}$*, the expected distance between* y_i *and its closest facility equals* $\alpha(\mathbf{y})$ *for* $1 \leq i \leq n$*. Moreover, for any component, there are at least two placements in* $\{P_1, P_2, P_3, P_4\}$ *where the component does not contain any facility (but facilities can be at the extremities of the component).*

Proof. Consider an agent $i \neq i^*(\mathbf{y}), j^*(\mathbf{y})$ where we recall that $i^*(\mathbf{y}), j^*(\mathbf{y})$ denote the two players whose reported locations are at minimum distance. In any placement, the closest facility is opened at distance $\alpha(\mathbf{y})$ from y_i. For agent $i^* = i^*(\mathbf{y})$, the distance from y_{i^*} to the closest facility is: 0 in P_1, $2\alpha(\mathbf{y})$ in P_2, $\alpha(\mathbf{y})$ in P_3 and P_4. Hence, the average is $\alpha(\mathbf{y})$, and similarly for agent $j^*(\mathbf{y})$.

Let T be the component containing the locations of agents $i^*(\mathbf{y})$ and $j^*(\mathbf{y})$. No facility is opened inside T under placements P_1 and P_2. Besides, by the definition of the mechanism, there are at least two placements in $\{P_1, P_2, P_3, P_4\}$ where a component does not contain a facility[1]. □

Now we prove the strategy-proofness of the mechanism. Suppose that an agent i strategically misreports its location as x'_i (while other agents' locations remain unchanged). Let $\mathbf{x}' = (x'_i, x_{-i})$, where $\mathbf{x} = (x_1, \ldots, x_n)$ is the true location profile. Define the parameters $2\alpha := 2\alpha(\mathbf{x}), i^* := i^*(\mathbf{x}), j^* := j^*(\mathbf{x})$. For every agent i, $\mathcal{N}(i, \mathbf{x})$ denotes the set of its neighbors in profile $\mathbf{x}$ ($\mathcal{N}(i, \mathbf{x})$ does not contain i). The strategy-proofness is due to the two following main lemmas.

Lemma 3. *No agent* i *has incentive to misreport its location as* x'_i *such that* $\mathcal{N}(i, \mathbf{x}) \neq \mathcal{N}(i, \mathbf{x}')$*.*

Proof. Suppose that $\mathcal{N}(i, \mathbf{x}) \neq \mathcal{N}(i, \mathbf{x}')$. In this case, the locations of agents in $\mathcal{N}(i, \mathbf{x})$ form a component T' of tree $\mathcal{T}$ with respect to profile $\mathbf{x}'$. By Lemma 2, with probability at least $1/2$, no facility is opened in T', i.e., in those cases agent i is serviced by a facility outside T'. Note that the distance from x_i to the location of any agent in $\mathcal{N}(i, \mathbf{x})$ is at least 2α. Therefore, the new cost of agent i is at least α, meaning i has no incentive to report x'_i. □

Lemma 4. *Agent* i *cannot strictly decrease its cost by reporting a location* $x'_i \neq x_i$ *such that* $\mathcal{N}(i, \mathbf{x}) = \mathcal{N}(i, \mathbf{x}')$*.*

Proof. As $\mathcal{N}(i, \mathbf{x}) = \mathcal{N}(i, \mathbf{x}')$, the path connecting x_i and x'_i contains no other agent's location. Hence, there is a component T'_i in the partition of $\mathcal{T}$ with respect to $\mathbf{x}'$ such that $x'_i \in T'_i$ and $x_i \in T'_i$. Let $2\alpha'$ be the minimum distance between two neighbors in $\mathbf{x}'$. Also let $e = d(x_i, x'_i)$.

[1] There are facilities in T under P_3 and P_4 but facilities are put on the extremities under placements P_1 and P_2. Notice that a component may never receive a facility if there are two components named $\{i, \infty\}$ and i is located at the intersection of two branches of the tree, see location 3 in Figure 2.

Case 1: Consider the case where, with the new location x'_i, i is neither $i^*(\mathbf{x}')$ nor $j^*(\mathbf{x}')$. Hence, $\alpha' \geq \alpha$. By Lemma 2, with probability at least 1/2, no facility is opened inside T'_i. In this case, the distance from x_i to the closest facility is at least $\min\{d(x_i, x'_i) + d(x'_i, F'_i), d(x_i, x_\ell) + d(x_\ell, F'_\ell)\}$ where: $\ell \in \mathcal{N}(i, \mathbf{x})$ and F'_ℓ is its associated facility; and F'_i is the facility opened at distance α' from x'_i, F'_i is in a component different from T'_i. In other words, this distance is at least $\min\{e + \alpha', 2\alpha\}$ since $d(x'_i, F'_i) = \alpha'$ and $d(x_i, x_\ell) \geq 2\alpha$. Besides, with probability at most 1/2, the closest facility to x_i is either F'_i (the facility opened in component T'_i at distance α' from x'_i) or some other facility F'_ℓ in T'_i for some $\ell \in \mathcal{N}(i, \mathbf{x})$. The former gives a distance $d(x_i, F'_i) \geq \max\{d(x'_i, F'_i) - d(x'_i, x_i), 0\} = \max\{\alpha' - e, 0\}$ (by triangular inequality). The latter gives a distance $d(x_i, F'_\ell) \geq \max\{d(x_i, x_\ell) - d(x_\ell, F'_\ell), 0\} \geq \max\{2\alpha - \alpha', 0\}$. Hence, the cost of agent i is at least

$$\frac{1}{2}\left(\min\{e + \alpha', 2\alpha\} + \min\{\max\{\alpha' - e, 0\}, \max\{2\alpha - \alpha', 0\}\}\right) \geq \alpha$$

where the inequality is due to $\alpha' \geq \alpha$. Indeed, this is immediate if $e + \alpha' \geq 2\alpha$. Otherwise, the cost is either at least $e + \alpha' + \alpha' - e = 2\alpha'$, or $e + \alpha' + 2\alpha - \alpha' \geq 2\alpha$. Hence, $c_i(\mathbf{x}') \geq c_i(\mathbf{x})$.

Case 2: Consider the case where with the new location x'_i agent $i = i^*(\mathbf{x}')$ (the case where $i = j^*(\mathbf{x}')$ is completely similar)[2]. Let $j = j^*(\mathbf{x}')$. Let d_1, d_2, d_3, d_4 be the distance from x_i to the closest facility in placements P_1, P_2, P_3, P_4 (in $\mathbf{x}'$), respectively. Let T' be the component in $\mathcal{T}$ with respect to $\mathbf{x}'$ that contains x'_i and x_j. By the triangle inequality, we know that

$$e + 2\alpha' = d(x_i, x'_i) + d(x'_i, x_j) \geq d(x_i, x_j) \geq 2\alpha \tag{1}$$

We study the two sub-cases and prove that $\sum_{t=1}^{4} d_t \geq 4\alpha$ always holds, meaning that agent i's deviation cannot be profitable since its cost is α when it reports its true location x_i.

(a) The true location x_i belongs to T'.
For each agent $\ell \neq i, j$, let F'_ℓ be its associated facility. The facility opened in the middle of $[x'_i, x_j]$ is denoted by $F^*(\mathbf{x}')$. We have:

$$d_1 = \min\{d(x_i, x'_i), d(x_i, F'_\ell)\} = \min\{e, d(x_i, x_\ell) + d(x_\ell, F'_\ell)\} \geq \min\{e, 2\alpha + \alpha'\} \tag{2}$$

$$d_2 = \min\{d(x_i, x_j), d(x_i, F'_\ell)\} \geq \min\{d(x_i, x_j), 2\alpha + \alpha'\} \geq 2\alpha \tag{3}$$

$$d_3 = \min\{d(x_i, F^*(\mathbf{x}')), d(x_i, F'_\ell)\} \geq \min\{2\alpha - \alpha', e + \alpha', 2\alpha + \alpha'\} \tag{4}$$

$$d_4 = \min\{d(x_i, F^*(\mathbf{x}')), d(x_i, F'_\ell)\} \geq \min\{2\alpha - \alpha', e + \alpha', 2\alpha + \alpha'\} \tag{5}$$

where $\ell \neq i, j$ is some agent in $\mathcal{N}(i, \mathbf{x}')$ (note that agents ℓ in the expressions above are not necessarily the same). The first equality in (2) is due to the fact that in placement P_1, agent i goes either to the facility opened at x'_i or

[2] Contrasting with Case 1, $\alpha' \leq \alpha$ does not necessarily hold.

to a facility (outside T') associated to some other agent. In placement P_2, agent i can either choose a facility opened at x_j or another one outside T', that is translated in the equality in (3). In placement P_3 and P_4, agent i can go either to facility $F^*(\mathbf{x}')$ opened in the midpoint connecting x'_i and x_j, or to the facility associated with some agent ℓ (inside and outside T' respectively).

If $e+\alpha' < 2\alpha-\alpha'$ then $\sum_{t=2}^{4} d_t \geq 2\alpha+2e+2\alpha' \geq 4\alpha$ (since $e+2\alpha' \geq 2\alpha$). In the sequel, assume $e+\alpha' \geq 2\alpha-\alpha'$. If $e \geq 2\alpha+\alpha'$ then $d_1+d_3 \geq 4\alpha$. Otherwise, $\sum_{t=1}^{4} d_t \geq e+\min\{d(x_i,x_j), 2\alpha+\alpha'\}+2\max\{2\alpha-\alpha',0\}$. Note that by the triangle inequality $e+d(x_i,x_j) = d(x'_i,x_i)+d(x_i,x_j) \geq d(x'_i,x_j) = 2\alpha'$. Therefore, $\sum_{t=1}^{4} d_t \geq \min\{2\alpha'+4\alpha-2\alpha', 2\alpha+\alpha'+2\alpha-\alpha'\} = 4\alpha$. Hence, the new cost of i is at least α.

(b) The true location x_i does *not* belong to T'.
Let T'_i be the component in $\mathcal{T}$ with respect to profile $\mathbf{x}'$ such that T'_i contains x_i and x'_i. Similar to the previous case, we have:

$$d_2 = \min\{d(x_i,x_j), d(x_i,F'_\ell)\} = \min\{d(x_i,x'_i)+d(x'_i,x_j), d(x_i,x_\ell)+d(x_\ell,F'_\ell)\}$$
$$\geq \min\{e+2\alpha', 2\alpha+\alpha'\} \geq 2\alpha \quad (6)$$
$$d_3 = \min\{d(x_i,F^*(\mathbf{x}')), d(x_i,F'_\ell)\}$$
$$\geq \min\{d(x_i,x'_i)+d(x'_i,F^*(\mathbf{x}')), d(x_i,x_\ell)-d(x_\ell,F'_\ell)\} \quad (7)$$
$$= \min\{e+\alpha', 2\alpha-\alpha'\} \quad (8)$$
$$d_4 = \min\{d(x_i,F^*(\mathbf{x}')), d(x_i,F'_\ell)\} \geq \min\{e+\alpha', 2\alpha+\alpha'\} \quad (9)$$

where $\ell \neq i,j$ is some agent in $\mathcal{N}(i,\mathbf{x}')$ (again agents ℓ in the expressions above are not necessarily the same). In placement P_2, agent i can choose either a facility opened at x_j or another one outside T'_i. The last inequality of (6) is due to $e+2\alpha' \geq 2\alpha$ (Inequality 1). In placement P_3 and P_4, agent i can go either to facility $F^*(\mathbf{x}')$ opened in the midpoint connecting x'_i and x_j, or some facilities associated with some agent ℓ.
If $e+\alpha' < 2\alpha-\alpha'$ then $\sum_{t=2}^{4} d_t \geq 2\alpha+2e+2\alpha' \geq 4\alpha$ (since $e+2\alpha' \geq 2\alpha$). Otherwise, $\sum_{t=2}^{4} d_t \geq \min\{e+4\alpha, 4\alpha\} \geq 4\alpha$. Again, the new cost of agent i is at least α.

In conclusion, no agent has incentive to strategically misreport its location. □

Theorem 2. *The mechanism is strategy-proof and it induces an $n/2$-approximation according to the utilitarian objective and a tight $3/2$-approximation according to the egalitarian objective.*

Proof. The mechanism is strategy-proof by previous lemmas. The cost of each agent is α, so in the utilitarian objective, the cost induced by the mechanism is $n\alpha$. An optimal placement is to open facilities at the locations of all agents but i^*, which induces a cost 2α. Hence, the mechanism is $n/2$-approximation for the utilitarian objective.

Consider the egalitarian objective. By the mechanism, in P_3 and P_4 the maximum cost of an agent is α, while in P_1 and P_2 it is 2α. The average maximum cost of the mechanism is $3\alpha/2$. An optimal solution is to open facilities at locations of agents other than i^*, j^* and open one facility at the midpoint of the path connecting x_{i^*} and x_{j^*}; that gives a cost α. So, the approximation ratio is 3/2 and this ratio is tight, i.e., no randomized strategy-proof mechanism can do better [9, Theorem 2.4]. □

3.3 Lower Bounds on the Approximation Ratio of SP Mechanisms

In this section, we consider only the utilitarian objective (as the tight bound for the egalitarian objective has been derived in the previous section). The proof of Proposition 2 is omitted.

Proposition 2. *No deterministic strategy-proof mechanism on a line metric space has an approximation ratio smaller than* 3/2.

The following proposition indicates that even with randomization, we cannot get an optimal strategy-proof mechanism for the utilitarian objective.

Proposition 3. *No randomized strategy-proof mechanism on a line metric space has an approximation ratio smaller than* $10 - 4\sqrt{5} \approx 1.055$.

Proof. Let f be a randomized strategy-proof mechanism with an approximation ratio strictly better than $1 + \varepsilon > 1$. Consider a profile $\mathbf{x}$ where the positions of the agents are $x_1 = A, x_2 = B, x_3 = C, x_4 = D$ (Figure 5). For any placement of three facilities, the total cost is at least 1. Hence, there exists an agent with (expected) cost at least 1/4. Without loss of generality, suppose that agent 1 (with $x_1 = A$) has cost $c_1(f, \mathbf{x}) \geq 1/4$.

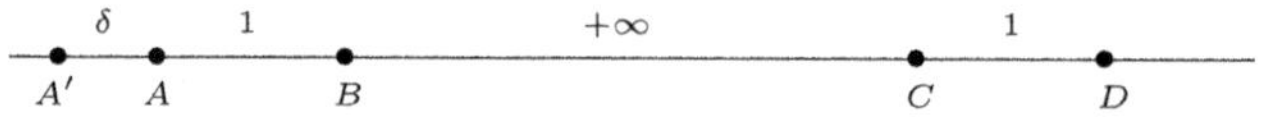

Fig. 5. Instance which gives the lower bound on the ratio of a randomized strategy-proof mechanism in a line metric space

Let $0 < \delta < 1/4$ be a constant to be defined later. Let $A' \notin [A, B]$ be a location at distance δ from A. Let $\mathbf{y}$ be the profile in which agent 1 is located at $y_1 = A'$ and the other agents' locations are the same as in $\mathbf{x}$. By strategy-proofness, $c_1(f, \mathbf{x}) \leq \delta + c_1(f, \mathbf{y})$. Hence, $c_1(f, \mathbf{y}) \geq 1/4 - \delta$. In $\mathbf{y}$, an optimal solution has cost 1 (e.g. place the facilities at the locations of the agents other than agent 4). As f is a $(1 + \varepsilon)$-approximation, the total cost of the solution returned by the mechanism is $c_1(f, \mathbf{y}) + c_2(f, \mathbf{y}) + c_3(f, \mathbf{y}) + c_4(f, \mathbf{y}) \leq 1 + \varepsilon$. Thus, $c_3(f, \mathbf{y}) + c_4(f, \mathbf{y}) \leq 3/4 + \varepsilon + \delta$.

In outcome $f(\mathbf{y})$, let p be the probability that the closest facility of agent 3 is also the closest facility of agent 4 (in other words, agents 3 and 4 share one facility with probability p; and with probability $(1 - p)$ there is at most one facility between A' and B). We have $c_3(f, \mathbf{y}) + c_4(f, \mathbf{y}) \geq p \cdot 1 = p$. Therefore, $p \leq 3/4 + \varepsilon + \delta$.

Besides, the social cost of $f(\mathbf{y})$ is at least $p + (1-p)(1+\delta) = 1 + \delta - p\delta$. This is lower bounded by $1 + \delta - (3/4 + \varepsilon + \delta)\delta$. Hence, $1 + \delta - (3/4 + \varepsilon + \delta)\delta \leq C(f, \mathbf{y}) \leq 1 + \varepsilon$. We deduce that $\varepsilon \geq \frac{\delta/4 - \delta^2}{1+\delta}$.

The function $\frac{\delta/4-\delta^2}{1+\delta}$ for $\delta \in (0, \frac{1}{4})$ attains maximal value $9 - 4\sqrt{5}$ for $\delta = \sqrt{5}/2 - 1$. Thus the approximation ratio is at least $1 + \varepsilon \geq 10 - 4\sqrt{5} \approx 1.055$.□

4 Discussion and Further Directions

The results presented in this paper are a first step toward handling the general case where one wishes to locate k facilities in a metric space with n agents (for $1 \leq k \leq n$). The general case is widely open since nothing on the performance of strategy-proof mechanisms is known. Any positive or negative results on the problem would be interesting. We suggest a mechanism based on the *Inversely Proportional Mechanism* in which the k facilities are put on reported locations. Starting with the n reported locations the mechanism would iteratively eliminate a candidate until k locations remain. We do not know whether this mechanism is strategy-proof. For restricted spaces such as line, cycle or tree metric spaces, there might be some specific strategy-proof mechanisms with guaranteed performance which exploits the structures of such spaces. Besides, some characterization of strategy-proof mechanisms (as done by Moulin [6] or Schummer and Vohra [10]), even not a complete characterization, would be helpful.

References

1. Alon, N., Feldman, M., Procaccia, A.D., Tennenholtz, M.: Strategyproof approximation of the minimax on networks. Math. Oper. Res. 35, 513–526 (2010)
2. Coppersmith, D., Doyle, P., Raghavan, P., Snir, M.: Random Walks on Weighted Graphs and Applications to On-line Algorithms. J. of ACM 40(3), 421–453 (1993)
3. Fotakis, D., Tzamos, C.: Winner-imposing strategyproof mechanisms for multiple facility location games. In: Saberi, A. (ed.) WINE 2010. LNCS, vol. 6484, pp. 234–245. Springer, Heidelberg (2010)
4. Koutsoupias, E.: The k-server problem. Comp. Science Rev. 3(2), 105–118 (2009)
5. Lu, P., Sun, X., Wang, Y., Zhu, Z.A.: Asymptotically optimal strategy-proof mechanisms for two-facility games. In: ACM Conf. on Electronic Com, pp. 315–324 (2010)
6. Moulin, H.: On strategy-proofness and single peakedness. Public Choice 35, 437–455 (1980)
7. Nguyen Kim, T.: On (Group) strategy-proof mechanisms without payment for facility location games. In: Saberi, A. (ed.) WINE 2010. LNCS, vol. 6484, pp. 531–538. Springer, Heidelberg (2010)
8. Pollatos, G.G., Telelis, O.A., Zissimopoulos, V.: On the social cost of distributed selfish content replication. In: Das, A., Pung, H.K., Lee, F.B.S., Wong, L.W.C. (eds.) NETWORKING 2008. LNCS, vol. 4982, pp. 195–206. Springer, Heidelberg (2008)
9. Procaccia, A.D., Tennenholtz, M.: Approximate mechanism design without money. In: ACM Conference on Electronic Commerce, pp. 177–186 (2009)
10. Schummer, J., Vohra, R.V.: Strategy-proof location on a network. Journal of Economic Theory 104 (2001)

Making Decisions in Multi Partitioning

Alain Guénoche

IML - CNRS, 163 Av. de Luminy, 13009 Marseille, France
guenoche@iml.univ-mrs.fr

Abstract. Starting from individual judgments given as categories (i.e., a profile of partitions on an X item set), we attempt to establish a collective partitioning of the items. For that task, we compare two combinatorial approaches. The first one allows to calculate a consensus partition, namely the median partition of the profile, which is the partition of X whose sum of distances to the individual partitions is minimum. Then, the collective classes are the classes of this partition. The second one consists in first calculating a distance D on X based on the profile and then in building an X-tree associated to D. The collective classes are then some of its subtrees. We compare these two approaches and more specifically study in what extent they produce the same decision as a set of collective classes.

Keywords: Categorization data, Consensus, Partitions, Tree representation.

1 Introduction

In this paper, we propose to compare two combinatorial methods to analyze categorical data. These data correspond to subjects also called experts who cluster items - photos, sounds, products - according to individual *categories* gathering close ones. We assume that an item is only classified once by each expert and so each subject expresses his judgment as a partition with any number of classes, thus carrying out a *free categorization*. Therefore, these data define a profile Π of partitions on the same X set. Such a situation is also encountered:

– when items are described by nominal variables, since each variable is a partition. As a particular case, binary data constitute a two-classes partition profile.
– when applying a partitioning method on the same X set according to bootstrapped data.

We then aim at classifying the elements of X, i.e. to go from the partition *profile* based on individual categories to a unique partition in collective classes also called *concepts* here. Staying in the Combinatorial Data Analysis frame, we compare two methods:

– The first one consists in building the median partition for Π, i.e. a partition whose sum of distances to the profile partitions is minimum. This partition best represents the set of individual categorizations and can be considered as the collective judgment of the experts.
– The second has been developed by Barthélemy (1991) and consists in calculating a distance D between items and to represent this distance in the form of an X-tree denoted A. This tree is such that the set of leaves is X and

R.I. Brafman, F. Roberts, and A. Tsoukiàs (Eds.): ADT 2011, LNAI 6992, pp. 82–95, 2011.

the other nodes are the roots of subtrees corresponding to classes. The distance on X, that takes into account all the partitions in the profile, enables to go from the individual to the collective categorization and some subtrees in A are regarded as concepts.

The point is to know if these two methods produce similar results on the same data. Rather than comparing concepts built on classical data (benchmark), we are going to establish a simulation protocol. From any given initial partition, a profile of more or less similar partitions is generated by effecting a fixed number of transfers from the initial one. For each profile, we build on the one hand, the consensus partition and on the other hand, a series of splits of the corresponding X-tree, making a artition. Then, we calculate indices whose mean values allow to measure the adequacy of both methods.

The rest of the paper is organized as follows: In Section 2 we describe how to calculate median partitions that are either optimal for limited size profiles or very close to the optimum for larger ones. In Section 3, we accurately review Barthélemy's method and give a way to determine the optimal partition in an X-tree. In Section 4 we describe the simulation process used to measure the adequacy of these methods. This process leads to conclude that the median consensus method has better ability to build concepts from categorical data than the X-tree procedure. All along this text, we illustrate the methodologies with categorization data, made of 16 pieces of music clustered by 17 musicians :

Example 1

Table 1. Categorizations of the 17 experts giving partitions[1]. Each row, corresponding to a musician, indicates the class number of the 16 pieces. For instance, Amelie, makes 8 classes, {7, 8, 14} are in the first one, {1, 5, 13} in the second one, and so on.

	1	2	3	4	5	6	7	8	9	10	11	12	13	14	15	16
Amelie	2	3	4	7	2	4	1	1	5	3	6	6	2	1	5	8
Arthur	1	4	4	2	1	1	4	4	3	4	2	5	1	5	3	4
Aurore	1	1	2	3	3	2	1	3	4	1	3	4	2	4	3	2
Charlotte	3	6	5	1	3	5	6	2	3	6	1	3	5	4	2	5
Clement	7	3	5	8	4	5	1	1	4	3	9	6	7	6	2	2
Clementine	1	1	2	3	1	2	1	5	4	1	3	2	2	5	4	2
Florian	2	5	6	8	1	6	7	7	5	3	4	3	2	4	7	7
Jean-Philippe	2	3	3	1	2	3	4	4	2	3	1	1	2	1	4	4
Jeremie	1	2	3	4	1	3	2	5	5	2	5	6	3	6	5	3
Julie	4	4	3	4	4	3	1	1	2	4	2	2	3	2	1	3
Katrin	1	2	2	2	1	3	3	3	1	2	3	4	2	3	2	3
Lauriane	2	1	1	3	2	1	4	4	3	2	1	3	2	4	4	1
Louis	3	1	3	3	3	1	3	2	2	1	2	3	1	2	2	1
Lucie	4	2	3	4	4	1	5	6	6	2	6	6	1	5	6	3
Madeleine	3	2	1	5	3	1	2	2	4	4	5	4	1	2	2	3
Paul	1	4	4	1	1	4	3	3	1	4	3	2	1	3	3	3
Vincent	5	2	2	1	1	2	3	3	4	2	3	4	5	3	4	3

[1] I would like to thanks P. Gaillard (Dept. Psychology, University of Toulouse, France, who provided these data.

2 Consensus Partition

A pioneer work about consensus of partitions is Régnier's paper (1965). Starting from the problem of partitioning items described by nominal variables, he introduced the concept of central or median partition, defined as the partition minimizing the sum of symmetric difference distances to the profile partitions.

2.1 Consensus Formalization

Let $X = \{x_1, x_2, \ldots x_n\}$ be a set of cardinality n. A partition of X is any collection of disjoint and non empty classes of elements of X whose union equals X. Hereafter, we denote by $\mathcal{P}$ the set of all the partitions of X and by $\Pi = (P_1, \ldots, P_m)$ a *profile* of m partitions in $\mathcal{P}$. Moreover, for any partition $P \in \mathcal{P}$ and any element $x_i \in X$, we denote by $P(i)$ the class of x_i in P. Then, for a given Π, finding the *consensus partition* consists in determining a partition $\pi \in \mathcal{P}$ as close as possible to Π for some criterion.

The criterion used in the sequel may be computed as follows. For any $(P, Q) \in \mathcal{P}^2$, we first define the similarity S between P and Q as the number of pairs of elements of X that are commonly joined or separated in both P and Q. So, S equals the non normalized Rand index, which is the complementary number of the symmetric difference cardinality. We then define the score of the partition P relatively to the profile Π as

$$S_\Pi(P) = \sum_{i=1,\ldots,m} S(P, P_i).$$

So, with respect to this criterion, the optimal partition is a median partition of Π. Actually, Régnier (1965) shows that maximizing S_Π is equivalent to maximize over $\mathcal{P}$ the quantity

$$W_\Pi(P) = \sum_{(i<j)\in J(P)} \left(T_{i,j} - \frac{m}{2}\right), \tag{1}$$

where $T_{i,j}$ denotes the number of partitions of Π in which x_i and x_j are joined and $J(P)$ is the set of every joined pairs in P.

The value $W_\Pi(P)$ has a very intuitive meaning. Indeed, it points out that a joined pair in P has a positive (resp. negative) contribution to the criterion as soon as its elements are gathered in more (resp. less) than half of the partitions of Π.

Example 2
Table 2 indicates twice the value of pair scores : $2w_{i,j} = 2T_{i,j} - m$. Pieces of music 1 and 2 being joined together in only 3 partitions (Aurore, Clementine and Julie) their score is 6 - 17 = -11. One can see that there are very few positive values underlined in bold.

Table 2. Score of pieces of music pairs according to the profile in Table 1

	1	2	3	4	5	6	7	8	9	10	11	12	13	14	15
2	-11														
3	-15	-5													
4	-9	-13	-13												
5	**9**	-13	-15	-5											
6	-15	-7	**9**	-17	-15										
7	-11	-5	-13	-15	-13	-15									
8	-17	-13	-15	-15	-15	-15	**5**								
9	-9	-15	-17	-13	-7	-17	-17	-11							
10	-9	**11**	-7	-13	-11	-9	-7	-15	-15						
11	-17	-15	-15	-5	-15	-13	-11	-3	-9	-17					
12	-13	-17	-13	-11	-13	-15	-15	-15	-3	-13	-9				
13	-1	-13	-3	-13	-7	**1**	-17	-17	-13	-11	-17	-15			
14	-17	-15	-17	-15	-17	-15	-3	-1	-11	-17	-3	-5	-17		
15	-17	-13	-15	-13	-15	-17	-5	**5**	-3	-15	-7	-13	-15	-9	
16	-15	-11	-1	-17	-15	-1	-5	-5	-17	-13	-9	-15	-5	-11	-9

Let $\mathbf{K}_n$ be the complete graph on X whose edges are weighted by

$$w_{i,j} = (T_{i,j} - \frac{m}{2}).$$

Thus, maximizing W_Π turns out to build a partition or equivalently, a set of disjoint cliques in $(\mathbf{K}_n, W)$ having maximal weight. This problem generalizes Zahn (1971)'s NP-hard problem to weighted graphs. Therefore, no polynomial algorithm leading to an optimal solution is known.

As mentioned in Régnier (1965), the consensus partition problem can be solved by integer linear programming. Given a partition P, with notations $\alpha_{ij} = 1$, iff items x_i and x_j belong to the same class, the W_Π criterion can be formulated as:

$$W_\Pi(\alpha) = \sum_{i<j} \alpha_{ij} w_{i,j}. \qquad (2)$$

The optimization problem is to determine a symmetric matrix α maximizing W_Π under constraints making P an equivalence relation on X.

$$\begin{cases} \forall (i<j), \alpha_{ij} \in \{0,1\} \\ \forall (i \neq j \neq k), \alpha_{ij} + \alpha_{jk} - \alpha_{ik} \leq 1 \end{cases}$$

There exist optimal resolution methods to find α, and so partition π, realizing the global maximum of function W_Π over $\mathcal{P}$. Several mathematical programming solutions have been proposed, begining with Grötschel & Wakabayashi (1989). We use the GLPK software (GNU Linear Programming Kit) to calculate maximal scores when possible. There are $\binom{n}{2}$ variables and $3\binom{n}{3}$ constraints. The set of constraints $\alpha_{ij} + \alpha_{jk} - \alpha_{ik} \leq 1$ makes a table indexed by constraints and by pairs $(i<j)$ of elements of X. For each triple $(i<j<k)$ there are 3 rows, one with coefficients $\alpha_{ij} = 1, \alpha_{jk} = 1, \alpha_{ik} = -1$, the second with $\alpha_{ij} = 1, \alpha_{jk} = -1, \alpha_{ik} = 1$

and the third with $\alpha_{ij} = -1, \alpha_{jk} = 1, \alpha_{ik} = 1$, the other coefficients being equal to 0. Consequently, there are $\frac{n(n-1)(n-2)}{2}$ linear constraints. For $n = 100$ it makes 4950 binary variables and 485,100 constraints. For a bipartition profile, $n = 20$ could generate intractable problems in reasonable time. These are the limits of our simulations, meaning that not all instances are computable, particularly for binary tables.

2.2 The Fusion-Transfer Method *FT*

A lot of heuristics have been proposed. Among them, Régnier's *transfer method* consists in affecting an element of an initial partition π to another class of π as long as the W_Π criterion increases. This optimization method achieves a local maximum of the score criterion. In the following, we propose a new heuristic leading to excellent results for the optimization of W_Π. It is based on average-linkage and transfer methods followed by a stochastic optimization procedure.

- Firstly, we apply an ascending hierarchical method that we call Fusion. Starting from the atomic partition P_0, we join, at each step, both classes maximizing the resulting partition score. These are the classes whose the between-class pair average weight is maximum. The process stops when no more fusion leads to increase the criterion. The obtained partition $\pi = (X_1, \ldots, X_p)$ is such that every partition π_{ij} obtained by gathering the classes X_i and X_j has a weaker score: $W_\Pi(\pi_{ij}) < W_\Pi(\pi)$; doing so, the number of classes is automatically determined.
- Secondly, we implement a transfer procedure. We begin with calculating the weight of the assignment of each element x_i to each class X_k of π by $K(i,k) = \sum_{x_j \in X_k} w(i,j)$. If x_i belongs to X_k, $K(i,k)$ denotes the contribution of x_i to its class, and to $W_\Pi(\pi)$. Otherwise, it corresponds to the weight of a possible assignment to another class $X_{k'}$ and the difference $K(i,k') - K(i,k)$ is the variation of the criterion due to the transfer of x_i from class X_k to class $X_{k'}$. Our procedure consists in selecting, at each step, the element x_i and class $X_{k'}$ maximizing this variation, then (unless $K(i,k') < 0$) in moving x_i from X_k to $X_{k'}$. Let us notice that $X_{k'}$ may be created, if there is no existing class to which x_i positively contributes. In this last case, the element becomes a singleton and has a null contribution to the score, thus increasing the criterion. From now on, we denote by π the partition obtained at the end of the process.
- Finally, we add a stochastic optimization procedure to the two aforementioned deterministic steps. Having observed that the transfer procedure is very fast, we decide to apply it to random partitions obtained from the best current one by swapping random elements taken in two classes. For that task, two parameters have to be defined: the maximum number of swaps to start transfers ($SwapMax$) and the maximum number of consecutive trials without improving W_Π (NbT).

Thanks to the simulation protocol given in section 4.1, allowing to generate profiles on which the optimal consensus partition can be calculated, we have shown (Guénoche, 2011) that the FT method provides results that are optimal in more than 80% of cases, up to $n = m = 100$, and always very near from optimum, even for very difficult problems. We have also compared FT to other heuristics such as improving by transfers a random partition or the partition belonging to the profile which is the central one, and also to the *méthode de Louvain* (Blondel et al., 2008) which can be applied to any complete graph with positive and negative weights. The Fusion-Transfer method performs better than the others in the average.

Example 3
In the median partition of the profile in Table 1 there are only small classes, 7 of them being reduced to a single element. The score of each class is indicated, and also a robustness coefficient, ρ equal to the percentage of judges joining pairs of this class. This partition, also given by the Fusion-Transfer algorithm, has the optimal score equal to 34.

- Class 1 : (1, 5) ($Score = 9, \rho = 0.765$)
- Class 2 : (2, 10) ($Score = 11, \rho = 0.824$)
- Class 3 : (3, 6) ($Score = 9, \rho = 0.765$)
- Class 4 : (7, 8, 15) ($Score = 5, \rho = 0.549$)
- Singletons : (4|9|11|12|13|14|16)

3 Tree Representation of Partitions

In the beginning of the nineties, in order to determine the collective categories corresponding to a partition profile, J.P. Barthélemy collaborating with D. Dubois came up with the idea of measuring a distance between items and of representing it in the form of an X-tree. An X-tree is a tree such that its leaves (external vertices) are the elements of X, its nodes (internal vertices) have at least degree 3 and its edges have a non negative length (Barthélemy & Guénoche, 1991). To each X-tree A is associated a tree distance D_A such that $D_A(x, y)$ is the path length in the tree between leaves x and y ; it is the sum of the edge lengths along this single path. So, for a given distance D between items, an X-tree A, whose tree distance D_A is as near as possible to D, is searched. This is an approximation problem.

To equip X with a metric allows to go from individual judgments to collective categories, via subtrees. An item is connected to a set of elements that form a subtree, not because it is nearer, as in a hierarchical tree, but because it is associated to the others elements of this subtree at the opposite of the pairs located outside this subtree. This is the notion of *score* developed by Sattah & Tversky (1977) which makes that a pair (x, y) is opposed in the tree to another pair (z, t) because:

$$D(x, y) + D(z, t) \leq \min\{D(x, z) + D(y, t), D(x, t) + D(y, z)\}. \tag{3}$$

It means that at least one edge separates pair (x, y) from pair (z, t).

This notion is different to that of score of the median consensus section so that we use the term *weight* in place of score in the sequel. Precisely, the weight of a pair (x, y) is the number of pairs (z, t) satisfying Equation 3. The Sattah & Tversky algorithm, ADDTREE, aims at gathering, at each step, the maximum weight pairs and then builds an X-tree associated to a distance D.

The problem of how to choose a metric on X based on a partition profile has been solved as follows: since partitions essentially consist of relations on the either joined or separated pairs of X, a natural distance between x and y is the number of partitions of the profile in which x and y are separated, that is the *split distance* D_s. With the notations of Section 2,

$$D_s(x_i, x_j) = \mid \{P(i) \neq P(j)\}_{P \in \Pi} \mid = m - T_{i,j}.$$

Example 4

Table 3. The split distance between pieces of music

	1	2	3	4	5	6	7	8	9	10	11	12	13	14	15	16
1	0	14	16	13	4	16	14	17	13	13	17	15	9	17	17	16
2	14	0	11	15	15	12	11	15	16	3	16	17	15	16	15	14
3	16	11	0	15	16	4	15	16	17	12	16	15	10	17	16	9
4	13	15	15	0	11	17	16	16	15	15	11	14	15	16	15	17
5	4	15	16	11	0	16	15	16	12	14	16	15	12	17	16	16
6	16	12	4	17	16	0	16	16	17	13	15	16	8	16	17	9
7	14	11	15	16	15	16	0	6	17	12	14	16	17	10	11	11
8	17	15	16	16	16	16	6	0	14	16	10	16	17	9	6	11
9	13	16	17	15	12	17	17	14	0	16	13	10	15	14	10	17
10	13	3	12	15	14	13	12	16	16	0	17	15	14	17	16	15
11	17	16	16	11	16	15	14	10	13	17	0	13	17	10	12	13
12	15	17	15	14	15	16	16	16	10	15	13	0	16	11	15	16
13	9	15	10	15	12	8	17	17	15	14	17	16	0	17	16	11
14	17	16	17	16	17	16	10	9	14	17	10	11	17	0	13	14
15	17	15	16	15	16	17	11	6	10	16	12	15	16	13	0	13
16	16	14	9	17	16	9	11	11	17	15	13	16	11	14	13	0

3.1 X-Trees and Subtrees

Initially, the tree has been built using the ADDTREE method (cf. Barthélemy & Guénoche, 1991). Let us remind that ADDTREE is a clustering (ascending) method such that at each iteration:

- the weight of each pair is evaluated by enumerating quartets;
- the maximal weight pair is joined and connected to a new node in the tree.
- the edge lengths are calculated (by formulae that are not displayed here);
- the dimension of the distance table is reduced, replacing the joined elements by their common adjacent node in the tree.

The main drawback of ADDTREE is its complexity (in $O(n^4)$ at each iteration). Therefore, the NJ method (Saitou & Ney, 1987) has subsequently been used in place of ADDTREE. Moreover, NJ tends to have more ability to fit tree distances and to recover known trees.

Unlike hierarchical trees, X-trees are not rooted so that the notion of subtree has to be clarified. Indeed, an X-tree is a set of bipartitions (splits), each of them being defined by an internal edge of the tree setting on both sides one class against the other. Since there are $n-3$ internal edges in a fully resolved tree, there are $2(n-3)$ possible classes or subtrees that are not reduced to 1 or $n-1$ elements.

Reading these X-trees is usually made by considering the length of the internal edges separating two subtrees: the longer is an edge, the more robust can be appreciated the corresponding subtree so that psychologist can interpret it as a collective category underlined by the length. Such long edges, with probably above average lengths indicate *well separated* classes chosen by the user according to the tree. But their number remains to define. For that task, we use the number of classes, with more than one element, of the consensus partition.

Example 5
The tree given by the NJ method applied to the split distance in Table 4 is represented in Figure 1. Classes $1, 5|2, 10|3, 6|7, 8, 15$ are subtrees, but .. As there are 4 classes with at least 2 elements in the consensus partition, we look for the 4 best separated subtrees :

- Length = 5.341 → Class 1 : (2, 10)
- Length = 3.766 → Class 2 : (1, 5)
- Length = 2.477 → Class 3 : (3, 6)
- Length = 2.034 → Class (3, 6, 13, 16) → is eliminated since it contains (3 , 6) previously retained,
- Length = 1.984 → Class 4 : (9, 12) (but its score is equal to -3 !)

- Length = 1.428 → Class 5 : (7, 8, 15) ... does not belong to the 4 best separated subtrees !

4 Adequacy of Both Methods

In order to assess if the two above mentioned methods are congruent, we have set up a simulation protocol and defined several criteria allowing to quantify their adequacy.

4.1 Generation of More or Less Scattered Random Profiles

We start from a partition of X in p balanced classes, which is the initial partition of the profile. Then, we generate $m-1$ partitions by applying t random transfers to the initial one. A transfer consists in assigning an element taken at random to a class of the current partition or to a new class. For the first transfer, one class between 1 and $p+1$ is selected at random ; for the second, one class between 1

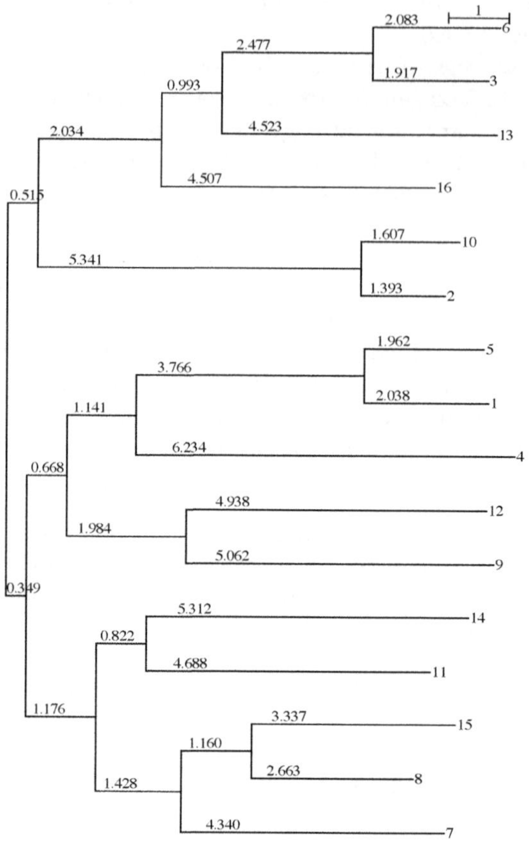

Fig. 1. The X-tree of the 16 pieces of music with the edge lengths

and $p+2$ is uniformly chosen if a new class has been created and so on. Therefore, generally the obtained partitions do not have the same number of classes.

For fixed n and m and according to the value of t, we obtain either homogeneous profiles for which the consensus partition is the initial one or very scattered profiles for which the consensus is, most of times the atomic partition. Varying the numbers of initial classes and transfers, we obtain either strong categorization problems around the classes of the initial partition or weak categorization problems with few joined pairs in most of partitions, leading to a consensus partition with high number of classes and low score.

4.2 Some Criteria

Thus, from each profile, we build the consensus partition (π) and the A tree that best approximate the split distance. We then calculate the score of each class of π and each subtree of A by the sum of scores of the joined pairs. This allows to compute two partitions from A only made with subtrees and eventually completed by singletons :

- P_A maximizing the W_Π score function;
- P_S made with the best separate subtrees. Indeed, the median partition indicates the optimal number of collective categories which is the number N_c of classes of π with more than one element. This leads to keep in P_S the subtrees with maximal score corresponding to the N_c splits with greatest edge lengths. That is what would do a user knowing in advance the number of classes to select. These N_c classes are used to measure the score $W_\Pi(P_S)$ of the best separated classes in A.

In Table 1, we display the values $W_\Pi(\pi)$, $W_\Pi(P_A)$ and $W_\Pi(S)$ as well as three criteria:

- One can compare, for each class of π containing at least two elements, the size of the class and the one of the smallest subtree including it. This criterion gives an idea of how similar the X-tree subtrees and the consensus partition classes are. The class and subtree sizes are generally very near or equal, so we indicate below the percentage τ_c of classes of the consensus partition that are identical to a subtree.
- The percentage of problems for which the score of partition P_A, built from the $2(n-3)$ subtrees, equals that of π. The former score is never greater than the latter but it is often equal.
- The percentage of problems for which the score of P_S equals that of the consensus partition.

4.3 Results

Let us recall that n is the number of classified items, m is the number of partitions, p is the class number of the initial partition of the profile and t is the number of transfers done from the initial partition in view to generate the profile. These results are the average values over 100 profiles.

n = m	p	t	$W_\Pi(\pi)$	$W_\Pi(P_A)$	$W_\Pi(S)$	τ_c	$\pi = P_A$	$\pi = S$
10	3	3	40.2	40.2	39.6	.98	.98	.79
10	2	5	33.9	33.2	28.8	.83	.80	.25
20	3	5	463.4	454.1	462.9	.99	.94	.92
20	5	10	33.0	32.8	-3.4	.92	.92	.01
20	3	15	11.8	11.2	-114.2	.83	.79	.04
50	5	10	4954.7	4954.7	4954.7	1.0	1.0	1.0
50	10	20	233.5	231.7	-10.9	.92	.66	.00
50	5	30	29.8	29.4	-1876.9	.86	.84	.00

Table 1 - Score of the consensus partition π, of the best partition P_A in the tree and of the best-separated classes S.

5 Conclusions

One first concluding remark is that the idea of looking for a consensus categorization via X-trees was pertinent. Whatever the hardness of the problem is,

X-trees include the consensus partition classes. Most of 90 % of the consensus classes are subtrees or they vary from at most one to two elements otherwise. Moreover, the best partitions of the trees into subtrees lead to scores close to optimal ones.

The second concluding remark is that it is not always easy to *read* these trees. The best subtrees, and consequently classes, do not necessarily correspond to the longest edges, and the score of the best separated classes is noticeably weaker than that of the consensus partition as soon as the problem gets harder.

The difficulty relies on the choice of the classes in the tree. Guénoche & Garretta (2001) attempted to appreciate the robustness of the internal edges by enumerating the number of quartets whose topology supports each edge, which can be counted comparing the three sums in Formula 3. This is a general measure for any distance based tree reconstruction, but in the case of distances between partitions, the score of classes corresponding to subtrees is a much better criterion.

For the psychologist who gathered the data, it is very disappointing to get a consensus partition with many singletons and only small classes. What could be the conclusion ? Is there no common opinion in this profile, or is the method not appropriate to detect it ? May be there are several opinions and, when they are merged, no consensus can appear. Extending the median approach, one can propose two complementary algorithms.

For a set of categorizations with no clear collective one, some extensions of the median consensus methodology allow to analyze a scattered profile giving either a weak consensus or several consensus partitions corresponding to divergent subgroups.

5.1 A Weak Consensus

If there are no majority pair, the atomic partition is the consensus partition. It is not informative, and it suggests that there is no valid class for this profile. However, the majority threshold ($m/2$) can be decreased, resulting in higher values in the complete weighted graph. Therefore, there will be more positive pairs. The consensus partition is no more a median, but it can always be interpreted as a common opinion, even if it is not supported by a majority. Instead of $w_{i,j} = T_{i,j} - m/2$ a threshold σ can be chosen and we pose :

$$w_{i,j} = T_{i,j} - \sigma.$$

When $\sigma < m/2$, the weights will be increased and larger classes with positive weight could appear.

Example 6

For the 17 judges profile in Table 1, the majority threshold is equal to 8.5. Fixing $\sigma = 6$, one get an optimal score partition and classes :

- Class 1 : 1, 5 ($Score = 14, \rho = 0.765$)
- Class 2 : 2, 10 ($Score = 16, \rho = 0.824$)

- Class 3 : 3, 6, 13, 16 ($Score = 30, \rho = 0.500$)
- Class 4 : 7, 8, 14, 15 ($Score = 22, \rho = 0.461$)
- Class 5 : 9, 12 ($Score = 2, \rho = 0.412$)
- Singletons : 4 | 11

Compared to the median partition, Classes 1 and 2 remain the same, Classes 3 and 4 are enlarged, and Class 5 appears with a robustness coefficient lower than .5 as for the new Class 4.

5.2 Subgroups of Experts

To cluster judges according to their opinions, the profile partitions are to be compared to put together close partitions, making appear homogeneous subgroups of experts. Comparing partitions is usually done with, *distance indices* between partitions (Rand, Jaccard, ..), which are similarity functions, with high values when partitions are close. But according to the partition neighborhood established with transfers, one can recommend the *Transfer distance.* For two partitions P and Q it counts the smallest number of transfers to pass from one to the other. This distance is implicit in the Régnier's article, clearly defined by Day (1981) with many other editing distances between partitions, and precisely analyzed by Denœud in her PhD and its article in 2008. To establish these subgroups, the class diameter seems to be natural.

Example 7
The hierarchy of bipartitions with minimum diameter (Guénoche et al. 1991), applied to the transfer distance, is represented in Figure 2. It clearly indicates two balanced groups of experts. Their consensus partitions, at the majority threshold, give different opinions :

- Group 1 (Amelie, Clement, Florian, Jean-Philippe, Katrin, Lauriane, Paul, Vincent)
  - Class 1 : 1, 5, 13 ($Score = 8, \rho = 0.667$)
  - Class 2 : 2, 3, 6, 10 ($Score = 10, \rho = 0.604$)
  - Class 4 : 7, 8, 14, 16 ($Score = 14, \rho = 0.646$)
  - Singletons : 4|9|11|12|15
- Group 2 (Arthur, Aurore, Charlotte, Clementine, Jeremie, Julie, Louis, Lucie, Madeleine)
  - Class 1 : 1, 5 ($Score = 7, \rho = 0.889$)
  - Class 2 : 2, 7, 10 ($Score = 9, \rho = 0.714$)
  - Class 3 : 3, 6, 13, 16 ($Score = 28, \rho = 0.833$)
  - Class 4 : 4, 11 ($Score = 1, \rho = 0.556$)
  - Class 5 : 8, 15 ($Score = 5, \rho = 0.778$)
  - Class 6 : 9,12 ($Score = 1, \rho = 0.556$)
  - Singletons : 14

A software corresponding to the Fusion-Transfer method to establish a consensus partition, at a chosen threshold, either from a set of partitions, or from a nominal variables array, can be loaded from http://bioinformatics.lif.univ-mrs.fr/.

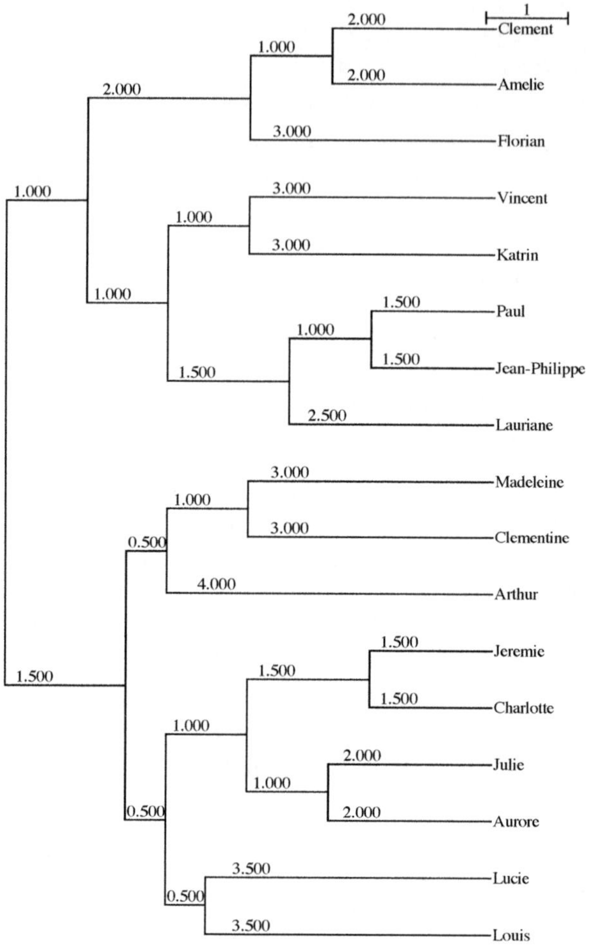

Fig. 2. Expert hierarchy from the transfer distance between their partitions

References

1. Barthélemy, J.P., Guénoche, A.: Trees and Proximity Representations. J. Wiley, London (1991)
2. Barthélemy, J.P.: Similitude, arbres et typicalités. In: Dubois, D. (ed.) Sémantique et cognition - Catégories, prototypes et typicalité. du CNRS, Paris (1991)
3. Blondel, V., Guillaume, J.L., Lambiotte, R., Lefebvre, E.: Fast unfolding of communities in large networks. Journal of Stat. Mechanics: Theory and Experiment, 10008 (2008)
4. Day, W.: The complexity of computing metric distances between partitions. Math. Soc. Sci. 1, 269–287 (1981)
5. Denœud, L.: Transfer distance between partitions. Advances in Data Analysis and Classification 2, 279–294 (2008)

6. Grötschel, M., Wakabayashi, Y.: A cutting plan algorithm for a clustering problem. Math. Program. 45, 59–96 (1989)
7. Guénoche, A., Hansen, P., Jaumard, B.: Efficient algorithms for divisive hierarchical clustering with the diameter criterion. Journal of Classification 8(1), 5–30 (1991)
8. Guénoche, A., Garreta, H.: Can We Have Confidence in a Tree Representation? In: Gascuel, O., Sagot, M.-F. (eds.) JOBIM 2000. LNCS, vol. 2066, pp. 45–53. Springer, Heidelberg (2001)
9. Guénoche, A.: Consensus of partitions: a constructive approach. Advances in Data Analysis and Classification (to appear, 2011)
10. Régnier, S.: Sur quelques aspects mathématiques des problèmes de classification automatique. Mathématiques et Sciences humaines 82, 13–29 (1983); reprint of I.C.C. bulletin 4, 175–191 (1965)
11. Sattah, S., Tversky, A.: Additive Similarity Trees. Psychometrica 42, 319–345 (1977)
12. Saitou, N., Nei, M.: The neighbor-joining method: a new method for reconstructing phylogenetic trees. Mol. Biol. Evol. 4, 406–425 (1987)
13. Zahn, C.T.: Graph-theoretical methods for detecting and describing gelstalt clusters. IEEE Trans. on Computers 20 (1971)

Efficiently Eliciting Preferences from a Group of Users

Greg Hines and Kate Larson

Cheriton School of Computer Science
University of Waterloo
Waterloo, Canada
{ggdhines,klarson}@cs.uwaterloo.ca.ca
www.cs.uwaterloo.ca/~ggdhines

Abstract. Learning about users' preferences allows agents to make intelligent decisions on behalf of users. When we are eliciting preferences from a group of users, we can use the preferences of the users we have already processed to increase the efficiency of the elicitation process for the remaining users. However, current methods either require strong prior knowledge about the users' preferences or can be overly cautious and inefficient. Our method, based on standard techniques from non-parametric statistics, allows the controller to choose a balance between prior knowledge and efficiency. This balance is investigated through experimental results.

Keywords: Preference elicitation.

1 Introduction

There are many real world problems which can benefit from a combination of research in both decision theory and game theory. For example, we can use game theory in studying the large scale behaviour of the Smart Grid [6]. At the same time, software such as Google's powermeter can interact with Smart Grid users on an individual basis to help them create optimal energy use policies.

Powermeter currently only provides people with information about their energy use. Future versions of powermeter (and similar software) could make choices on behalf of a user, such as how much electricity to buy. This would be especially useful when people face difficult choices involving risk; for example, is it worth waiting until tomorrow night to run my washing machine if there is a 10% chance that the electricity cost will drop by 5%? To make intelligent choices, we need to elicit preferences from each household by asking them a series of questions. The fewer questions we need to ask, the less often we need to interrupt a household's busy schedule.

In preference elicitation, we decide whether or not to ask additional questions based on a measure of confidence in the currently selected decision. For example, we could be 95% confident that waiting until tomorrow night to run the washing machine is the optimal decision. If our confidence is too low, then we need to ask additional questions to confirm that we are making the right decision.. Therefore, to maximize efficiency, we need an accurate measurement of confidence.

R.I. Brafman, F. Roberts, and A. Tsoukiàs (Eds.): ADT 2011, LNAI 6992, pp. 96–107, 2011.

Confidence in a decision is often measured in terms of *regret*, or the loss in utility the user would experience if the decision in question was taken instead of some (possibly unknown) optimal decision. Since the user's preferences are private, we cannot calculate the actual regret. Instead, we must estimate the regret based on our limited knowledge.

Regret estimates, or measures, typically belong to one of two models. The first measure, *expected regret*, estimates the regret by assuming that the user's utility values are drawn from a known prior distribution [2]. However, there are many settings where it is challenging or impossible to obtain a reasonably accurate prior distribution. The second measure, *minimax regret*, makes no assumptions about the user's utility values and provides a worst-case scenario for the amount of regret [7]. In many cases, however, the actual regret may be considerably lower than the worst-case regret. This difference may result in needless querying of the user.

In this paper, we propose a new measure of regret that achieves a balance between expected regret and minimax regret. As with expected regret, we assume that all users' preferences are chosen according to some single probability distribution [2]. We assume no knowledge, however, as to what this distribution is. Instead, we are allowed to make multiple *hypotheses* as to what the distribution may be. Our measure of regret is then based on an aggregation of these hypotheses.

Our measurement of regret will never be higher than minimax regret, and in many cases we can provide a considerably lower estimate than minimax regret. As long as one of the hypotheses is correct, even without knowing which is the correct hypothesis, we can show that our estimate is a proper upper bound on the actual regret. Since our measure allows for any number of hypotheses, this flexibility gives the controller the ability to decide on a balance between speed (with fewer hypotheses) and certainty (with more hypotheses). Furthermore, when we have multiple hypotheses, our approach is able to gather evidence to use in rejecting the incorrect hypotheses. Thus, the performance of our approach can improve as we process additional users.

Although our approach relies on standard techniques from non-parametric statistics, we never assign hypotheses a probability of correctness. This makes our method non-Bayesian. While a Bayesian approach might be possible, we discuss why our method is simpler and more robust.

2 The Model

Consider a set of possible outcomes $\mathbb{X} = [x_\perp, \ldots, x_\top]$. A user exists with a private utility function u. The set of all possible utility functions is $\mathbb{U} = [0,1]^{|\mathbb{X}|}$. There is a finite set of decisions $D = [d_1, \ldots, d_n]$. Each decision induces a probability distribution over $\mathbb{X}$, *i.e.*, $\Pr_d(x_i)$ is the probability of the outcome x_i occurring as a result of decision d. We assume the user follows *expected utility theory* (EUT), *i.e.*, the overall expected utility for a decision d is given by

$$EU(d, u) = \sum_{x \in \mathbb{X}} \Pr_d(x) u(x).$$

Since expected utility is unaffected by positive affine transformations, without loss of generality we assume that
$u : \mathbb{X} \rightarrow [0, 1]$ with $u(x_\perp) = 0$ and $u(x_\top) = 1$.

Since the user's utility function is private, we represent our limited knowledge of her utility values as a set of constraints. For the outcome x_i, we have the constraint set

$$[C_{\min}(x_i), C_{\max}(x_i)],$$

which gives the minimum and maximum possible values for $u(x_i)$, respectively. The complete set of constraints over $\mathbb{X}$ is $C \subseteq \mathbb{U}$.

To refine C, we query the user using *standard gamble queries* (SGQs) [3]. SGQs ask the user if they prefer the outcome x_i over the gamble $[1 - p; x_\perp, p; x_\top]$, *i.e.*, having outcome $x_\top$ occur with probability p and otherwise having outcome $x_\perp$ occur. By EUT, if the user says yes, we can infer that $u(x_i) > p$. Otherwise, we infer that $u(x_i) \leq p$.

2.1 Types of Regret

Regret, or loss of utility, can be used to help us choose a decision on the user's behalf. We can also use regret as a measure of how good our choice is. There are two main models of regret which we describe in this section.

Expected Regret. Suppose we have a known family of potential users and a prior probability distribution, P, over $\mathbb{U}$ with respect to this family. In this case, we can sample from P, restricted to C, to find the expected utility for each possible decision. We then choose the decision d^* which maximizes expected utility. To estimate the regret from stopping the elicitation process and recommending d^* (instead of further refining C), we calculate the *expected regret* as [2]

$$\int_C [EU(d^*(u), u) - EU(d^*, u)] P(u) du, \quad (1)$$

where $d^*(u)$ is the decision which maximizes expected utility given utility values u.

The disadvantage of expected regret is that we must have a reasonable prior probability distribution over possible utility values. This means that we must have already dealt with many previous users whom we know are drawn from the same probability distribution as the current users. Furthermore, we must know the exact utility values for these previous users. Otherwise, we cannot calculate $P(u)$ in Equation 1.

Minimax Regret. When there is not enough prior information about users' utilities to accurately calculate expected regret, and in the extreme case where we have no prior information, an alternative measure to expected regret is *minimax regret.* Minimax regret minimizes the worst-case regret the user could experience and makes no assumptions about the user's utility function.

To define minimax regret, we first define *pairwise maximum regret* (PMR) [7]. The PMR between decisions d and d' is

$$PMR(d, d', C) = \max_{u \in C} \{EU(d', u) - EU(d, u)\}. \quad (2)$$

Table 1. A comparison of the initial minimax and actual regret for users with and without the monotonicity constraint

Regret	Nonmonotonic	Monotonic
Minimax	0.451	0.123
Actual	0.052	0.008

The PMR measures the worst-case regret from choosing decision d instead of d'. The PMR can be calculated using linear programming. PMR is used to find a bound for the actual regret, $r(d)$, from choosing decision d, *i.e.*,

$$r(d) \leq MR(d, C) = \max_{d' \in D} PMR(d, d', C), \tag{3}$$

where $MR(d, C)$ is the *maximum regret* for d given C. For a given C, the *minimax decision* d^* guarantees the lowest worst-case regret, *i.e.*,

$$d^*(C) = \arg\min_{d \in D} MR(d, C). \tag{4}$$

The associated *minimax regret* is [7]

$$MMR(C) = \min_{d \in D} MR(d, C). \tag{5}$$

Wang and Boutilier argue that in the case where we have no additional information about a user's preferences, we should choose the minimax decision [7].

The disadvantage of minimax regret is that it can overestimate the actual regret, which can result in unnecessary querying of the user. To investigate this overestimation, we created 500 random users, each faced with the same 20 outcomes. We then picked 10 decisions at random for each user. Each user was modeled with the utility function

$$u(x) = x^{\eta}, \; x \in \mathbb{X} \tag{6}$$

with η picked uniformly at random between 0.5 and 1 and $\mathbb{X}$ some set of nonnegative outcomes. Equation 6 is commonly used to model peoples' utility values in experimental settings [5]. Table 1 shows the mean initial minimax and actual regret for these users. Since Equation 6 guarantees that each users' utility values are monotonically increasing, one possible way to reduce the minimax regret is to add a monotonicity constraint to the utility values in Equation 2. Table 1 also shows the mean initial minimax and actual regret when the monotonicity constraint is added. Without the monotonicity constraints, the minimax regret is, on average, 8.7 times larger than the actual regret. With the monotonicity constraints, while the minimax regret has decreased in absolute value, it is now 15.4 times larger than the actual regret.

It is always possible for the minimax regret and actual regret to be equal. The proof follows directly from calculating the minimax regret and is omitted for brevity. This means that despite the fact that the actual regret is often considerably less than the minimax regret, we cannot assume this to always be the case. Furthermore, even if we knew that the actual regret is less than the minimax regret, to take advantage of this knowledge, we need a quantitative measurement of the difference. For example, suppose we are in a situation where the minimax regret is 0.1. If the maximum actual regret we can tolerate is 0.01, can we stop querying the user? According to the results in Table 1, the minimax regret could range from being 8.7 times to 15.4 times larger than the actual regret. Based on these values, the actual regret could be as large as 0.0115 or as small as 0.006. In the second case, we can stop querying the user and in the first case, we cannot. Therefore, a more principled approach is needed.

2.2 Elicitation Heuristics

Choosing the optimal query can be difficult. A series of queries may be more useful together then each one individually. Several heuristics have been proposed to help. The *halve largest-gap* (HLG) heuristic queries the user about the outcome x which maximizes the utility gap $C_{\max}(x) - C_{\min}(x)$ [1]. Although HLG offers theoretical guarantees for the resulting minimax regret after a certain number of queries, other heuristics may work better in practice. One alternative is the *current solution* (CS) heuristic which weights the utility gap by $|\Pr_{d^*}(x) - \Pr_{d^a}(x)|$, where d^a is the "adversarial" decision that maximizes the pairwise regret with respect to d^* [1].

3 Hypothesis-Based Regret

We now consider a new method for measuring regret that is more accurate than minimax regret but weakens the prior knowledge assumption required for expected regret. We consider a setting where we are processing a group of users one at a time. For example, we could be processing a sequence of households to determine their preferences for energy usage. As with expected regret, we assume that all users' preferences are chosen i.i.d. according to some single probability distribution [2]. However, unlike expected regret, we assume the distribution is completely unknown and make no restrictions over what the distribution could be. For example, if we are processing households, it is possible that high income households have a different distribution than low income households. Then the overall distribution would just be an aggregation of these two.

Our method is based on creating a set of hypotheses about what the unknown probability distribution could be. Suppose we knew the correct hypothesis $\mathcal{H}^*$. Then for any decision d, we could calculate the *cumulative probability distribution* (cdf) $F_{d,\mathcal{H}^*|_C}(r)$ for the regret from choosing decision d restricted to the utility constraints C. We can calculate $F_{d,\mathcal{H}^*|_C}(r)$ using a Monte Carlo method. In this setting, we define the *probabilistic maximum regret* (PrMR) as

$$PrMR(d, \mathcal{H}^*|_C, p) = F^{-1}_{d,\mathcal{H}^*|_C}(p), \qquad (7)$$

for some probability p. That is, with probability p the maximum regret from choosing d given the hypothesis $\mathcal{H}^*$ and utility constraints C is $PrMR(d, \mathcal{H}^*|_C, p)$. The *probabilistic minimax regret* (PrMMR) is next defined as

$$PrMMR(\mathcal{H}^*|_C, p) = \min_{d \in d} PrMR(d, \mathcal{H}^*|_C, p).$$

Since we do not know the correct hypothesis, then we need to make multiple hypotheses. Let $\mathbb{H} = \{\mathcal{H}_1, \ldots\}$ be our set of possible hypotheses. With multiple hypotheses, we generalize our definition of PrMR and PrMMR to

$$PrMR(d, \mathbb{H}|_C, p) = \max_{\mathcal{H} \in \mathbb{H}} PrMR(d, \mathcal{H}|_C, p). \tag{8}$$

and

$$PrMMR(\mathbb{H}|_C, p) = \min_{d} PrMR(d, \mathbb{H}|_C, p), \tag{9}$$

respectively.

We can control the balance between speed and certainty by deciding which hypotheses to include in $\mathbb{H}$. The more hypotheses we include in $\mathbb{H}$ the fewer assumptions we make about what the correct hypothesis is. However, additional hypotheses can increase the PrMMR and may result in additional querying.

Since the PrMMR calculations take into account both the set of possible hypotheses and the set of utility constraints, the PrMMR will never be greater than the MMR. As our experimental results show, in many cases the PrMMR may be considerably lower than the MMR. At the same time PrMMR still provides a valid bound on the actual regret:

Proposition 1. *If $\mathbb{H}$ contains $\mathcal{H}^*$, then*

$$r(d) \leq PrMR(d, \mathbb{H}|_C, p) \tag{10}$$

with probability of at least p.

Proof. Proof omitted for brevity.

3.1 Rejecting Hypotheses

The correctness of Proposition 1 is unaffected by incorrect hypotheses in $\mathbb{H}$. However, the more hypotheses we include, the higher the calculated regret values will be. Therefore, we need a method to reject incorrect hypotheses.

Some hypotheses can never be rejected with certainty. For example, it is always possible that a set of utility values was chosen uniformly at random. Therefore, the best we can do is to reject incorrect hypotheses with high probability while minimizing the chances of accidentally rejecting the correct hypothesis.

After we have finished processing user i, we examine the utility constraints from that user and all previous users to see if there is any evidence against each of the hypotheses. Our method relies on the Kolmogorov-Smirnov (KS) one-sample test [4]. This is a standard test in non-parametric statistics. We use the KS test to compare the regret values we would see if a hypothesis $\mathcal{H}$ was true against the regret values we see in practice. The test statistic for the KS test is

$$T^{\mathcal{H}}_{d,i} = \max_{r} |F_{d,\mathcal{H}}(r) - \hat{F}_{d,i}(r)|, \tag{11}$$

where $\hat{F}_{d,i}(r)$ is an *empirical distribution function* (edf) given by

$$\hat{F}_{d,i}(r) = \frac{1}{i} \sum_{j \leq i} I(r_j(d) \leq r), \tag{12}$$

where

$$I(A \leq B) = \begin{cases} 1 & \text{if } A \leq B \\ 0 & \text{otherwise,} \end{cases}$$

and $r_j(d)$ is the regret calculated according to user j's utility constraints.

If $\mathcal{H}$ is correct, then as i goes to infinity,

$$\sqrt{i} \cdot T^{\mathcal{H}}_{d,i}$$

converges to the *Kolmogorov distribution* which does not depend on $F_{d,\mathcal{H}}$. Let K be the cumulative distribution of the Kolmogorov distribution. We reject $\mathcal{H}$ if

$$\sqrt{i} \cdot T^{\mathcal{H}}_{d,i} \geq K_\alpha, \tag{13}$$

where K_α is such that

$$\Pr(K \leq K_\alpha) = 1 - \alpha.$$

Unfortunately, we do not know $r_j(d)$ and therefore, cannot calculate $\hat{F}$. Instead we rely on Equation 3 to provide an upper bound for $r_j(d)$ which gives us a lower bound for $\hat{F}$, *i.e.*

$$\hat{F}_{d,i}(r) \geq L_{d,i}(r) := \frac{1}{i} \sum_{j \leq i} I(MR(d, C_j) \leq r), \tag{14}$$

where C_j is the utility constraints found for user j. We assume the worst case by taking equality in Equation 14. As a result, we can give a lower bound to Equation 11 with

$$T^{\mathcal{H}}_{d,i} \geq \max\{0, \max_r (L_{d,i}(r) - F_{d,\mathcal{H}}(r))\}. \tag{15}$$

This statistic is illustrated in Figure 1. Since $L_{d,i}(r)$ is a lower bound, if $L_{d,i}(r) < F_{d,\mathcal{H}}(r)$, we can only conclude that $T^{\mathcal{H}}_{d,i} \geq 0$.

If $\mathcal{H}$ is true, then the probability that we incorrectly reject $\mathcal{H}$ based on $T^{\mathcal{H}}_{d,i}$ for a specific decision d is at most α. However, since we examine $T^{\mathcal{H}}_{d,i}$ for every decision, the probability of incorrectly rejecting $\mathcal{H}$ is much higher. (This is known as the *multiple testing problem.*) Our solution is to use the *Bonferroni Method* where we reject $\mathcal{H}$ if [8]

$$\max_{d \in D} \sqrt{i} \cdot T^{\mathcal{H}}_{d,i} \geq K'_\alpha,$$

where

$$\Pr(K \leq K'_\alpha) = \frac{1 - \alpha}{|D|}.$$

Using this method, the probability of incorrectly rejecting $\mathcal{H}$ is at most α.

3.2 Heuristics for Rejecting Hypotheses

A major factor in how quickly we can reject incorrect hypotheses is how accurate the utility constraints are for the users we have processed. In many cases, it may be beneficial in the long run to spend some extra time querying the initial users for improved

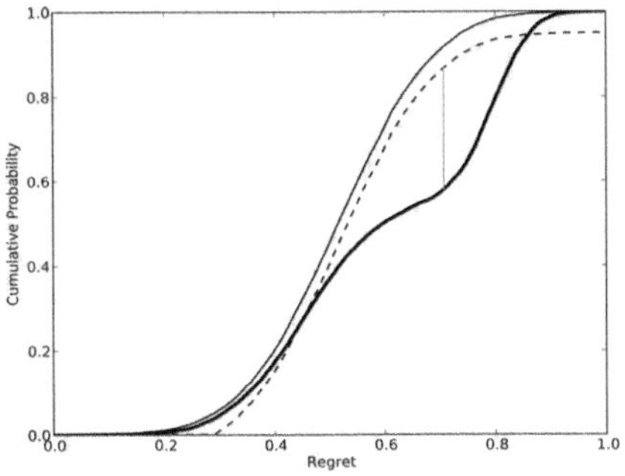

Fig. 1. An example of the KS one sample test. Our goal is to find evidence against the hypothesis $\mathcal{H}$. The KS test (Equation 11) focuses on the maximum absolute difference between the cdf $F_{d,\mathcal{H}}(r)$ (the thick lower line) and the edf $\hat{F}_{d,i}(r)$ from Equation 12 (the thin upper line). However, since we cannot calculate $\hat{F}_{d,i}(r)$, we must rely on Equation 14 to give the lower bound $L_{d,i}(r)$ shown as the dashed line. As a result, we can only calculate the maximum positive difference between $L_{d,i}(r)$ and $F_{d,\mathcal{H}}(r)$. This statistic, given in Equation 15, is shown as the vertical line. We reject the hypothesis $\mathcal{H}$ if this difference is too big, as according to Equation 13.

utility constraints. To study these tradeoffs between short term and long term efficiency we used a simple heuristic, $R(n)$. With the $R(n)$ heuristic, we initially query every user for the maximum number of queries. Once we have rejected n hypotheses, we query only until the PrMMR is below the given threshold. While this means that the initial users will be processed inefficiently, we will be able to quickly reject incorrect hypotheses and improve the long term efficiency over the population of the users.

4 Experimental Results

For our experiments, we simulated helping a group of households choose optimal policies for buying electricity on the Smart Grid. In this market, each day people pay a lump sum of money for the next day's electricity. We assume one aggregate utility company that decides on a constant per-unit price for electricity which determines how much electricity each person receives. We assume a competitive market where there is no profit from speculating.

A person's decision, c, is how much money to pay in advance. For simplicity, we consider only a finite number of possible amounts. There is uncertainty both in terms of how much other people are willing to pay and how much capacity the system will have the next day. However, based on historical data, we can estimate, for a given amount of payment, the probability distribution for the resulting amount of electricity. Again, for simplicity, we consider only a finite number of outcomes. Our goal is to process a set of Smart Grid users and help them each decide on their optimal decision.

Each person's overall utility function is given by

$$u(c, E) = u_{\text{elect}}(E) - c,$$

where E is the amount of electricity they receive.

All of the users' preferences were created using the probability distribution:

$\mathcal{H}^*$: The values for u_{elect} are given by

$$u_{\text{elect}}(E) = E^{\eta}, \tag{16}$$

where $0 \leq \eta \leq 1$ is chosen uniformly at random for each user. We are interested in utility functions of the form in given in Equation 16 since it is often used to describe peoples' preferences in experimental settings [5].

To create a challenging experiment, we studied the following set of hypotheses which are feasible with respect to $\mathcal{H}^*$.

$\mathcal{H}_1$: The values for u_{elect} are chosen uniformly at random, without a monotonicity constraint.

$\mathcal{H}_2$: The values for u_{elect} are chosen according to Equation 16, where $0 \leq \eta \leq 1$ is chosen according to a Gaussian distribution with mean 0.7 and standard deviation 0.1.

$\mathcal{H}_3$: The values for u_{elect} are chosen according to

$$u_{\text{elect}}(E) = E^{\eta} + \epsilon$$

where $0 \leq \eta \leq 1$ is chosen uniformly at random and ϵ is chosen uniformly at random between -0.1 and 0.1.

For these experiments we created 200 users whose preferences were created according to $\mathcal{H}^*$. Each user had the same 15 possible cost choices and 15 possible energy outcomes. We asked each user at most 100 queries. Our goal was to achieve a minimax regret of at most 0.01. We rejected hypotheses when $\alpha < 0.01$. (This is typically seen as very strong evidence against a hypothesis [8].) For all of our experiments, we chose p in Equation 7 to be equal to 1.

As a benchmark, we first processed users relying just on minimax regret (with and without the monotonicity constraint). The average number of queries needed to solve each user is shown in Table 2. We experimented with both the HLG and CS elicitation heuristics. Without the monotonicity constraint, the average number of queries was 42.0 using HLG and 66.7 using CS. With the monotonicity constraint, the average was 22.7 using HLG and 53.6 using CS. Table 2 also shows the results using hypothesis-based regret with $\mathbb{H} = \{\mathcal{H}^*\}$, *i.e.* what would happen if we knew the correct distribution. In this case, using HLG the average number of queries is 2.4 and using CS the average is 13.3. These results demonstrate that the more we know about the distribution, the better the performance is.

Our next experiments looked at the performance of hypothesis-based regret using the $R(0)$ heuristic with the following sets for $\mathbb{H}$: $\{\mathcal{H}^*, \mathcal{H}_1\}$, $\{\mathcal{H}^*, \mathcal{H}_2\}$, and $\{\mathcal{H}^*, \mathcal{H}_3\}$. Since, as shown in Table 2, the HLG elicitation strategy outperforms the CS strategy for our model, we used the HLG strategy for the rest of our experiments. The average number of queries needed, shown in Table 3, was 23.7, 2.4 and 12.9 for $\mathcal{H}_1$, $\mathcal{H}_2$, and $\mathcal{H}_3$, respectively. Both $\mathcal{H}_1$ and $\mathcal{H}_3$ overestimate the actual regret, resulting in an increase in the number of queries needed. While $\mathcal{H}_2$ is not identical to $\mathcal{H}^*$, for our simulations,

Table 2. The mean number of queries needed to process a user using either the HLG or CS strategy based on different models of regret. Unless otherwise noted, all users were solved. The averages are based on only those users we were able to solve, *i.e.* obtain a regret of at most 0.01.

Regret	HLG	CS
Minimax	42.0	66.7 (135 users not solved)
Minimax with monotonicity	22.7	53.6 (143 users not solved)
Hypothesis-based regret with $\mathbb{H} = \{\mathcal{H}^*\}$	2.4	13.3

Table 3. Average number of queries using $R(0)$ heuristic for different hypotheses sets

$\mathbb{H}$	Mean
$\{\mathcal{H}^*, \mathcal{H}_1\}$	24.7
$\{\mathcal{H}^*, \mathcal{H}_2\}$	2.4
$\{\mathcal{H}^*, \mathcal{H}_3\}$	12.9

the regret estimates provided by these two hypotheses are close enough that there is no increase in the number of queries when we include $\mathcal{H}_2$ in $\mathbb{H}$. We were unable to reject any of the incorrect hypotheses using $R(0)$.

We next experimented with the $R(1)$ heuristic the HLG elicitation strategy. We tested the same sets of hypotheses for $\mathbb{H}$ and the results are shown in Table 4. We were able to reject $\mathcal{H}_1$ after 5 users, which reduced the overall average number of queries to 7.4 when $\mathbb{H} = \{\mathcal{H}^*, \mathcal{H}_1\}$. Thus, we can easily differentiate $\mathcal{H}_1$ from $\mathcal{H}^*$ and doing so improves the overall average number of queries. With the additional querying in $R(1)$, we were able to quickly reject $\mathcal{H}_2$. However, since including $\mathcal{H}_2$ did not increase the average number of queries, there is no gain from rejecting $\mathcal{H}_2$ and as a result of the initial extra queries, the average number of queries rises to 8.29. It took 158 users to reject $\mathcal{H}_3$. As a result, the average number of queries increased to 80.0. This means it is relatively difficult to differentiate $\mathcal{H}_3$ from $\mathcal{H}^*$. In this case, while including $\mathcal{H}_3$ in $\mathbb{H}$ increases the average number of queries, we would be better off not trying to reject $\mathcal{H}_3$ when processing only 200 users.

Finally, we experimented with $\mathbb{H} = \{\mathcal{H}^*, \mathcal{H}_1, \mathcal{H}_2, \mathcal{H}_3\}$ using $R(n)$ with different values of n. The results are shown in Table 5. With $n = 0$ we are unable to reject any of the incorrect hypotheses, however the average number of queries is still considerably lower than for minimax regret results shown in Table 2. With $n = 1$, we are able to quickly reject $\mathcal{H}_1$ and, as a result, the average number of queries decreases to 15.0. For $n = 2$, we are able to also reject $\mathcal{H}_2$. However, $\mathcal{H}_2$ takes longer to reject and since $\mathcal{H}_2$ does not increase the number of queries, for $R(2)$, the average number of queries rises to 18.5. Finally, with $n = 3$, we are able to reject $\mathcal{H}_3$ as well as $\mathcal{H}_1$ and $\mathcal{H}_2$. While having $\mathcal{H}_3$ in $\mathbb{H}$ increases the number of queries, rejecting $\mathcal{H}_3$ is difficult enough that the average number of queries rises to 80.0.

These experiments show how hypothesis-based regret outperforms minimax regret. While this is most noticeable when we are certain of the correct hypothesis, our

Table 4. Average number of queries using $R(1)$ heuristic for different hypotheses sets

$\mathbb{H}$	Mean	Number of users needed to reject hypothesis
$\{\mathcal{H}^*, \mathcal{H}_1\}$	7.4	5
$\{\mathcal{H}^*, \mathcal{H}_2\}$	8.3	11
$\{\mathcal{H}^*, \mathcal{H}_3\}$	80.0	158

Table 5. Mean number of queries and number of users not solved for $\mathbb{H} = \{\mathcal{H}^*, \mathcal{H}_1, \mathcal{H}_2, \mathcal{H}_3\}$ using the $R(n)$ heuristic for different values of n. NR stands for not rejected.

$n =$	Mean	Number of users needed to reject $\mathcal{H}_1$,$\mathcal{H}_2$,$\mathcal{H}_3$
0	26.0	NR,NR,NR
1	15.0	5,NR,NR
2	18.5	5,11,NR
3	80.0	5,11,158

approach continues to work well with multiple hypotheses. The $R(n)$ heuristic can be effective at rejecting hypotheses, improving the long term performance of hypothesis-based regret.

5 Using a Bayesian Approach with Probabilistic Regret

An alternative method could use a Bayesian approach. In this case we start off with a prior estimate of the probability of each hypothesis being correct. As we processed each user, we would use their preferences to update our priors. A Bayesian approach would help us ignore unlikely hypotheses which might result in a high regret.

Unfortunately, there is no simple guarantee that the probabilities would ever converge to the correct values. For example, if we never queried any of the users and as a result, we only had trivial utility constraints for each user, the probabilities would never converge. However, finding some sort of guarantee of eventual convergence is not enough. We need to provide each individual user with some sort of guarantee. An individual user does not care whether we will eventually be able to choose the right decision, each user only cares whether or not we have chosen the right decision for them specifically. Therefore, for each user we need to bound how far away the current probabilities can be from the correct ones. We would also need to give a way of bounding the error introduced into our regret calculations from the difference between the calculated and actual probabilities. Again, these bounds depend on more than just the number of users we have processed. Given these complications of trying to apply a Bayesian approach, we argue that our approach is simpler and more robust.

6 Conclusion

In this paper we introduced hypothesis-based regret, which bridges expected regret and minimax regret. Furthermore, hypothesis-based regret allows the controller to decide on the balance between accuracy and necessary prior information. We also introduced a method for rejecting incorrect hypotheses which allows the performance of hypothesis-based regret to improve as we process additional users.

While the $R(n)$ heuristic is effective it is also simple. We are interested in seeing whether other heuristics are able to outperform $R(n)$. One possibility is create a measure of how difficult it would be to reject a hypothesis. We are also interested in using $\mathcal{H}$ to create better elicitation heuristics.

References

1. Boutilier, C., Patrascu, R., Poupart, P., Schuurmans, D.: Constraint-based optimization and utility elicitation using the minimax decision criterion. Artificial Intelligence 170, 686–713 (2006)
2. Chajewska, U., Koller, D., Parr, R.: Making rational decisions using adaptive utility elicitation. In: Proceedings of the National Conference on Artificial Intelligence (AAAI), Austin, TX, pp. 363–369 (2000)
3. Keeney, R., Raiffa, H.: Decisions with multiple objectives: Preferences and value tradeoffs. Wiley, New York (1976)
4. Pratt, J.W., Gibbons, J.D.: Concepts of Nonparametric Theory. Springer, Heidelberg (1981)
5. Tversky, A., Kahneman, D.: Advances in prospect theory: Cumulative representation of uncertainty. Journal of Risk and Uncertainty 5(4), 297–323 (1992), `http://ideas.repec.org/a/kap/jrisku/v5y1992i4p297-323.html`
6. Vytelingum, P., Ramchurn, S.D., Voice, T.D., Rogers, A., Jennings, N.R.: Trading agents for the smart electricity grid. In: Proceedings of the 9th International Conference on Autonomous Agents and Multiagent Systems (AAMAS 2010), pp. 897–904. International Foundation for Autonomous Agents and Multiagent Systems, Richland, SC (2010), `http://portal.acm.org/citation.cfm?id=1838206.1838326`
7. Wang, T., Boutilier, C.: Incremental utility elicitation with the minimax regret decision criterion. In: Proceedings of the 18th International Joint Conference on Artificial Intelligence (IJCAI 2003), Acapulco, Mexico, pp. 309–318 (2003)
8. Wasserman, L.: All of Statistics. Springer, Heidelberg (2004)

Risk-Averse Production Planning

Ban Kawas[1], Marco Laumanns[1], Eleni Pratsini[1], and Steve Prestwich[2]

[1] IBM Research – Zurich, 8803 Rueschlikon, Switzerland
{kaw,mlm,pra}@zurich.ibm.com
[2] University College Cork, Ireland
s.prestwich@cs.ucc.ie

Abstract. We consider a production planning problem under uncertainty in which companies have to make product allocation decisions such that the risk of failing regulatory inspections of sites - and consequently losing revenue - is minimized. In the proposed decision model the regulatory authority is an adversary. The outcome of an inspection is a Bernoulli-distributed random variable whose parameter is a function of production decisions. Our goal is to optimize the conditional value-at-risk (CVaR) of the uncertain revenue. The dependence of the probability of inspection outcome scenarios on production decisions makes the CVaR optimization problem non-convex. We give a mixed-integer nonlinear formulation and devise a branch-and-bound (BnB) algorithm to solve it exactly. We then compare against a Stochastic Constraint Programming (SCP) approach which applies randomized local search. While the BnB guarantees optimality, it can only solve smaller instances in a reasonable time and the SCP approach outperforms it for larger instances.

Keywords: Risk Management, Compliance Risk, Adversarial Risk Analysis, Conditional Value-at-Risk, Production Planning, Combinatorial Optimization, MINLP.

1 Introduction

More and more regulations are enforced by government authorities on companies from various sectors to ensure good business practices that will guarantee quality of services and products and the protection of consumers. For example, pharmaceutical companies must follow current Good Manufacturing Practices (cGMPs) enforced by the Food and Drug Administration (FDA) [1]. In the financial sector, investment banks and hedge funds must comply with regulations enforced by the U.S. Securities and Exchange Commission (SEC), and in the Information, Technology, and Communication sector, companies must adhere to the Federal Communications Commission (FCC) rules. As a consequence, companies are increasingly faced by non-compliance risks, i.e., risks arising from violations and non-conformance with given regulations. Risk here is defined as the potential costs that can come in the form of lost revenues, lost market share, reputation damage, lost customers' trust, or personal or criminal liabilities. Not all of these risks are easily quantified.

R.I. Brafman, F. Roberts, and A. Tsoukiàs (Eds.): ADT 2011, LNAI 6992, pp. 108–120, 2011.

Due to the high costs, companies try to achieve maximum compliance and use different means to achieve that. Generally, they employ a system to manage all their risks (non-compliance included) [2], [3],[4],[5]. Some companies use "governance, risk, and compliance" (GRC) software, systems, and services [6], the total market of which in 2008 was estimated at $52.1 billion [7]. Within these systems, necessary measures are taken to ensure compliance, and an internal inspection policy is sometime instituted to make sure that those measures have the desired effect. A recent paper [8] explores the use of technology and software in managing non-compliance risk and considers its consequences.

To quantify the exposure of a company to non-compliance risk, [9] proposes the use of causal networks based on a mixture of data and expert-driven modeling and illustrates the approach in pharmaceutical manufacturing processes and IT systems availability. In [10], a quantitative model was developed using statistical approaches to measure non-conformace risks of a company from historical data. The resulting risk indices are then used as input data for an optimization model that not only minimizes a company's risk exposure and related costs but also maximizes its revenue. In [11], the authors give a quantitative risk-based optimization model that allows a company to dynamically apply the optimal set of feasible measures for achieving an adequate level of compliance.

In this paper, we investigate non-compliance risks in the planning stage of a business process. In particular, we focus on production planning and resource allocation [12], [13]. An exhaustive literature survey of models for production planning under uncertainty can be found in [14]. This survey identifies the need for the development of new models to address additional types of uncertainty since the main focus for most models is on demand uncertainty. In a recent paper [15], a production planning model addressing compliance uncertainties was considered and a mixed integer program (MIP) was formulated for two risk measures, the expected and the worst-case return. We consider a similar production planning model but optimize for the conditional value-at-risk (CVaR) of a company's return instead.

Conditional value-at-risk – also known as the average value-at-risk or expected shortfall – is a risk measure that is widely used in finacial risk management [16], [17], [18]. For a confidence level $\alpha \in (0, 1)$, the CVaR of the loss or profit associatecd with a decision $x \in \mathbb{R}^n$ is defined as the mean of the α or $(1-\alpha)$-tail distribution of the loss or profit function, respectively. The popularization of CVaR is due to its coherence characteristcs – coherency in the sense of Artzner et al. [19] – and the introduction of efficient convex linear fomulations by Rockafellar and Uryasev [20], [21]. In the latter, the authors consider general loss functions $z = f(x, y)$, where $x \in \mathbb{R}^n$ is the decision vector and $y \in \mathbb{R}^m$ represents the random future values of a number of variables with known probability distributions. Their key results on convexity of CVaR and the use of linear programming formulations rely on the assumption that the probability measure governing the random vector y is *independent* of the decision vector x. When this is not the case, the proposed CVaR optimization problem is not necessarily convex, even if the function $f(x, y)$ is itself convex.

In this work, a risk-averse one period setting decision model is analyzed. An authoritative inspection agency is considered an adversary with full information and unlimited budget. This agency inspects all production sites of a company for regulatory compliance. Moreover, it is assumed that at each site *only* the most hazardous product is inspected. If a site fails inspection, all revenue generated at it is lost. The company's objective is to allocate its products to the sites such that the CVaR of the net-revenue is maximized. The inspection outcome of a site is a Bernoulli random variable with success and failure probabilities that are *dependent* on the company's allocation decisions. Hence, the resulting CVaR maximization problem is nonlinear, and more importantly, nonconvex. We give a mixed-integer nonlinear program (MINLP) and devise a branch-and-bound (BnB) to solve it exactly. The results of which are compared against a Stochastic Constrained Programming (SCP) [22] approach that is based on a simple randomized local search. While the latter generally outperforms in terms of CPU times, the former provides bounds on the optimal solution for larger instances of the problem and optimality guarantees for smaller ones.

The main contribution of this paper is a general framework to address non-compliance uncertainties in an adversarial-setting decision model with a focus on a well-known and widely used risk measure (CVaR). The devised solution techniques can easily be generalized to other problems and applications with decision dependent probability measures and for which the CVaR of a preference functional is to be optimized.

The remaining sections are organized as follows: in Sect. 2, we introduce the adversarial problem along with the notations. We then give the MINLP formulation of the CVaR maximization problem in Sect. 3 followed by the devised BnB algorithm in Sect. 4. The SCP approach is described in Sect. 5 and the numerical results of both the BnB and the SCP are given in Sect. 6. The paper is then concluded in Sect. 7.

2 Problem Setup

In this section, we describe the aggressive adversarial problem in which the adversary is the inspection agency that has full information and unlimited budget. The inspected company has P products and S production sites. Each product $p \in \mathcal{P} = \{1, \cdots, P\}$ generates a net-revenue of r_p and can be produced at any of the sites $s \in \mathcal{S} = \{1, \cdots, S\}$. However, a product cannot be produced at more than one site. Furthermore, products have an associated site-specific risk hazard $h_{p,s} \in [0, 1]$. An adversarial authoritative agency regularly inspects the company's production sites to make sure that regulatory measures are being maintained. We assume that *only* the most hazardous product at each site is inspected. If a site fails inspection, the company loses all revenues generated at that site. Given the safety-hazards $h_{p,s}, \forall p, s$, and the revenues generated by each product r_p, the company's objective is to allocate products to sites in a way that will maximize the CVaR of its expected revenue, because maximizing the expected worst-case-scenarios of future revenues gives some guarantees that realized revenues will not be below a certain threshold with some probability α.

The following section presents the probability distribution governing the process of inspections and gives the aforementioned MINLP formulation for maximizing the CVaR of a preference functional, a company's net-revenue.

3 The CVaR of the Net-Revenue of a Company under Non-compliance Risks

CVaR has been commonly defined for loss functions, because it is mostly used in managing financial losses. In this work, we focus on the CVaR of a company's net-revenue to control non-compliance risks. Hence, we conveniently redefine CVaR to represent the average $(1-\alpha)$-tail distribution of revenue.

Let $f(x,y)$ be the revenue function where $x \in \mathbb{R}^n$ is a decision vector and the vector $y \in \mathbb{R}^m$ represents the random future outcome of the adversarial agency inspections. We consider the discrete probability space $(\Omega, \mathcal{F}, \mathbb{P})$ and assume that $f(x,y)$ is $\mathcal{F}$-measurable in $y \in \mathbb{R}^m \subseteq \Omega$. Since the sampling space Ω is discrete with a finite number of scenarios I and the probability function is assumed to be a stepwise right-continuous, then the random revenue $f(x,y)$ for a fixed x can be represented as an ordered set $F = \{f(x,y^i), \mathbb{P}(y^i)\}_{i=1,\ldots,I}$ where $f(x,y^i)$ is the i-th smallest revenue scenario. The $(1-\alpha)$-quantile will then be the value $f(x,y^{i^*})$, where i^* is the unique index such that the sum of probabilities of scenarios $\{1,\ldots,i^*-1\}$ is strictly less than $1-\alpha$ and of scenarios $\{1,\ldots,i^*\}$ is greater than or equal $1-\alpha$. Accordingly, the CVaR for a given α and decision x is given by:

$$CVaR(x,\alpha) = \begin{cases} f(x,y^{i^*}), & \text{if } i^* = 1, \\ \dfrac{1}{1-\alpha}\left(\displaystyle\sum_{i=1}^{i^*-1} \mathbb{P}(y^i) f(x,y^i) + f(x,y^{i^*})\left(1-\alpha-\sum_{i=1}^{i^*-1}\mathbb{P}(y^i)\right)\right), & \text{o.w.} \end{cases} \tag{1}$$

equivalently [20], [21],

$$CVaR(x,\alpha) = \max_V \; V - \frac{1}{1-\alpha}\sum_{i=1}^{I} \mathbb{P}(y^i)\left(\max\{0, V - f(x,y^i)\}\right) \tag{2}$$

For a fixed decision $x \in \mathbb{R}^n$ and known probabilities $\mathbb{P}(y^i)$, $\forall i \in \{1,\ldots,I\}$, (2) is a convex linear optimization problem. If $f(x,y^i)$ is a concave function in its arguments, then the maximization of (2) with respect to $x \in \mathbb{R}^n$ is also a convex problem. As will be shown below, if the probabilities are not known independently of the decision vector $x \in \mathbb{R}^n$, then the optimization problem in x is nonlinear and noncovex and will require special solution techniques to be solved exactly. Moreover, the sampling of the space Ω to deal with complexity when solving large instances of the problem will not be possible. Hence, all scenarios in Ω are to be considered, the number of which increases exponentially with the size of vector $y \in \mathbb{R}^m$.

3.1 Probability Distribution of the Inspection Process

With the assumption of an unlimited budget adversary, we are also assuming that all production sites are inspected. This means that there are 2^S different scenarios of inspections results. Let f_s denote the maximum safety-hazard at site s, i.e., $f_s = \max_p\{h_{p,s}\, x_{ps}\}$, where $x_{ps} \in \{0,1\}, \forall p \in \mathcal{P}, s \in \mathcal{S}$ are binary decision variables indicating if a product p is allocated at site s ($x_{ps} = 1$) or not ($x_{ps} = 0$). The process of a site inspection follows a Bernoulli probability distribution with f_s as the probability of a success event – a site failing inspection – and $(1 - f_s)$ is the probability of a site passing inspection. As mentioned above, if a site fails inspection, all revenues generated at it will be lost. Thus, the total revenue of a company is directly associated with inspection results, and the probability distribution of the revenue function is multivariate Bernoulli:

$$\mathbb{P}\{X_1^i = k_1^i, \cdots, X_S^i = k_S^i\} = \mathbb{P}\left\{\cap_{s=1}^S [X_s^i = k_s^i]\right\}, \; \forall i \in I = \{1, \cdots, 2^S\} \tag{3}$$

where X_s^i is a Bernoulli random variable representing the event of inspecting site s in scenario $i \in I$, and $k_s^i \in \{0,1\}$ is an indicator that has a value of 1 if site s passes inspection in scenario i and 0 otherwise. We assume that inspection results for each site s is independent of other sites, hence, the value of the probability in (3) is simply given by:

$$\prod_{s=1}^S \mathbb{P}\{X_s^i = k_s^i\} = \prod_{s=1}^S \left[f_s^{(1-k_s^i)}.(1 - f_s)^{k_s^i}\right], \; \forall i \in I = \{1, \cdots, 2^S\} \tag{4}$$

the expression in (4) represents the probability of scenario i of inspection results.

3.2 MINLP Formulation

After enumerating all scenarios of inspection results $I = \{1, ..., 2^S\}$, we use (2) along with (4) to formulate the production planning problem with the objective of maximizing the CVaR of net-revenues:

$$\begin{aligned}
\max_{x,u,f,v,V} \; & \frac{1}{1-\alpha} \sum_{i \in I} u_i . \prod_{s=1}^S \left[f_s^{(1-k_s^i)}.(1 - f_s)^{k_s^i}\right] + V \\
\text{s.t. } & u_i \leq \sum_{s=1}^S k_s^i \, v_s - V, \; \forall i, \\
& v_s \leq \sum_{p=1}^P r_p \, x_{p,s}, \; \forall s, \\
& h_{p,s}\, x_{p,s} \leq f_s \leq 1, \; \forall s, p, \\
& \sum_{s=1}^S x_{p,s} \leq 1, \; \forall p, \\
& x_{p,s} \in \{0,1\}, \; \forall p, s, \;\; v_s \geq 0, \; \forall s, \;\; u_i \leq 0, \; \forall i.
\end{aligned} \tag{5}$$

where u_i, $\forall\, i$, and v_s, $\forall\, s$ are auxilliary variables. Note that if the probabilities f_s were independent of decision variables x_{ps} and are known a prior, (5) would be a MIP and can be solved using any MIP solver, such as CPLEX. However, this is not the case and (5) is a non-convex MINLP that requires special solution techniques. We have attempted to solve this problem using COUENNE (a solver for non-convex MINLP problems) [23], but even for a small problem size of 2 sites and 3 products, the solver did not arrive to a solution. Hence, we developed a problem-specific BnB algorithm that builds upon the idea that failure probability of sites f_s, $\forall\, s$, is to be kept at a minimum. We describe the algorithm in the following section.

4 Branch-and-Bound Algorithm (BnB)

To solve the MINLP in (5) exactly we devise a BnB utilizing many of the basic techniques of BnB algorithms in the literature [24], [25] and drawing from the structure of our problem. The general idea of the algorithm is to fix the variables f_s, $\forall\, s$ in (5) and solve the LP-relaxation of the resulting MIP. At each branch, the algorithm fixes some of the decision variables x_{ps} and finds the corresponding worst- and best-case values of failure probabilities f_s, $\forall\, s$, denoted f_s^{WC} and f_s^{BC}, respectively. The worst-case values are an overestimation of f_s and when used as constants in the objective of (5), the resulting MIP is a lower bound. Similarly, best-case values are an underestimation and when used in the objective, the resulting MIP after relaxing the constraints ($f_s \geq h_{ps}\, x_{ps}, \forall\, s, p$) is an upper bound. We solve the LP-relaxation of both the worst- and the best-case MIPs. The resulting solutions are an upper bound to their respective MIPs. For prunning, we utilize a heuristic, described below, that gives a feasible solution to the original problem (5).

At the root node of the BnB tree, the worst-case f_s^{WC}, $\forall\, s$ is the maximum hazard value amongst all products ($f_s^{WC} = \max_{p \in \mathcal{P}}\{h_{p,s}\}$) and the best-case value is the minimum ($f_s^{BC} = \min_{p \in \mathcal{P}}\{h_{p,s}\}$). At each node, we start branching by allocating a candidate product to the different sites. At each branch, when allocating product $\hat{p}$ to site $\hat{s}$ ($x_{\hat{p}\hat{s}} = 1$), its hazard value at other sites is not considered when evaluating f_s^{BC}, f_s^{WC}, $\forall\, s \in \mathcal{S} \backslash \{\hat{s}\}$, consequently the allocation of $\hat{p}$ can have the following effects on the current values of f_s^{BC}, f_s^{WC}, $\forall\, s$:

1. If $h_{\hat{p}\hat{s}}$ is greater than the value of $f_{\hat{s}}^{BC}$, then $f_{\hat{s}}^{BC} = h_{\hat{p}\hat{s}}$
2. If product $\hat{p}$ is strictly the most hazardous product for site $s \in \mathcal{S} \backslash \{\hat{s}\}$ ($h_{\hat{p}s} > h_{ps}, \forall\, p \in \mathcal{P} \backslash \hat{p}$), then the value of f_s^{WC} will decrease ($f_s^{WC} = \max_{p \in \mathcal{P} \backslash \hat{p}}\{h_{ps}\}$).
3. If product $\hat{p}$ is strictly the least hazardous product for site $s \in \mathcal{S} \backslash \{\hat{s}\}$ ($h_{\hat{p}s} < h_{ps}, \forall\, p \in \mathcal{P} \backslash \hat{p}$), then the value of f_s^{BC} will increase ($f_s^{BC} = \min_{p \in \mathcal{P} \backslash \hat{p}}\{h_{ps}\}$)

After obtaining f_s^{BC}, f_s^{WC}, $\forall\, s$ for the current branch, we solve the LP-relaxation of the best- and worst-case MIPs. If the branch is not pruned, then we record the best-case objective value and analyze the resulting solutions. If the solution of the worst-case problem is binary feasible, then we compare its objective value against the objective of the best known feasible solution and update the latter when the worst-case objective is better. On the other hand, if the worst-case solution is binary infeasible, then we populate the list of candidate products for branching with the ones that are associated with non-binary variables x_{ps}. The pruning and branching rules of the algorithm are as follows:

Pruning Rule. We prune a branch from the tree, if the optimal objective of the LP-relaxation of the best-case MIP is lower than the best known feasible solution.

Branching Rule. From the list of candidate problems, we start with the one that has the highest best-case objective. We then rank candidate products according to the sum of their hazards across all sites and we branch on the most hazardous one. The idea behind this is to force early prunning, because a more hazardous product will have more effects on the values of f_s^{BC}, f_s^{WC}, $\forall s$.

Going down the search tree, by allocating more and more products, the worst- and best-case bounds become closer and closer until the gap is closed and we reach optimality. We use two search directions. One is a breadth-first (BF) that gives tighter upper bounds and the other is a depth-first (DF) that gives tighter lower bounds as will be shown in the numerical experiments in Sect. 6.

Heurisitc. To improve the prunning process of both BnB algorithms, we derive a very simple and intuitive heuristic that only requires solving a single MIP. The basic idea is similar to the premise of the BnB, we fix the probabilities in (5) and then solve the resulting MIP. Intuitively, all site hazards f_s should be kept at a minimum. For each product, the heuristic finds the least hazardous site (i.e. $\min_s\{h_{ps}\}, \forall p$) and assumes that the product will be allocated to it. Then for each site s, it sets f_s to the maximum hazard amongst those products that has their minimum hazard at s. This heuristic is very simple and always guarantees a feasible solution to be used in the pruning process of the devised BnB. The peformance of the heuristic is dependent on input data, sometimes it gives optimal or close to optimal solutions and other times it perfomrs poorly.

5 Stochastic Constraint Programming (SCP)

Stochastic Constraint Programming (SCP) is an extension of Constraint Programming (CP) designed to model and solve complex problems involving uncertainty and probability, a direction of research first proposed in [22]. SCP is closely related to SP, and bears roughly the same relationship to CP as SP does to MIP. A motivation for SCP is that it should be able to exploit the more expressive constraints used in CP, leading to more compact models and the use of powerful filtering algorithms. *Filtering* is the process of removing values from the domains of variables that have not yet been assigned values during search, and is the main CP method for pruning search trees. If all values have been pruned from an unassigned variable then the current partial assignment cannot be extended to a solution, and backtracking can occur.

An m-stage *Stochastic Constraint Satisfaction Problem* (SCSP) is defined as a tuple $(V, S, D, P, C, \theta, L)$ where V is a set of decision variables, S a set of stochastic variables, D a function mapping each element of $V \cup S$ to a domain of values, P a function mapping each variable in S to a probability distribution, C a set of constraints on $V \cup S$, θ a function mapping each constraint in C to a threshold value $\theta \in (0, 1]$, and $L = [\langle V_1, S_1 \rangle, \ldots, \langle V_m, S_m \rangle]$ a list of *decision stages* such that the V_i partition V and the S_i partition S. Each constraint must contain at least one V variable, a constraint with threshold $\theta(h) = 1$ is a *hard constraint*, and one with $\theta(h) < 1$ is a *chance constraint*.

To solve an SCSP we must find a *policy tree* of decisions, in which each node represents a value chosen for a decision variable, and each arc from a node represents the value assigned to a stochastic variable. Each path in the tree represents a different possible *scenario* and the values assigned to decision variables in that scenario. A *satisfying policy tree* is a policy tree in which each chance constraint is satisfied with respect to the tree. A chance constraint $h \in C$ is satisfied with respect to a policy tree if it is satisfied under some fraction $\phi \geq \theta(h)$ of all possible paths in the tree.

An objective function to be minimized or maximized may be added, transforming the SCSP into a *Stochastic Constrained Optimization Problem* (SCOP). We also add two further features that are non-standard. Firstly, we allow stochastic variable distributions to be dependent on earlier decisions. This feature, which we refer to as *conditional stochastic variables*, lies outside both SCP and SP but is common in Stochastic Dynamic Programming. We implement it by allowing the probabilities associated with stochastic variable domain values to be represented by decision variables. This motivates the second SCP extension: decision variables may have real-valued domains. These must be functionally dependent on the values of already-assigned variables.

An SCOP model for our problem is shown in Figure 1. A decision $x_p = s$ means that product p is made at site s. The y_s are real-valued decision variables that are functionally dependent on the x_p. The o_s are conditional stochastic variables whose probability distributions are given by the y_s, which are written in brackets after the values they are associated with (1 for inspection success, 0 for failure). Each probability y_s represents the greatest hazard among products made at site s. We need dummy hazards $h_{0,p} = 1$ ($\forall p \in \mathcal{P}$): we allow dummy site 0 to be inspected, but the dummy hazards force these inspections to fail. Note that this is a very compact model, largely because of the use of conditional stochastic variables.

To solve this problem we need an SCP algorithm. As will be seen below, using our problem-specific branch-and-bound algorithm, we are unable to find good solutions to large instances in a reasonable time. There are various complete methods that have been proposed for solving SCP problems (see [26] for a short survey) but we do not believe that these would be any more scalable. Instead we shall apply an incomplete search method based on local search: an SCP solver, to exploit the high-level modeling capabilities of SCP. This solver is much like that described in [26,27] and will be fully described in a future paper, but we summarize it here.

Objective:
$\max\left(\text{CVaR}_\alpha\left\{\sum_{p\in\mathcal{P}} o_{x_p} r_p\right\}\right)$
Subject to:
$y_s = \max_{p\in\mathcal{P}}\left\{h_{s,p}\cdot \text{reify}(x_p = s)\right\} \quad (\forall s \in \mathcal{S}\cup\{0\})$
Decision variables:
$x_p \in \mathcal{S}\cup\{0\} \quad (\forall p \in \mathcal{P})$
$y_s \in [0,1] \quad (\forall s \in \mathcal{S}\cup\{0\})$
Stochastic variables:
$o_s \in \{0(y_s), 1(1-y_s)\} \quad (\forall s \in \mathcal{S}\cup\{0\})$
Stage structure:
$L = [\langle\{x,y\},\{o\}\rangle]$

Fig. 1. SCP model for CVaR case

We transform the problem of finding a satisfying policy tree to an unconstrained optimization problem. Define a variable at each policy tree node, whose values are the domain values for the decision variable at that node. Then a vector of values for these variables represents a policy tree. We can now apply a metaheuristic search algorithm to find a vector corresponding to a satisfying policy tree via *penalty functions*, which are commonly used when applying genetic algorithms or local search to problems with constraints [28]. For each constraint $h \in C$ define a penalty x_h in each scenario, which is 0 if h is satisfied and 1 if it is violated in that scenario. Then the objective function for a vector $\mathbf{v}$ is:

$$f(\mathbf{v}) = \sum_{h\in C}(\mathbb{E}\{x_h\} - \theta(h))^+$$

where $(.)^+$ denotes $\max\{.,0\}$. We compute each $\mathbb{E}\{x_h\}$ by performing a complete search of the policy tree, and checking at each leaf whether constraint h is satisfied. If it is then that scenario contributes its probability to $\mathbb{E}\{x_h\}$. If $f(\mathbf{v}) = 0$ then each constraint h is satisfied with probability at least that of its satisfaction threshold $\theta(h)$ so $\mathbf{v}$ represents a satisfying policy tree. We can now apply metaheuristic search to the following unconstrained optimization problem: minimize $f(\mathbf{v})$ to 0 on the space of vectors $\mathbf{v}$. We handle an objective function by computing its value f when traversing the policy tree, and modifying the penalty to include an extra term $(f - f_{best})^+$ for minimization and $(f - f_{best})^+$ for maximization, where f_{best} is the objective value of the best solution found so far. By solving a series of SCSPs with improving values of f_{best} we hope to converge to an optimal satisfying policy tree.

However, instead of treating hard constraints as chance constraints with threshold 1, we can do better. We simply *enforce* any hard constraints when traversing the policy tree, backtracking if they are violated (or if filtering indicates that this will occur). If we have chosen a poor policy then this traversal will be incomplete, and we penalize this incompleteness by adding another penalty term. This enables a poor policy to be evaluated more quickly, because less of the policy tree is traversed. Moreover, if filtering indicates that the value specified by our policy will lead to backtracking, then we can instead choose another value, for example the cyclically-next value in the variable's domain. Thus a policy that would be incorrect *if we treated hard constraints as chance constraints* might become correct using this method, making it easier to find a satisfying policy.

It remains to choose a meta-heuristic, and we obtained good results using *randomized hill climbing*: at each step, mutate the policy and evaluate its penalty: if it has not increased, or with a small probability (we use 0.005), accept the mutation, otherwise reject it. This very simple heuristic outperformed a genetic algorithm, indicating that the search space is not very rugged.

6 Numerical Experiments

In what follows, we show a comparison of performance in terms of CPU times between the BnB and the SCP approach. We consider 4 different sizes of the problem and solve 10 random instances per each size. For the SCP, we run each instance 6 times and record the median result. A cut-off time of 2^S is used for larger sizes, after which the SCP approach does not improve much.

Figure 2 gives time-evolution graphs for two problem instances per size; the two that took the shortest (in the left column) and the longest time for the SCP to reach a stagnate state (i.e. a state without many improvements on the solution). As can be seen, computational times increases exponentially with the size of the problem. The results show that the SCP outperforms the BnB, but does not give optimality guarantees for smaller problem sizes nor provide bounds for larger sizes as the BnB do. The solutions of the SCP improves rapidly at the beginning and then reaches a plateau. With fixed running times, the bounds provided by the BnB generally become weaker and the optimality gap increases as the problem becomes larger. Table 1 gives the optimality gap of both approaches for each of the instances in Fig. 2. The optimality gap for the SCP is found using the upper bound provided by the BnB. Note that the running times of the different sizes are different and one cannot conclude that the optimality gap improves with the size of the problem.

Table 1. Optimality Gap (%) for each of the instances in Fig. 2. L:=left position in the figure, R:=right position.

	BnB	SCP
S=4,P=8, L+R	0.0000	0.0000
S=6,P=12, L	16.6022	11.6749
S=6,P=12, R	15.6860	13.2577
S=8,P=16, L	9.3365	7.7201
S=8,P=16, R	13.4124	6.4949
S=10,P=20, L	5.5466	4.9225
S=10,P=20, R	11.2761	7.5745

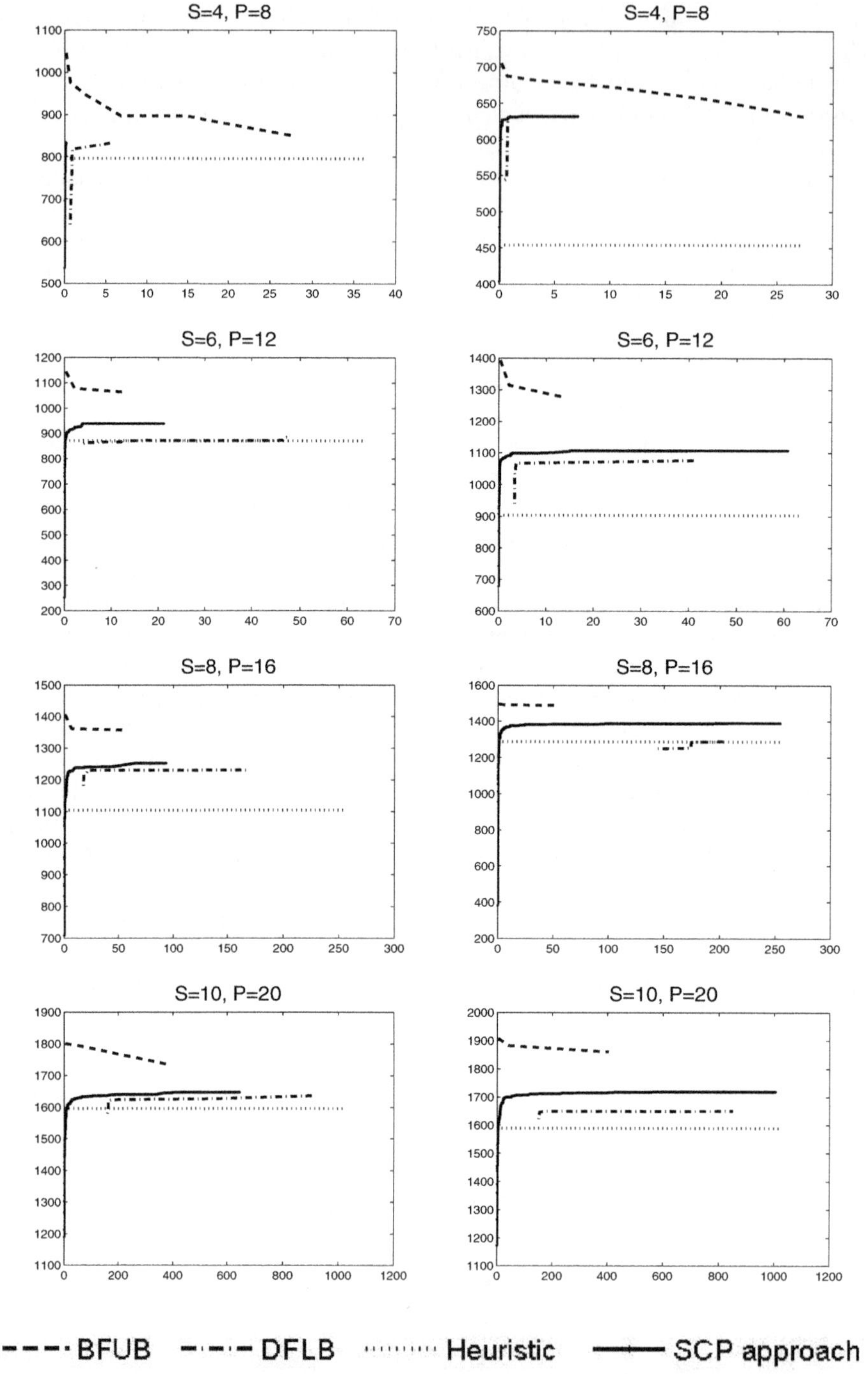

Fig. 2. Objective value (y-axis) vs CPU time in seconds for two instances per problem size (S:= number of sites and P:= number of products, BFUB:= upper bound of the breadth-first BnB, DFLB:= lower bound of the depth-first BnB)

7 Conclusions

This paper provides a general framework to address non-compliance risks in production planning. A risk-averse one-period adversarial decision model is given in which regulatory agencies are considered adversaries. A widely used coherent risk measure, the conditional value-at-risk (CVaR), is optimized. We show that the CVaR optimization problem is nonconvex and nonlinear when the probability measure is dependent on the decision variables and solving it require special solution techniques. We give a MINLP formulation and devise a branch-and-bound algorithm to solve it exactly. A comparison in terms of CPU times with a Stochastic Constraint Programming approach is given. The results show that both approaches have unique advantages. The BnB provides bounds and optimality guarantees and the SCP provides better solutions in less CPU time. This suggest the use of hybrid techniques that builds on the strengths of both approaches. One of our current research directions is to develop such hybrid techniques that can be tailored to the specific needs of applications, i.e., if an application requires fast solutions that are an ϵ away from optimality, then one would use SCP and monitor its solutions with the bounds provided by the BnB. If another application requires precise and very close to optimal solutions, then one would use a BnB algorithm that utilizes SCP solutions within the pruning and branching procedures to improve its peformance. Other current research directions are to investigate more risk measures that can be used in controlling non-compliance risks and to address input data uncertainty by utilizing robust optimization techniques within the current framework.

References

1. Facts about current good manufacturing practices (cGMPs), U.S. Food and Drug Administration, `http://www.fda.gov/Drugs/DevelopmentApprovalProcess/Manufacturing/ucm169105.htm`
2. Abrams, C., von Kanel, J., Muller, S., Pfitzmann, B., Ruschka-Taylor, S.: Optimized Enterprise Risk Management. IBM Systems Journal 46(2), 219–234 (2007)
3. Beroggi, G.E.G., Wallace, W.A.: Operational Risk Management: A New Paradigm for Decision Making. IEEE Transactions on Systems, Man, and Cypernetics 24(10), 1450–1457 (1994)
4. McNeil, A.J., Frey, R., Embrechts, P.: Quantitative Risk Management: Concepts, Techniques, and Tools. Princeton University Press, Princeton (2005)
5. Liebenbergm, A.P., Hoyt, R.E.: The Determinants of Enterprise Risk Management: Evidence From the Appointment of Chief Risk Officers. Risk Management and Insurance Review 6, 37–52 (2003)
6. Frigo, M.L., Anderson, R.J.: A Strategic Framework for Governance, Risk, and Compliance. Strategic Finance 44, 20–61 (2009)
7. Rasmussen, M.: Corporate Integrity: Strategic Direction for GRC, 2008 GRC Drivers, Trends, and Market Directions (2008)
8. Bamberger, K.A.: Technologies of Compliance: Risk and Regulation in a Digital Age. Texas Law Review 88, 670–739 (2010)

9. Elisseeff, A., Pellet, J.-P., Pratsini, E.: Causal Networks for Risk and Compliance: Methodology and Applications. IBM Journal of Research and Development 54(3), 6:1–6:12 (2010)
10. Pratsini, E., Dea, D.: Regulatory Compliance of Pharmaceutical Supply Chains. In: ERCIM News, no. 60
11. Muller, S., Supatgiat, C.: A Quantitative Optimization Model for Dynamic risk-based Compliance Management. IBM Journal of Research and Development 51, 295–307 (2007)
12. Silver, E.A., Pyke, D.F., Peterson, R.: Inventory Management and Production Planning and Scheduling, 3rd edn. John Wiley and Sons, Chichester (1998)
13. Graves, S.C.: Manufacturing Planning and Control. In: Resende, M., Paradalos, P. (eds.) Handbook of Applied Optimization, pp. 728–746. Oxford University Press, NY (2002)
14. Mula, J., Poler, R., Garcia-Sabater, J.P., Lario, F.C.: Models for Production Planning Under Uncertainty: A Review. International Journal of Production Economics 103, 271–285 (2006)
15. Laumanns, M., Pratsini, E., Prestwich, S., Tiseanu, C.-S.: Production Planning for Pharmaceutical Companies Under Non-Compliance Risk (submitted) (2010)
16. Acerbi, C.: Coherent Measures of Risk in Everday Market Practice. Quantitative Finance 7(4), 359–364 (2007)
17. Acerbi, C., Tasche, D.: Expected shortfall: A Natural Coherent Alternative to Value at Risk. Economic Notes 31(2), 379–388 (2002)
18. Alexander, G.J., Baptista, A.M.: A Comparison of VaR and CVaR Constraints on Portfolio Selection with the Mean-Variance Model. Management Science 50(9), 1261–1273 (2004)
19. Artzner, P., Delbaen, F., Eber, J.-M., Heath, D.: Coherent Measures of Risk. Mathematical Finance 3, 203–228 (1999)
20. Rockafellar, R.T., Uryasev, S.P.: Optimization of Conditional Value-at-Risk. The Journal of Risk 2, 21–41 (2000)
21. Rockafellar, R.T., Uryasev, S.P.: Conditional Value-at-Rsk for a General Loss Distribtion. Journal of Banking and finance 26, 1443–1471 (2002)
22. Walsh, T.: Stochastic Constraint Programming. In: 15th European Conference on Artificial Intelligence (2002)
23. Belotti, P., Lee, J., Liberti, L., Margot, F., Wachter, A.: Branching and Bounds Tightening Techniques, for Non-Convex MINLP. Optimization Methods and Software 24(4-5), 597–634 (2009)
24. Clausen, J.: Branch and Bound Algorithms - Principles and Examples. Parallel Computing in Optimization (1997)
25. Gendron, B., Crainic, T.G.: Parallel Branch-And-Bound Algorithms: Survey and Synthesis. Operations Research 42(6), 1042–1066 (1994)
26. Prestwich, S.D., Tarim, S.A., Rossi, R., Hnich, B.: Evolving Parameterised Policies for Stochastic Constraint Programming. In: Gent, I.P. (ed.) CP 2009. LNCS, vol. 5732, pp. 684–691. Springer, Heidelberg (2009)
27. Prestwich, S.D., Tarim, S.A., Rossi, R., Hnich, B.: Stochastic Constraint Programming by Neuroevolution With Filtering. In: Lodi, A., Milano, M., Toth, P. (eds.) CPAIOR 2010. LNCS, vol. 6140, pp. 282–286. Springer, Heidelberg (2010)
28. Craenen, B., Eiben, A.E., Marchiori, E.: How to Handle Constraints with Evolutionary Algorithms. In: Chambers, L. (ed.) Practical Handbook of Genetic Algorithms, pp. 341–361 (2001)

Minimal and Complete Explanations for Critical Multi-attribute Decisions

Christophe Labreuche[1], Nicolas Maudet[2], and Wassila Ouerdane[3]

[1] Thales Research & Technology
91767 Palaiseau Cedex, France
christophe.labreuche@thalesgroup.com
[2] LAMSADE, Université Paris-Dauphine
Paris 75775 Cedex 16, France
maudet@lamsade.dauphine.fr
[3] Ecole Centrale de Paris
Chatenay Malabry, France
wassila.ouerdane@ecp.fr

Abstract. The ability to provide explanations along with recommended decisions to the user is a key feature of decision-aiding tools. We address the question of providing minimal and complete explanations, a problem relevant in critical situations where the stakes are very high. More specifically, we are after explanations with minimal cost supporting the fact that a choice is the weighted Condorcet winner in a multi-attribute problem. We introduce different languages for explanation, and investigate the problem of producing minimal explanations with such languages.

1 Introduction

The ability to provide explanations along with recommended decisions to the user is a key feature of decision-aiding tools [1,2]. Early work on expert systems already identified it as one of the main challenge to be addressed [3], and the recent works on recommender systems face the same issue, see *e.g.* [4]. Roughly speaking, the aim is to increase the user's acceptance of the recommended choice, by providing supporting evidence that this choice is justified.

One of the difficulties of this question lies on the fact that the relevant concept of an explanation may be different, depending on the problem at hand and on the targeted audience. The objectives of the explanations provided by an online recommender system are not necessarily the same as the ones of a pedagogical tool. To better situate our approach, we emphasize two important distinctive dimensions:

– *data vs. process*—following [5], we first distinguish explanations that are based on the *data* and explanations that are based on the *process*. Explanations based on the data typically focus on a "relevant" subset of the available data, whereas those based on the process make explicit (part of) the mathematical model underlying the decision.

R.I. Brafman, F. Roberts, and A. Tsoukiàs (Eds.): ADT 2011, LNAI 6992, pp. 121–134, 2011.

– *complete vs. incomplete explanations*—as opposed to incomplete explanations, complete explanations support the decision unambiguously, they can be seen as proofs supporting the claim that the recommended decision is indeed the best one. This is the case for instance in critical situations (*e.g.* involving safety) where the stakes are very high.

In this paper we shall concentrate on complete explanations based on the data, in the context of decisions involving multiple attributes from which, associating a preference model, we obtain *criteria* upon which options can be compared. Specifically, we investigate the problem of providing simple but complete explanations to the fact that a given option is a weighted Condorcet winner (WCW). An option is a WCW if it beats any other options in pairwise comparison, considering the relative weights of the different criteria. Unfortunately, a WCW may not necessarily exists. We focus on this case because (i) when a WCW exists it is the unique and uncontroversial decision to be taken, (ii) when it does not many decision models can be seen as "approximating" it, and (iii) the so-called outranking methods (based on the Condorcet method) are widely used in multi-criteria decision aiding, (iv) even though the decision itself is simple, providing a *minimal* explanation may not be.

In this paper we assume that the problem involves two types of *preferential information* (PI): preferential information regarding the importance of the criteria, and preferential information regarding the ranking of the different options.

To get an intuitive understanding of the problem, consider the following example.

Example 1. There are 6 options $\{a, b, c, d, e, f\}$ and 5 criteria $\{1, \cdots, 5\}$ with respective weights as indicated in the following table. The (full) orderings of options must be read from top (first rank) to bottom (last rank).

criteria	1	2	3	4	5
weights	0.32	0.22	0.20	0.13	0.13
ranking	c	b	f	d	e
	a	a	e	f	b
	e	f	a	b	d
	d	e	c	a	f
	b	d	d	c	a
	f	c	b	e	c

In this example, the WCW is a. However this option does not come out as an obvious winner, hence the need for an explanation. Of course a possible explanation is always to explicitly exhibit the computations of every comparison, but even for moderate number of options this may be tedious. Thus, we are seeking explanations that are minimal, in a sense that we shall define precisely below. What is crucial at this point is to see that such a notion will of course be dependent on the language that we have at our disposal to produce explanations. A tentative "natural" explanation would be as follows:

"First consider criteria 1 and 2, a is ranked higher than e, d, and f in both, so is certainly better. Then, a is preferred over b on criteria 1 and

> 3 (which is almost as important as criterion 2). Finally, it is true that c is better than a on the most important criteria, but a is better than c on all the other criteria, which together are more important."

The aim of this paper is not to produce such natural language explanation, but to provide the theoretical background upon which such explanations can later be generated.

This abstract example may be instantiated in the following situations. In the first one, a decision-maker presents a choice recommendation regarding a massive investment before funding agency. The decision was based on a multi-criteria analysis during which criteria and preferences were elicited. In the second one, a committee (where members have different voting weights) just proceeded to a vote on a critical issue, and the chairman is now to explain why a given option was chosen as a result. The reason why we take these two concrete examples is that beyond their obvious similarity (members of the committee play the role of the criteria in the funding example), they share the necessity to produce a complete explanation. The type of explanation we seek for is relevant when the voters (for the committee example) are not anonymous, which is often the case in committee.

The remainder of this paper is as follows. In the next section, we provide the necessary background notions, and introduce in particular the languages we shall use for formulating explanations. Section 3 defines minimal complete explanations. Section 4 and Section 5 deal with languages allowing to express the preferences on the rankings of options only, starting with the language allowing basic statements, then discussing a more refined language allowing to "factor" statements. Finally, Section 6 discusses connections to related works, in particular argumentation theory.

2 Background and Basic Definitions

2.1 Description of the Choice Problem

We assume a finite set of options O, and a finite set of criteria $H = \{1, 2, \dots, m\}$. The options in O are compared thanks to a weighted majority model based on some preferential information (PI) composed of preferences and weights. Preferences are linear orders, that is, complete rankings of the options in O, and $a \succ_i b$ stands for the fact that a is strictly preferred over b on criterion i. Weights are assigned to criteria, and W_i stands for the weight of criterion i. Furthermore, they are normalized in the sense that they sum up to 1. An instance of the choice problem, denoted by ρ, is given by the full specification of this PI. The decision model over O given ρ is defined by $b \succ_\rho c$ iff $\sum_{b \succ_i c} W_i > \sum_{c \succ_i b} W_i$.

Definition 1. *An option $a \in O$ is called* weighted Condorcet winner w.r.t. ρ *(noted* WCW(ρ)*) if for all $b \in O^\star := O \setminus \{a\}$, $a \succ_\rho b$.*

We shall also assume throughout this paper the existence of a weighted Condorcet winner labeled $a \in O$.

2.2 Description of the Language for the Explanation

Following the example in the introduction, the simplest language on the partial preferences is composed of terms of the form $[i : b \succ c]$, with $i \in H$ and $b, c \in O$, meaning that b is strictly preferred to c on criterion i. Such terms are called *basic preference statements.* In order to reduce the length of the explanation, they can also be factored into terms of the form $[I : b \succ P]$, with $I \subseteq H$, $b \in O$ and $P \subseteq O \setminus \{b\}$, meaning that b is strictly preferred to all options in P on all criteria in I. Such terms are called *factored preference statements.* The set of all subsets of basic preference statements (resp. factored preference statements) that correspond to a total order over O on each criterion is denoted by $\mathcal{S}$ (resp. $\widehat{\mathcal{S}}$). For $K \in \mathcal{S}$, we denote by $K^{\uparrow}$ the set of statements of the form $[I : b \succ P]$ with $I \subseteq H$ and $P \subseteq O$ such that for all $i \in I$ and $c \in P$, $[i : b \succ c] \in K$. Conversely, for $\widehat{K} \in \widehat{\mathcal{S}}$, let $\widehat{K}^{\downarrow} = \{[i : b \succ c] \ : \ \exists [I : b \succ P] \in \widehat{K}$ s.t. $i \in I$ and $c \in P\}$ be the atomization of the factored statements $\widehat{K}$. Now assuming that a is the WCW, it is useful to distinguish different types of statements:

- positive statements, of the form $[I : a \succ P]$
- neutral statements, of the form $[I : b \succ P]$ with $a \notin P$
- negative statements, of the form $[I : b \succ P]$ with $a \in P$.

We note that in the case of basic statements, negative statements are "purely" negative since $P = \{a\}$.

Example 2. The full ranking of actions, on criterion 1 only, yields the following basic statements:

- $[1 : c \succ a]$ (negative statement),
- $[1 : c \succ e], [1 : c \succ d], [1 : c \succ b], [1 : c \succ f], [1 : e \succ d], [1 : e \succ b, [1 : e \succ f], [1 : d \succ b], [1 : d \succ f], [1 : b \succ f]$ (neutral statements),
- $[1 : a \succ e], [1 : a \succ d], [1 : a \succ b], [1 : a \succ f]$ (positive statements).

Regarding factored statements, the following examples can be given:

- $[1, 2 : e \succ d]$ is a neutral statement;
- $[1 : c \succ a, e]$ is a negative statement;
- $[1, 2 : a \succ d, e, f]$ is a positive statement.

The explanation shall also mention the weights in order to be complete. We assume throughout this paper that the values of weights can be shown to the audience. This is obvious in voting committee where the weights are public. This is also a reasonable assumption in a multi-criteria context when the weights are elicited, as the constructed weights are validated by the decision-maker and then become an important element of the explanation [6]. The corresponding language on the weights is simply composed of statements (called *importance statements*) of the form $[i : \alpha]$ with $i \in H$ and $\alpha \in [0, 1]$ meaning that the weight of criterion i is α. Let $\mathcal{W}$ (the set of normalized weights) be the set of sets $\{[i : w_i] \ : \ i \in H\}$ such that $w \in [0, 1]^H$ satisfies $\sum_{i \in H} w_i = 1$. For $W \in \mathcal{W}$ and $i \in H$, $W_i \in [0, 1]$ is the value of the weight on criterion i, that is that $[i : W_i] \in W$. A set $A \subseteq H$ is called a *winning coalition* if $\sum_{i \in A} W_i > \frac{1}{2}$.

2.3 Cost Function over the Explanations

An *explanation* is a pair composed of an element of $\widehat{\mathcal{S}}$ (note that $\mathcal{S} \subset \widehat{\mathcal{S}}$) and an element of $\mathcal{W}$. We seek for minimal explanations in the sense of some cost function. For simplicity, the cost of an element of $\widehat{\mathcal{S}}$ or $\mathcal{W}$ is assumed to be the sum of the cost of its statements. A difficult issue then arises: how should we define the cost of a statement?

Intuitively, the cost should capture the simplicity of the statement, the easiness for the user to understand it. Of course this cost must depend in the end of the basic pieces of information transmitted by the statement. The statements are of various complexity. For instance $[1,2,5,7,9 : a \succ b,c,g,h]$ looks more complex to grasp than $[1 : a \succ b]$, so that factored preference statements are basically more complex than basic preference statements.

Let us considered the case of preference statements. At this point we make the following assumptions:

- *neutrality*— the cost is insensitive to the identity of both criteria and options, i.e. $cost([I : b \succ P])$ depends only on $|I|$ and $|P|$ and is noted $C(|I|,|P|)$,
- *monotony*— the cost of a statement is monotonic w.r.t. criteria and to options, i.e. function C is non-decreasing in its two arguments. Neutrality implies that all basic statements have the same cost $C(1,1)$.

Additionally to the previous properties, the cost may be sub-additive in the sense that $cost(I \cup I', P) \leq cost(I,P) + cost(I',P)$ and $cost(I, P \cup P') \leq cost(I,P) + cost(I,P')$, or super-additive if the converse inequalities hold. Finally, we assume the cost function can be computed in polynomial time.

3 Minimal Complete Explanations

Suppose now that the PI of choice problem is expressed in the basic language as a pair $\langle S,W\rangle \in \mathcal{S} \times \mathcal{W}$. Explaining why a is the Condorcet winner for $\langle S,W\rangle$ amounts to simplifying the PI (data-based approach [5]). We focus in this section on explanations in the language $\mathcal{S} \times \mathcal{W}$. The case of the other languages will be considered later in the paper.

A subset $\langle K,L\rangle$ of $\langle S,W\rangle$ is called a *complete explanation* if the decision remains unchanged regardless of how $\langle K,L\rangle$ is completed to form an element of $\mathcal{S} \times \mathcal{W}$. The completeness of the explanation is thus ensured. The pairs are equipped with the ordering $\langle K,L\rangle \sqsubseteq \langle K',L'\rangle$ if $K \subseteq K'$ and $L \subseteq L'$. More formally, we introduce the next definition.

Definition 2. *The set of* complete explanations *for language* $\mathcal{S} \times \mathcal{W}$ *is:*

$$
\begin{aligned}
Ex_{\mathcal{S},\mathcal{W}} := \{\langle K,L\rangle \sqsubseteq \langle S,W\rangle :\\
\forall K' \in \mathcal{S}(K)\ \forall L' \in \mathcal{W}(L) \quad \mathrm{WCW}(K',L') = \{a\}\},
\end{aligned}
$$

where $\mathcal{S}(K) = \{K' \in \mathcal{S} : K' \supseteq K\}$ *and* $\mathcal{W}(L) = \{L' \in \mathcal{W} : L' \supseteq L\}$.

Example 3. The explanation $K_1 = [1,2 : a \succ d,e,f], [1,3 : a \succ b], [2,3 : a \succ c]$ is not complete, since it does not provide enough evidence that a is preferred over c. Indeed, $H_{K_1}(a,c) < 0$ (since $0.42 - 0.58 = -0.16$). On the other hand, $[1 : a \succ e,d,b,f], [2 : a \succ f,e,d,c], [3 : a \succ b,c,d], [4 : a \succ c,e], [5 : a \succ c]$ is complete but certainly not minimal, since (for instance) exactly the same explanation without the last statement is also a complete explanation whose cost is certainly lower (by monotonicity of the cost function). Now if the cost function is sub-additive, then a minimal explanation cannot contain (for instance) both $[1,2 : a \succ d,e]$ and $[1,2 : a \succ f]$. This is so because then it would be possible to factor these statements as $[1,2 : a \succ d,e,f]$, all other things being equal, so as to obtain a new explanation with a lower cost.

In the rest of the paper, complete explanations will be called simply explanations when there is no possible confusion. One has $\langle S, W\rangle \in Ex_{\mathcal{S},\mathcal{W}}$ and $\langle \emptyset, \emptyset\rangle \notin Ex_{\mathcal{S},\mathcal{W}}$. As shown below, adding more information to a complete explanation also yields a complete explanation.

Lemma 1. *If $\langle K, L\rangle \in Ex_{\mathcal{S},\mathcal{W}}$ then $\langle K', L'\rangle \in Ex_{\mathcal{S},\mathcal{W}}$ for all K', L' with $K \subseteq K' \subseteq S$ and $L \subseteq L' \subseteq W$.*

Proof : Clear since $\mathcal{S}(K) \supseteq \mathcal{S}(K')$ when $K \subseteq K'$, and $\mathcal{W}(L) \supseteq \mathcal{W}(L')$ when $L \subseteq L'$. ■

We will assume in the rest of the paper that there is no simplification regarding the preferential information W. Indeed the gain of displaying less values of the weights is much less significant than the gain concerning S. This comes from the fact that $|W| = m$ whereas $|S| = \frac{1}{2} m\, p\,(p-1)$, where $m = |H|$ and $p = |O|$. Only the information about the basic statements $S \in \mathcal{S}$ is simplified. We are thus interested in the elements of $Ex_{\mathcal{S},\mathcal{W}}$ of the form $\langle K, W\rangle$. Hence we introduce the notation $Ex_{\mathcal{S}} = \{K \in \mathcal{S} \,:\, \langle K, W\rangle \in Ex_{\mathcal{S},\mathcal{W}}\}$.

4 Simple Language for $\mathcal{S}$

We consider in this section explanations with the basic languages $\mathcal{S}$ and $\mathcal{W}$. In this section, the PI is expressed as $\langle S, W\rangle$. The aim of this section is to characterize and construct minimal elements of $Ex_{\mathcal{S}}$ w.r.t. the cost.

We set $H_K(a,b) := \sum_{i\,:\,[i:a\succ b]\in K} W_i - \sum_{i\,:\,[i:a\succ b]\notin K} W_i$ for $K \subseteq S$ and $b \in O^\star$. This means that $K \subseteq S$ is completed only with negative preference statements (in other words, what is not explicitly provided in the explanation is assumed to be negative).

Lemma 2. $Ex_{\mathcal{S}} = \{K \subseteq S \,:\, \forall b \in O^\star \quad H_K(a,b) > 0\}$.

Proof : We have $\mathrm{WCW}(K', W) = \{a\}$ $\forall K' \in \mathcal{S}(K)$ iff $\mathrm{WCW}(K', W) = \{a\}$ for $K' = K \cup \{[i : b \succ a] \,:\, b \in O^\star \text{ and } [i : a \succ b], [i : b \succ a] \notin K\}$ iff $H_K(a,b) > 0$ $\forall b \in O^\star$. ■

A consequence of this result is that neutral statements can simply be ignored since they do not affect the expression $H_K(a,b)$. The next lemma shows furthermore that the minimal explanations are free of negative statements.

Lemma 3. *Let $K \in Ex_{\mathcal{S}}$ minimal w.r.t. the cost. Then K does not contain any negative or neutral preference statement.*

Proof : $K \in Ex_{\mathcal{S}}$ cannot minimize the cost if $[i : b \succ a] \in K$ since then $H_{K'}(a,b) = H_K(a,b)$ and thus $K' \in Ex_{\mathcal{S}}$, with $K' = K \setminus \{[i : b \succ a]\}$. It is the same if $[i : b \succ c] \in K$ with $b, c \neq a$. ■

Then we prove that we can replace a positive basic statement appearing in a complete explanation by another one, while having still a complete explanation, if the weight of the criterion involved in the first statement is not larger than that involved in the second one.

Lemma 4. *Let $K \in Ex_{\mathcal{S}}$, $[i : a \succ b] \in K$ and $[j : a \succ b] \in S \setminus K$ with $W_j \geq W_i$. Then $(K \setminus \{[i : a \succ b]\}) \cup \{[j : a \succ b]\} \in Ex_{\mathcal{S}}$.*

Proof : Let $K' = (K \setminus \{[i : a \succ b]\}) \cup \{[j : a \succ b]\}$. We have $H_{K'}(a,b) = H_K(a,b) + 2(W_j - W_i) > 0$. Hence $K' \in Ex_{\mathcal{S}}$. ■

We define $\Delta_i^S(a,b) = +1$ if $[i : a \succ b] \in S$, and $\Delta_i^S(a,b) = -1$ if $[i : b \succ a] \in S$. For each option $b \in O^\star$, we sort the criteria in H by a permutation π_b on H such that $W_{\pi_b(1)} \Delta_{\pi_b(1)}^S(a,b) \geq \cdots \geq W_{\pi_b(m)} \Delta_{\pi_b(m)}^S(a,b)$.

Proposition 1. *For each $b \in O^\star$, let p_b the smallest integer such that $H_{K_{p_b}^b}(a,b) > 0$, where $K_{p_b}^b = \{[\pi_b(1) : a \succ b], [\pi_b(2) : a \succ b], \ldots, [\pi_b(p_b) : a \succ b]\}$. Then $\{[\pi_b(j) : a \succ b] \; : \; b \in O^\star$ and $j \in \{1, \ldots, p_b\}\}$ is a minimal element of $Ex_{\mathcal{S}}$ w.r.t. the cost.*

Proof (Sketch): Let $Ex_{\mathcal{S}}(b) = \{K \subseteq S_b \; : \; H_K(a,b) > 0\}$, where S_b is the set of statements of S involving option b. The existence of p_b follows from the fact that a is a WCW. Now let $j \in \{1, \ldots, p_b - 1\}$. From the definition of p_b, $K_{p_b-1}^b \notin Ex_{\mathcal{S}}(b)$. This, together with $W_{\pi_b(j)} \geq W_{\pi_b(p_b)}$ and Lemma 4, implies that $K_{p_b}^b \setminus \{[\pi_b(j) : a \succ b]\} \notin Ex_{\mathcal{S}}(b)$. Hence $K_{p_b}^b$ is minimal in $Ex_{\mathcal{S}}(b)$ in the sense of $\subseteq$. It is also apparent from Lemma 4 that there is no element of $Ex_{\mathcal{S}}(b)$ with a strictly lower cardinality and thus lower cost (since, from Section 2.3, the cost of a set of basic statements is proportional to its cardinality). Finally, $\cup_{b \in O^\star} K_{p_b}$ minimizes the cost in $Ex_{\mathcal{S}}$ since the conditions on each option $b \in O^\star$ are independent. ■

This proposition provides a polynomial computation of a minimal element of $Ex_{\mathcal{S}}$. This is obtained for instance by the following greedy Algorithm 1. The complexity of this algorithm is $O(m \cdot p \cdot \log(p))$ (where $m = |H|$ and $p = |O|$).

```
Function Algo(W, Δ) :
  K = ∅;
  For each b ∈ O* do
    Determine a ranking π_b of the criteria according to W_j Δ_j^S(a, b) such
    that W_{π_b(1)} Δ^S_{π_b(1)}(a, b) ≥ ··· ≥ W_{π_b(m)} Δ^S_{π_b(m)}(a, b);
    K_b = {[π_b(1) : a > b]}; k = 1;
    While (H_{K_b}(a, b) ≤ 0) do
      k = k + 1; K_b = K_b ∪ {[π_b(k) : a > b]};
    done
    K = K ∪ K_b;
  end For
  return K;
End
```

Algorithm 1. Algorithm for the determination of a minimal element of Ex_S. The outcome is K.

We illustrate this on our example.

Example 4. Consider the iteration regarding option b. The ranking of criteria for this option is 1/3/4/5/2. During this iteration, the statements $[1 : a \succ b], [3 : a \succ b]$ are added to the explanation. In the end the explanation produced by Algorithm 1 is $[1 : a \succ b], [3 : a \succ b], [2 : a \succ c], [3 : a \succ c], [4 : a \succ c], [1 : a \succ d], [2 : a \succ d], [1 : a \succ e], [2 : a \succ e], [1 : a \succ f], [2 : a \succ f]$. Note that criterion 5 is never involved in the explanation.

5 Factored Language for *S*

The language used in the previous section is simple but not very intuitive. As illustrated in the introduction, a natural extension is to allow more compact explanations by means of factored statements. We thus consider in this section explanations with the factored language $\widehat{\mathcal{S}}$ and the basic language $\mathcal{W}$. As in previous section, all weight statements in $W \in \mathcal{W}$ are kept. The explanations for $\widehat{\mathcal{S}}$ are:

$$Ex_{\widehat{\mathcal{S}}} = \left\{ \widehat{K} \subseteq S^{\uparrow} \ : \ \forall K \in \mathcal{S}(\widehat{K}^{\downarrow}) \quad \mathrm{WCW}(K, W) = \{a\} \right\}.$$

Similarly to what was proved for basic statements, it is simple to show that minimal explanation must only contain positive statements.

Lemma 5. *Let $\widehat{K} \in Ex_{\widehat{\mathcal{S}}}$ minimal w.r.t. the cost. Then $\widehat{K}$ only contains positive preference statements.*

Proof : Similar to the proof of Lemma 3. ■

A practical consequence of this result is that it is sufficient to represent the PI as a binary matrix, for a, where an entry 1 at coordinates (i, j) represents the fact that the option i is less preferred than a on criteria j. Doing so, we do not encode the preferential information expressed by neutral statements.

This representation is attractive because factored statements visually correspond to (combinatorial) rectangles. Informally, looking for an explanation amounts to find a "cheap" way to "sufficiently" cover the 1's in this matrix. However, an interesting thing to notice is that a minimal explanation with factored statements does not imply that factored statements are non overlapping. To put it differently, it may be the case that some preferential information is repeated in the explanations. Consider the following example:

Example 5. There are 5 criteria of equal weight and 6 options, and a is the weighted Condorcet winner. As for the cost of statements, it is constant whatever the statement.

	1	2	3	4	5
	0.2	0.2	0.2	0.2	0.2
b	1	1	0	0	1
c	1	1	0	1	0
d	1	1	1	0	0
e	0	1	1	0	1
f	0	1	1	1	0

There are several minimal explanations involving 4 statements, but all of them result in a covering in the matrix, like for instance $[1,2 : a \succ b,c,d]$, $[2,3 : a \succ d,e,f]$, $[4 : a \succ c,f][5 : a \succ b,e]$, where the preferential information that $a \succ_2 d$ is expressed twice (in the first and second statement).

The previous section concluded on a simple algorithm to compute minimal explanations with basic statements. Unfortunately, we will see that the additional expressive power provided by the factored statements comes at a price when we want to compute minimal explanations.

Proposition 2 (Min. explanations with factored statements). *Deciding if (using factored statements $S^{\uparrow}$) there exists an explanation of cost at most k is NP-complete. This holds even if criteria are unweighted and if the cost of any statement is a constant.*

Proof (Sketch): Membership is direct since computing the cost of an explanation can be done in polynomial time. We show hardness by reduction from the BICLIQUE EDGE COVER (BEC), known to be NP-complete (problem [GT18] in [7]). In BEC, we are given a finite bipartite graph $G = (X, Y, E)$ and positive integer k'. A biclique is a complete bipartite subgraph of G, *i.e.*, a subgraph induced by a subset of vertices such that any vertex is connected to a vertex of the other part. The question is whether there exists a collection of bicliques covering edges of G of size at most k'.

Let $I = (X, Y, E)$ be an instance of BEC. From I, we build an instance I' of the explanation problem as follows. The set O of actions contains $O_1 = \{o_1, \ldots, o_n\}$ corresponding to the elements in X, and a set O_2 of dummy actions consisting of $n+3$ actions $\{o'_1, \ldots, o'_{n+3}\}$. The set H of criteria contains $H_1 = \{h_1, \ldots, h_n\}$ corresponding to the elements in Y, and a set H_2 of dummy criteria consisting of $n+3$ criteria $\{h'_1, \ldots, h'_{n+3}\}$. First, for each $(x_i, y_j) \in E$, we build a statement $[h_i : a \succ o_j]$. Let S_{O_1,H_1} be this set of statements. Observe that a factored statement $[I : a \succ P]$ with $I \subseteq H_1$ and $P \subseteq O_1$ correspond to a biclique in I. But a may not be a Condorcet winner. Thus for each action $o \in O_1$, we add $(n+2) - |\{[h_i : a \succ o] \in O_1\}|$ statement(s) $[h'_j : a \succ o]$. Let S_{O_1,H_2} be this set of statements. Note that at this point, a is preferred to any other $o \in O_1$ by $n+2$ criteria. Next $\forall (h'_i, o'_j) \in (H_2 \times O_2)$ such that $i \neq j$ we add the following statement: $[h'_i : a \succ o'_j]$. There are $n+2$ such statements, hence a is preferred to any other $o \in O_2$ by a majority of exactly $n+2$ criteria. Let S_{O_2,H_2} be this set of statements. We claim that I admits a biclique vertex partition of at most $k - (n+3)$ subsets iff I' admits an explanation $\widehat{K}_*$ of cost at most k using factored statements. Take ($\Leftarrow$). By construction, all the basic statements must be "covered", *i.e.* $\widehat{K}_*^{\downarrow} = S_{O_1,H_1} \cup S_{O_1,H_2} \cup S_{O_2,H_2}$. We denote by $cov(.)$ the cost of covering a set of basic statements of $S_{O,H}$ (this is just the number of factored statements used, as the cost of statements is constant). Furthermore, as there are no statements using actions from O_2 and criteria from H_1, no factored statement can cover at the same time statements from S_{O_1,H_1} and S_{O_2,H_2}. Hence $cost(\widehat{K}_*) = cov(S_{O_1,H_1} \cup S') + cov(S_{O_2,H_2} \cup S'')$, such that $S' \cup S'' = S_{O_1,H_2}$.

But now observe that $cov(S_{O_2,H_2}) = cov(S_{O_2,H_2} \cup S_{O_1,H_2}) = n+3$, so $cost(\widehat{K}_*)$ boils down to $n+3+cov(S_{O_1,H_1} \cup S')$. By monotony wrt. criteria, $cov(S_{O_1,H_1} \cup S')$ is minimized when $S' = \emptyset$, and this leads to the fact $cov(S_{O_1,H_1}) \leq k - (n+3)$. The ($\Rightarrow$) direction is easy. ■

The previous result essentially shows that when the cost function implies to minimize the number of factored statements, no efficient algorithm can determine minimal explanations (unless P=NP). But there may be specific class(es) of cost functions for which the problem may turn out to be easy. As shown in the next lemma, when the cost function is super-additive, then it is sufficient to look for basic statements.

Lemma 6. *If the cost function is super-additive, then* $\min_{\widehat{K} \in Ex_{\widehat{\mathcal{S}}}} cost(\widehat{K}) = \min_{K \in Ex_{\mathcal{S}}} cost(K)$.

Proof : Let $\widehat{K} \in Ex_{\widehat{\mathcal{S}}}$. We know that $\widehat{K}^{\downarrow} \in Ex_{\mathcal{S}}$. By super-additivity, $cost(\widehat{K}) = \sum_{[I:b \succ P] \in \widehat{K}} cost([I : b \succ P]) \geq \sum_{[I:b \succ P] \in \widehat{K}} \sum_{i \in I\,,\, c \in P} cost([i : b \succ c]) \geq \sum_{[i:b \succ c] \in \widehat{K}^{\downarrow}} cost([i : b \succ c]) = cost(\widehat{K}^{\downarrow})$. ■

Yet, the cost is expected to be sub-additive. Relations (1) and (2) below give examples of sub-additive cost functions. In this case, factored statements are less costly (e.g. the cost of $[\{1,2\} : a \succ b]$ should not be larger than the cost of $[1 : a \succ b], [2 : a \succ b]$) and factored explanations become very relevant.

When the cost function is sub-additive, an intuitive idea could be to restrict our attention to statements which exhibit winning coalitions. For that purpose, let us assign to any subset $P \subseteq O^\star$ defended by a winning coalition the cost of using such statement. A practical way to do this is to build $T : 2^{O^\star} \to 2^H$ such that for all subsets $P \subseteq O^\star$, $T(P)$ is the largest set of criteria for which $[T(P) : a \succ P] \in S^\uparrow$. We have $T(P) = \cap_{b \in P} T(\{b\})$, where $T(\{b\}) := \{i \in H : [i : a \succ b] \in S\}$. Then subsets P of increasing cardinality are considered (but those supported by non-winning coalitions are discarded). The cost $C(\alpha, |P|)$ is finally assigned, where α is the size of the smallest winning coalition contained in $T(P)$. Then, the problem can be turned into a weighted set packing, for which the direct ILP formulation would certainly be sufficient in practice for reasonable values of $|O|$ and $|H|$.

Example 6. On our running example, the different potential factors would be $T(\{b\}) = \{1,3\}$ with $C(2,1)$, $T(\{c\}) = \{2,3,4,5\}$ with $C(4,1)$, $T(\{d\}) = \{1,2,3\}$ with $C(3,1)$, $T(\{e\}) = \{1,2,4\}$ with $C(3,1)$, $T(\{f\}) = \{1,2\}$ with $C(2,1)$, $T(\{b,d\}) = \{1,3\}$ with $C(2,2)$, etc. Depending on the cost function, two possible explanations remain: $\widehat{K}_1 = \{[1,3 : a \succ b], [2,3,4,5 : a \succ c], [1,2 : a \succ d,e,f]\}$ for a cost of $C(2,1) + C(4,1) + C(2,3)$, and $\widehat{K}_2 = \{[1,3 : a \succ b,d], [2,3,4,5 : a \succ c], [1,2 : a \succ e,f]$ for a cost of $C(2,2) + C(4,1) + C(2,2)$. The cost function

$$C(i,j) = i^\alpha j^\beta \tag{1}$$

(which is sub-additive when $\alpha \leq 1$ and $\beta \leq 1$) would select $\widehat{K}_1$. Note that criteria 4 or 5 will be dropped from the statement $[T(\{c\}) : a \succ c]$.

Now, considering only factored statements with winning coalitions may certainly prevent from reaching optimal factored explanations, as we illustrate below.

Example 7. We have 4 criteria and 3 options. Assume that a is preferred to b on criteria $1, 2$, and 3; that a is preferred to c on criteria $1, 2$, and 4 and that any coalition of at least 3 criteria is winning. The previous approach based on T gives $\widehat{K}_1 = \{[1,2,3 : a \succ b], [1,2,4 : a \succ c]\}$, with $cost(\widehat{K}_1) = 2\,C(3,1)$. Algorithm 1 gives $\widehat{K}_2 = (\widehat{K}_1)^\downarrow$ with $cost(\widehat{K}_2) = 6\,C(1,1)$. Another option is to consider $\widehat{K}_3 = \{[1,2 : a \succ b,c], [3 : a \succ b][4 : a \succ c]\}$, with $cost(\widehat{K}_3) = C(2,2) + 2\,C(1,1)$. Let us consider the following cost function[1]

$$C(i,j) = i \log(j+1). \tag{2}$$

Function C is sub-additive, since $C(i+i',j) = C(i,j) + C(i',j)$ and, from relation $j + j' + 1 \leq (j+1)(j'+1)$, we obtain $C(i, j+j') \leq C(i,j) + C(i,j')$. Then we have $cost(\widehat{K}_3) < cost(\widehat{K}_1) = cost(\widehat{K}_2)$ so that the explanation with the smallest cost is $\widehat{K}_3$.

Enforcing complete explanations implies a relatively large number of terms in the explanation. However, in most cases, factored statements allow to obtain small explanations. For instance, when all criteria have the same weight, the minimal elements of Ex_S contain exactly $(p-1)\,n$ basic statements (where $p = |O|$,

[1] Capturing that factoring over the criteria is more difficult to handle than factoring over the options.

$m = |H|$ and $m = 2n - 1$ if m is odd, and $m = 2n - 2$ if m is even. Indeed, one needs $p - 1$ terms to explain that a is globally preferred over b, for all $b \in O^\star$, and the minimal elements of $Ex_{\widehat{S}}$ contain at most $p - 1$ factored statements (factoring with winning coalitions for each $b \in O^\star$).

A current matter of investigation is to determine the class of cost functions for which the minimal explanation is not given either by trivial atomization or by factoring with winning coalitions only, thus requiring dedicated algorithms.

6 Related Work and Conclusion

The problem of producing explanations for complex decisions is a long-standing issue in Artificial Intelligence in general. To start with, it is sometimes necessary to (naturally) explain that no satisfying option can be found because the problem is over-constrained [8,9]. But of course it is also important to justify why an option is selected among many other competing options, as is typically the case in recommendations. Explanations based on the data seek to focus on a small subpart of the data, sufficient to either convince or indeed prove the claim to the user. Depending on the underlying decision model, this can turn out to be very challenging.

In this paper we investigate the problem of providing minimal and complete explanations for decisions based on a weighted majority principle, when a Condorcet winner exists. A first contribution of this paper is to set up the framework allowing to analyze notions of minimal explanations, introducing in particular different languages to express the preferential information. We then characterize minimal explanations, and study their computational properties. Essentially, we see that producing minimal explanations is easy with basic statements but may be challenging with more expressive languages.

Much work in argumentation set up theoretical systems upon which various types of reasoning can be performed, in particular argument-based decision-making has been advocated in [10]. The perspective taken in this paper is different in at least two respects: (i) the decision model is not argumentative in itself, the purpose being instead to generate arguments explaining a multiattribute decision model (weighted majority) issued from decision-theory; and (ii) the arguments we produce are complete (so, really *proving* the claim), whereas in argumentation the defeasible nature of the evidence put forward is a core assumption [11]. Regarding (ii), our focus on complete arguments has been justified in the introduction. Regarding (i), we should emphasize that we make no claim on the relative merits of argument-based vs. decision-theoretic models. But in many organizations, these decision models are currently in use, and although it may be difficult to change the habits of decision-makers for a fully different approach, adding explanatory features on top on their favorite model can certainly bring much added-value. This approach is not completely new, but previous proposals are mainly heuristic and seek to generate natural arguments [1] that are persuasive in practice. An exception is the recent proposal of [6] which provides solid theoretical foundations to produce explanations for a range of decision-theoretic weight-based models, but differs in (ii) since explanations are based on (defeasible) argument schemes. Our focus on complete explanations is a further motivation to build on solid theoretical grounds (even though weaker incomplete arguments may prove more persuasive in practice).

Recently, the field of computational social choice has emerged at the interface of AI and social choice, the study of various computational of various voting systems being one of the main topic in this field. There are connections to our work (and indeed one of the motivating example is a voting committee): for instance, exhibiting the smallest subsets of votes such that a candidate is a *necessary winner* [12] may be interpret as a minimal (complete) explanation that this candidate indeed wins. However, the typical setting of voting (*e.g.* guaranteeing the anonymity of voters) would not necessarily allow such explanations to be produced, as it implies to identify voters (to assign weights). An interesting avenue for future research would be to investigate what type of explanations would be acceptable in this context, perhaps balancing the requirements of privacy and the need to support the result. We believe our approach could be relevant. Indeed, two things are noteworthy: first, the proposed approach already preserves some privacy, since typically only parts of the ballots need to be exhibited. Secondly, in many cases it would not be necessary to exactly identify voters, at least when their weights are sufficiently close. Take again our running example: to explain that a beats b we may well say "the most important voter 1 is for a, and among 2 and 3 only one defends b".

We conclude by citing some possible extensions of this work. The first is to improve further the language used for explanations. The limitations of factored statements is clear when the following example is considered:

Example 8. In the following example with 6 alternatives and 5 criteria (with the same weight), the factored statements present in any minimal explanation contain at least 3 criteria or alternatives (for instance, $[1,2,3 : a \succ e, f]$, $[3,4,5 : a \succ b, c]$, $[1,2,4 : a \succ d]$)

1	2	3	4	5
0.2	0.2	0.2	0.2	0.2
b	c	d	e	f
a	a	a	a	a
c	d	e	f	b
d	e	f	b	c
e	f	b	c	d
f	b	c	d	e

However, an intuitive explanation that comes directly to mind is as follows: "a is only beaten by a different option on each criteria".

To take a step in the direction of such more natural explanations, the use of "except" statements allowing to assert that an option is preferred over any other option *except* the ones explicitly cited should be taken into account. (In fact, the informal explanation of our example makes also use of such a statement, since it essentially says that a is better than c on all criteria except 1). In that case, minimal explanations may cover larger sets of basic statements than strictly necessary (since including more elements of the PI may allow to make use of an except statement). Another extension would be to relax the assumption of neutrality of the cost function, to account for situations where some information is exogenously provided regarding criteria to be used preferably in the explanation (this may be based on the profile of the decision-maker, which may be more sensible to certain types of criteria).

Acknowledgments. We would like to thank Yann Chevaleyre for discussions related to the topic of this paper. The second author is partly supported by the ANR project ComSoc (ANR-09-BLAN-0305).

References

1. Carenini, G., Moore, J.: Generating and evaluating evaluative arguments. Artificial Intelligence 170, 925–952 (2006)
2. Klein, D.: Decision analytic intelligent systems: automated explanation and knowledge acquisition. Lawrence Erlbaum Associates, Mahwah (1994)
3. Buchanan, B.G., Shortliffe, E.H.: Rule Based Expert Systems: The Mycin Experiments of the Stanford Heuristic Programming Project. Addison-Wesley, Boston (1984)
4. Symeonidis, P., Nanopoulos, A., Manolopoulos, Y.: MoviExplain: a recommender system with explanations. In: Proceedings of the Third ACM Conference on Recommender Systems (RecSys 2009), pp. 317–320. ACM, New York (2009)
5. Herlocker, J.L., Konstan, J.A., Riedl, J.: Explaining collaborative filtering recommendations. In: Proceedings of the ACM Conference on Computer Supported Cooperative Work (CSCW 2000), pp. 241–250. ACM, New York (2000)
6. Labreuche, C.: A general framework for explaining the results of a multi-attribute preference model. Artificial Intelligence 175, 1410–1448 (2011)
7. Garey, M., Johnson, D.: Computers and intractability. A guide to the theory of NP-completeness. Freeman, New York (1979)
8. Junker, U.: QUICKXPLAIN: Preferred explanations and relaxations for over-constrained problems. In: McGuinness, D.L., Ferguson, G. (eds.) Proceedings of the Nineteenth AAAI Conference on Artificial Intelligence (AAAI 2004), pp. 167–172. AAAI Press, Menlo Park (2004)
9. O'Sullivan, B., Papadopoulos, A., Faltings, B., Pu, P.: Representative explanations for over-constrained problems. In: Proceedings of the Twenty-Second AAAI Conference on Artificial Intelligence (AAAI 2007), pp. 323–328. AAAI Press, Menlo Park (2007)
10. Amgoud, L., Prade, H.: Using arguments for making and explaining decisions. Artificial Intelligence 173, 413–436 (2009)
11. Loui, R.P.: Process and policy: Resource-bounded nondemonstrative reasoning. Computational Intelligence 14, 1–38 (1998)
12. Konczak, K., Lang, J.: Voting procedures with incomplete preferences. In: Brafman, R., Junker, U. (eds.) Proceedings of the IJCAI 2005 Workshop on Advances in Preference Handling, pp. 124–129 (2005)

Vote Elicitation with Probabilistic Preference Models: Empirical Estimation and Cost Tradeoffs

Tyler Lu and Craig Boutilier

Department of Computer Science, University of Toronto, Toronto, Canada
{tl,cebly}@cs.toronto.edu

Abstract. A variety of preference aggregation schemes and voting rules have been developed in social choice to support group decision making. However, the requirement that participants provide full preference information in the form of a complete ranking of alternatives is a severe impediment to their practical deployment. Only recently have incremental elicitation schemes been proposed that allow winners to be determined with partial preferences; however, while minimizing the amount of information provided, these tend to require repeated rounds of interaction from participants. We propose a probabilistic analysis of vote elicitation that combines the advantages of incremental elicitation schemes—namely, minimizing the amount of information revealed—with those of full information schemes—single (or few) rounds of elicitation. We exploit distributional models of preferences to derive the ideal *ranking threshold* k, or number of top candidates each voter should provide, to ensure that either a winning or a high quality candidate (as measured by max regret) can be found with high probability. Our main contribution is a general empirical methodology, which uses preference profile samples to determine the ideal ranking threshold for many common voting rules. We develop probably approximately correct (PAC) sample complexity results for one-round protocols with *any voting rule* and demonstrate the efficacy of our approach empirically on one-round protocols with Borda scoring.

Keywords: social choice, voting, preference elicitation, probabilistic rankings.

1 Introduction

Researchers in computer science have increasingly adopted preference aggregation methods from social choice, typically in the form of *voting rules*, for problems where a *consensus* decision or recommendation must be made for a group of users. The availability of abundant preference data afforded by search engines, recommender systems, and related artifacts, has accelerated the need for good computational approaches to social choice. One problem that has received little attention, however, is that of effective *preference elicitation* in social choice. Many voting schemes require users or *voters* to express their preferences over the entire space of options or *alternatives*, something that is not only onerous, but often extracts more information than is strictly necessary to determine a good consensus option, or *winner*. Reducing the amount of preference information elicited is critical to easing cognitive and communication demands on users and mitigating privacy concerns.

R.I. Brafman, F. Roberts, and A. Tsoukiàs (Eds.): ADT 2011, LNAI 6992, pp. 135–149, 2011.

Winners can't be determined in many voting schemes without a large amount of information in the worst case [2,3]. Nonetheless, the development of elicitation schemes that work well in practice has been addressed very recently. Lu and Boutilier [10] use the notion of *minimax regret* for vote elicitation: this measure not only allows one to compute worst-case bounds on the quality of a proposed winner given *partial* voter preferences, it can also be used to drive incremental elicitation. Kalech et al. [6] develop several heuristic strategies for vote elicitation, including one scheme that proceeds in *rounds* in which voters provide larger "chunks" of information. This offers an advantage over the Lu-Boutilier schemes, where each voter query is conditioned on all previous responses of the other voters. Unfortunately, Kalech et. al's approach does not admit approximation (with quality guarantees), and no principles are provided to select an appropriate chunk size.

In this work, we develop an approach to vote elicitation that exploits distributional information over voter preferences to simultaneously reduce the amount of information elicited from voters and the number of *rounds* (a notion defined formally below) of elicitation. Indeed, these factors can be explicitly traded off against one another. Our model also supports approximation, using minimax regret, to further minimize the amount of information elicited, the number of rounds, or both. In this way, we provide the first framework that allows the design of vote elicitation schemes that address the complicated three-way tradeoff between approximation quality, total information elicited, and the number of rounds of elicitation.

Developing analytical bounds depends, of course, on the specific distributional assumptions about the preferences and the voting rule in question. While we make some suggestions regarding the types of results one might derive along these lines, our primary contribution is an empirical methodology that allows a designer to assess these tradeoffs and design elicitation schemes for *any preference distribution*, and *any voting rule* that can be interpreted using some form of scoring. To illustrate the use of both our general elicitation framework and our empirical methodology, we analyze one-round vote elicitation protocols. We develop general PAC sample complexity bounds for such one-round protocols. We then analyze these protocols empirically using Mallows models of preferences distributions [11,12] and Borda scoring as the voting protocol. Our results suggest that good, even optimal, results can be obtained in one-round protocols even when only a small portion of the preferences of the voters is elicited.

2 Background

We begin with a brief overview of relevant background on social choice, vote elicitation, and preference distributions.

2.1 Voting Rules

We first define our basic social choice setting (see [5,1] for further background). We assume a set of *agents* (or *voters*) $N = \{1, \dots, n\}$ and a set of *alternatives* $A = \{a_1, \dots, a_m\}$. Alternatives can represent any outcome space over which the voters have preferences (e.g., product configurations, restaurant dishes, candidates for office, public projects, etc.) and for which a single collective choice must be made. Let Γ_A be the set of *rankings* (or *votes*) over A (i.e., permutations over A). Voter ℓ's preferences are represented by a ranking $v_\ell \in \Gamma_A$. Let $v_\ell(a)$ denote the *rank* of a in v_ℓ. Then ℓ prefers a_i to a_j, denoted $a_i \succ_{v_\ell} a_j$, if $v_\ell(a_i) < v_\ell(a_j)$. We refer to a collection of votes $\mathbf{v} = \langle v_1, \ldots, v_n \rangle \in \Gamma_A^n$ as a *preference profile*. Let $\mathcal{V}$ be the set of all such profiles.

Given a preference profile, we consider the problem of selecting a *consensus alternative*, requiring the design of a *social choice function* or *voting rule* $r : \mathcal{V} \rightarrow A$ which selects a "winner" given voter rankings/votes. *Plurality* is one of the most common rules: the alternative with the greatest number of "first place votes" wins (various tie-breaking schemes can be adopted). Plurality does not require that voters provide rankings; however, this "elicitation advantage" means that it fails to account for relative voter preferences for any alternative other than its top choice. Other schemes produce winners that are more sensitive to relative preferences, among them, the Borda rule, Copeland, single-transferable vote (STV), the Kemeny consensus, maximin, Bucklin, and many others. We outline the *Borda rule* since we use it extensively below: let $B(i) = m - i$ be the Borda score for each rank position i; the *Borda count or score* of alternative a given profile $\mathbf{v}$, is $s_B(a, \mathbf{v}) = \sum_\ell B(v_\ell(a))$. The winner is the a with the greatest Borda score.

Notice that both the Borda and plurality schemes *explicitly scores all alternatives* given voter preferences, implicitly defining "societal utility" for each alternative. Indeed, many (though not all) voting rules r can be interpreted as maximizing a "natural" *scoring function* $s(a, \mathbf{v})$ that defines some measure of the quality of an alternative a given a profile $\mathbf{v}$. We assume in what follows that our voting rules are *score-consistent* in this sense: $r(\mathbf{v}) \in \text{argmax}_{a \in A}\, s(a, \mathbf{v})$. some "natural" scoring function $s(a, \mathbf{v})$.[1]

2.2 Vote Elicitation

One obstacle to the widespread use of voting schemes that require full rankings is the informational and cognitive burden imposed on voters, and concomitant ballot complexity. Elicitation of sufficient, but still *partial* information about voter rankings could alleviate some of these concerns. We will assume in what follows that the partial information about any voter's ranking can be represented as a collection of pairwise comparisons. Specifically, let the *partial vote* p_ℓ of voter ℓ be a partial order over A, or equivalently (the transitive closure of) a collection of *pairwise comparisons* of the form $a_i \succ a_j$. Let $\mathbf{p}$ denote a partial profile, and $C(\mathbf{p})$ the set of consistent extensions of $\mathbf{p}$ to full ranking profiles. Let $\mathcal{P}$ denote the set of partial profiles.

If our aim is to determine the winner given a partial profile, theoretical worst-case results are generally discouraging, with the communication complexity of several common voting protocols (e.g., Borda) being $\Theta(nm \log m)$, essentially requiring communication of full voter preferences in the worst-case [3]. Despite its theoretical complexity, practical schemes for elicitation have been developed recently.

Lu and Boutilier [10] use *minimax regret (MMR)* to determine winners given partial profiles, and also to guide elicitation. Intuitively, one measures the quality of a proposed

[1] We emphasize that *natural* measures of quality are the norm; trivially, any rule can be defined as score consistent using a simple indicator function.

winner a given $\mathbf{p}$ by considering how far from optimal a could be in the worst case, given any completion of $\mathbf{p}$; this is a's *maximum regret* $MR(a, \mathbf{p})$. The *minimax optimal solution* is any alternative that is nearest to optimal in the worst case, i.e., with minimum max (minimax) regret. More formally:

$$Regret(a, \mathbf{v}) = \max_{a' \in A} s(a', \mathbf{v}) - s(a, \mathbf{v}) = s(r(\mathbf{v}), \mathbf{v}) - s(a, \mathbf{v}) \quad (1)$$

$$MR(a, \mathbf{p}) = \max_{\mathbf{v} \in C(\mathbf{p})} Regret(a, \mathbf{v}) \quad (2)$$

$$MMR(\mathbf{p}) = \min_{a \in A} MR(a, \mathbf{p}) \; ; \qquad a^*_{\mathbf{p}} \in \operatorname*{argmin}_{a \in A} MR(a, \mathbf{p}) \; . \quad (3)$$

This gives us a form of robustness in the face of vote uncertainty: every alternative has worst-case error at least as great as that of $a^*_{\mathbf{p}}$. Notice that if $MMR(\mathbf{p}) = 0$, then the minimax winner $a^*_{\mathbf{p}}$ is optimal in any completion $\mathbf{v} \in C(\mathbf{p})$. MMR can be computed in polytime for several common voting rules, including Borda [10].

MMR can also be used to determine (pairwise or top-k) queries that quickly reduce minimax regret; indeed, in a variety of domains, regret-based elicitation finds (optimal) winners with small amounts of voter preference information, and can find near-optimal candidates (with bounded maximum regret) with even less. However, these elicitation methods implicitly condition the choice of a voter-query pair on *all past responses*. Specifically, the choice any query is determined by first solving the minimax regret optimization (Eq. (3)) w.r.t. the responses to all prior queries. Hence each query must be posed in a separate *round*, making it impossible to "batch" multiple queries for a specific user.

Kalech et al. [6] develop two elicitation algorithms for winner determination with score-based rules (e.g., Borda, range voting) in which voters are asked for kth-ranked candidates in decreasing order of k. Their first method proceeds in fine-grained rounds much like the MMR-approach above, until a *necessary winner* [8,16] is discovered. Their second method proceeds for a predetermined number of rounds, asking each voter at each stage for fixed number of positional rankings (e.g., the top k candidates, or the next k' candidates, etc.). Since termination is predetermined, necessary winners may not be discovered; instead possible winners are returned. Tradeoffs between the number of rounds and amount of information per round are explored empirically. One especially attractive feature of this approach is the explicit batching of queries: voters are only queried a fixed (ideally small) number of times (though each query may request a lot of information), thus minimizing interruption, waiting time, etc. However, no quality guarantees are provided, nor is a theoretical basis provided for selecting the amount of information requested at any round.

2.3 Probabilistic Models of Population Preferences

Probabilistic analysis in social choice has often focused on the *impartial culture model*, which asserts that all preference orderings are equally likely. However, the plausibility of this assumption, and the relevance of theoretical results based on it, have been seriously called into question by behavioral social choice theorists [14]. More realistic probabilistic models of preferences, or parameterized families of distributions over rankings, have been proposed in statistics, econometrics and psychometrics. These models typically reflect some process by which people rank, judge or compare alternatives. Many models are unimodal, based on a "reference ranking" from which user rankings are seen as noisy perturbations. A commonly used model, adopted widely in machine learning—and one we exploit below—is the *Mallows ϕ-model* [11]. It is parameterized by a modal or *reference ranking* σ and a *dispersion parameter* $\phi \in (0, 1]$; and for any ranking r we define: $P(r; \sigma, \phi) = \frac{1}{Z}\phi^{d(r,\sigma)}$, where d is the Kendall-tau distance and Z is a normalization constant. When $\phi = 1$ we obtain the uniform distribution over rankings, and as $\phi \to 0$ we approach the distribution that concentrates all mass on σ. A variety of other models have been proposed that reflect different interpretations of the ranking process (e.g., Plackett-Luce, Bradley-Terry, Thurstonian, etc.); we refer to [12] for a comprehensive treatment. Mixtures of such models, which offer additional modeling flexibility (e.g., by admitting multimodal preference distributions), have also been investigated (e.g., [13,9]).

Sampling rankings from specific families of distributions is an important task that we also rely on below. The *repeated insertion model (RIM)*, introduced by Doignon et al. [4], is a generative process that can be used to sample from certain distributions over rankings and provides a practical way to sample from a Mallows model. A variant of this model, known as the *generalized repeated insseartion model (GRIM)*, offers more flexibility, including the ability to sample from conditional Mallows models [9].

3 A Regret-Based Model of Probabilistic Vote Elicitation

We begin by developing a general model of vote elicitation that allows one to make explicit tradeoffs between the number of rounds of elicitation, the amount of information provided by each voter, and approximation quality. Let a *query* refer to a "single" request for information from a voter. Types of queries include simple pairwise comparisons (e.g., "Do you prefer a to b?"); sets of such comparisons; more involved partial requests (e.g., "Who are your top k candidates?"); or requests for entire rankings. Different queries have different "costs"—both in terms of voter cognitive effort and communication costs (which range from 1 to roughly $m \log m$ bits)—and provide varying degrees of information.

Given a particular class of queries $\mathcal{Q}$, informally, a *multi-round voting protocol* selects, at each round, a subset of voters, and one query per selected voter. The *voter-query (VQ) pairs* selected at round t can be conditioned on the responses to all previous queries. More formally, let I_{t-1} be the *information set* available at round t (i.e., responses to queries at rounds $1, \ldots, t-1$). We represent this information set as a partial profile $\mathbf{p}^{t-1}$, or a set of pairwise comparisons for each voter.[2] A *protocol* then consists of: (a) a *querying function* π, i.e., a sequence of mappings $\pi^t : \mathcal{P} \mapsto (N \mapsto \mathcal{Q} \cup \{0\})$, selecting for each voter a single query at stage t given the current information set; and (b) a *winner selection function* $\omega : \mathcal{P} \mapsto A \cup \{0\}$, where $\omega(\mathbf{p})$ denotes the winner given partial profile $\mathbf{p}$. If $\omega(\mathbf{p}^t) = 0$, no winner is declared and the protocol proceeds

---

[2] Most natural constraints, including responses to many natural queries (e.g., pairwise comparison, top-k, etc.), can be represented in this way. One exception: arbitrary positional queries of the form "what candidate is in rank position k?" induce disjunctive constraints, unless positions k are queried in (ascending or descending) order.

to round $t+1$; otherwise the protocol terminates with the chosen winner at round t. If $\pi^t(\mathbf{p}^{t-1})(\ell) = 0$, then no query is posed to voter ℓ at round t.

Suppose we have a distribution P over complete voter profiles. Given a protocol $\Pi = (\pi, \omega)$, we have an induced distribution over runs of Π, which in turn gives us a distribution over various properties reflecting the cost and performance of Π. There are three general properties of interest to us:

(a) *Quality of the winner*: if Π terminates with information set $\mathbf{p}$ and winner a, we can measure quality using either expected regret, $\sum_{\mathbf{v}} Regret(a, \mathbf{v}) P(\mathbf{v}|\mathbf{p})$, or maximum regret, $MR(a, \mathbf{p})$. If Π is an *exact protocol* (always determining a true winner), both measures will be zero. We focus here on max regret, which provides worst-case guarantees on winner quality. In some settings, expected regret might be more suitable.
(b) *Amount of information elicited*: this can be measured in various ways (e.g., equivalent number of pairwise comparisons or bits).
(c) *Number of rounds* of elicitation.

There is a clear tradeoff between these factors. A greater degree of approximation in winner selection can be used to reduce informational requirements, rounds, or both [10]. For any fixed quality threshold, the number of rounds and the amount of information elicited can also be traded off against one another. At one extreme, optimal outcomes can clearly be found in one round if we ask each voter for full rankings. At the other extreme, optimal policies minimizing expected elicited information can always be constructed (though this will likely come at great computational expense) by selecting a single VQ-pair at each round, where each query carries very little information (e.g., a simple pairwise comparison), at a dramatic cost in terms of number of rounds. How one addresses these tradeoffs depends on the costs associated with each of these factors. For example, the cost of elicited information might reflect the number and type of queries asked of voters, while the cost associated with rounds might reflect interruption and delay experienced by voters as they "wait" for other voters to answer queries before receiving their own next query.[3]

Computing optimal protocols for specific voting rules, query classes, distributions over preferences, and cost models is a very important problem that can be addressed explicitly using our framework. The framework supports both Bayesian and PAC-style (probably approximately correct) analysis. We illustrate its use by considering a specific type of protocol using a PAC-style analysis in the next section.

4 Probably Approximately Correct One-Round Protocols

Imagine we require a *one-round protocol*, where each voter can be asked, exactly once, to list their top-k candidates. A natural question is: what is the minimum value k^* for which such top-k queries ensure that the resulting profile $\mathbf{p}$ has low minimax regret,

[3] We're being somewhat informal, since some voters may only be queried at subset of the rounds. If a (conditional) sequence of queries is asked of a single voter ℓ without any interleaving queries to another voter j, we might count this as a single "session" or round for ℓ. These distinctions won't be important in what follows.

$MMR(\mathbf{p}) \leq \varepsilon$, with high probability, at least $1 - \delta$? We call ε and δ the *minimax regret accuracy* and *confidence* parameters, respectively. Obviously, such a k^* exists: with $k = m - 1$, we elicit each voter's full ranking, always ensuring $MMR(\mathbf{p}) = 0$. This question is of interest when, for example, more than one round of elicitation is infeasible or very costly, an approximate solution (with tolerance ε) is suitable, and some small probability δ of a poor solution is acceptable.

Let $\mathbf{p}[k]$ denote the restriction of profile $\mathbf{v} = (v_1, \ldots, v_n)$ to the subrankings consisting of each voter's top k candidates. For any distribution P over voter preferences $\mathbf{v}$, $MMR(\mathbf{p}[k])$ is a random variable. Let $q_k = P(MMR(\mathbf{p}[k]) \leq \varepsilon)$. We would like to find $k^* = \min\{k : q_k \geq 1 - \delta\}$. Even if we assume P has a particular form, computing k^* might be analytically intractable; or the analytically derived upper bounds may too loose to be of practical use. If one can instead sample vote profiles from the true distribution—without necessarily knowing what P is—a simple empirical methodology can be used to determine a small $\hat{k}$ that, with high probability, has the desired MMR accuracy with near the desired MMR confidence (see Theorem 1 below). Specifically, we take the following steps:

(a) Specify the following parameters: MMR accuracy $\varepsilon > 0$, MMR confidence $\delta > 0$, *sampling accuracy* $\xi > 0$, and *sampling confidence* $\eta > 0$.
(b) Obtain t i.i.d. samples of vote profiles $S = (\mathbf{v}_1, \ldots, \mathbf{v}_t)$ where

$$t \geq \frac{1}{2\xi^2} \ln \frac{2(m-2)}{\eta}. \tag{4}$$

(c) Output $\hat{k}$, the smallest k for which

$$\hat{q}_k \equiv \frac{|\{i \leq t : MMR(\mathbf{p}_i[k]) \leq \varepsilon\}|}{t} > 1 - \delta - \xi \; .$$

The parameters ξ and η are required to account for sampling randomness, and are incorporated as part of the statistical guarantee on the algorithm's success (see Theorem 1). In summary, the approach is to estimate q_k (which is usually intractable to derive analytically) using $\hat{q}_k$, and take the smallest $\hat{k}$ that, accounting for sampling error, is highly likely to have the true probability, $q_{\hat{k}}$, lie close to the desired MMR confidence threshold $1 - \delta$. The larger the sample size t, the better the estimates, resulting in smaller ξ and η. Using a sample set specified as in the algorithm, one can obtain a PAC-style guarantee [15] on the quality of one-round, top-$\hat{k}$ elicitation:

Theorem 1. *Let* $\varepsilon, \delta, \eta, \xi > 0$. *If the sample size t satisfies Eq. (4), then for any preference profile distribution P, with probability* $1 - \eta$ *over i.i.d. samples* $\mathbf{v}_1, \ldots, \mathbf{v}_t$, *we have: (a)* $\hat{k} \leq k^*$; *and (b)* $P[MMR(\mathbf{p}[\hat{k}]) \leq \varepsilon] > 1 - \delta - 2\xi$.

Proof. For any $k \leq m - 2$ (for $k = 0$, minimax regret is $n(m-1)$ and for $k \geq m - 1$ minimax regret is 0, so we are not interested in these cases), the indicator random variables $\mathbf{1}[MMR(\mathbf{p}_i[k]) \leq \epsilon]$ for $i \leq t$ are i.i.d. By the Hoeffding bound, we have

$$\Pr_{S \sim P^t} [|\hat{q}_k - q_k| \geq \xi] \leq 2\exp(-2\xi^2 t).$$

If we choose t such that $\frac{\eta}{m-2} \leq 2\exp(-2\xi^2 t)$ we obtain Inequality (4) and

$$\begin{aligned}
&\Pr_{S\sim P^t}\left((|\hat{q}_1 - q_1| \leq \xi) \wedge (|\hat{q}_2 - q_2| \leq \xi) \wedge \ldots \wedge (|\hat{q}_{m-2} - q_{m-2}| \leq \xi)\right) \\
&= 1 - \Pr_{S\sim P^t}\left[\bigcup_{k=1}^{m-2} |\hat{q}_k - q_k| > \xi\right] \\
&\geq 1 - (m-2)\cdot\frac{\eta}{m-2} \qquad (5) \\
&= 1 - \eta,
\end{aligned}$$

where Inequality (5) follows from the union bound. Thus with probability at least $1-\eta$, uniform convergence holds, and we have $\hat{q}_{k^*} > q_{k^*} - \xi > 1 - \delta - \xi$. Since $\hat{k}$ is the smallest k with $\hat{q}_k > 1 - \delta - \xi$ we have $\hat{k} \leq k^*$. Furthermore, $q_{\hat{k}} > \hat{q}_{\hat{k}} - \xi > (1 - \delta - \xi) - \xi = 1 - \delta - 2\xi$, which shows part (2). ■

We note several significant features of this result. First, it is distribution-independent—we need t i.i.d. samples from P, where t depends only on ξ, η and m, and *not* on any property of P. Of course, depending on the nature of the distribution, the required sample size may be larger than necessary (e.g., if P is highly concentrated). Second, note that an algorithm that outputs $k = m - 1$ guarantees $MMR = 0$, but is effectively useless to the elicitor; hence we desire an algorithm that proposes a k that is not much larger than the optimal k^*. Our scheme guarantees $\hat{k} \leq k^*$. Third, while the true probability $q_{\hat{k}}$ of the estimated $\hat{k}$ satisfying the regret accuracy requirement may not meet the confidence threshold, it lies within some small tolerance of that threshold. This is unavoidable in general. For instance, if we have $q_{k^*} = 1 - \delta$, there is potentially a significant probability that $\hat{q}_{k^*} < 1 - \delta$ for any finite sample; but our result ensures that there is only a small probability that $\hat{q}_{k^*} < 1 - \delta - \xi$. Fourth, part (b) of Theorem 1 remains valid if the sum $\delta + \xi$ is fixed (and in some sense, this sum can be interpreted as our ultimate confidence); but variation in δ and ξ does impact sample size (and part (a)). One can reduce the required sample size by making ξ larger and reducing δ correspondingly, maintaining the same "total" degree of confidence, but the guarantee in part (a) becomes weaker since k^* generally increases as δ decreases. This is a subtle tradeoff that should be accounted for in the design of an elicitation protocol.

We can provide no *a priori* guarantees on how small k^* might be, since this depends crucially on properties of the distribution; in fact, it might be quite large (relative to m) for, say, the impartial culture model (as we see below). But our theorem provides a guarantee on the size of $\hat{k}$ w.r.t. the optimal k^*.

An analogous result can easily be obtained if one is interested in determining the smallest k for a one-round protocol that has small *expected MMR*. However, using expectation does not preclude MMR from being greater than a desired threshold with significant probability. Hence, expected MMR may be ill-suited to choosing k in many voting settings. The techniques above can also be used in a Bayesian fashion, where instead of using minimax regret to determine robust winners, one uses *expected regret* (i.e., expected loss relative to the optimal candidate given uncertainty over completions the partial profile). We defer treatment of expected regret to another article.

Our empirical methodology can also be used in a more heuristic fashion, *without* derivation of precise confidence bounds. One can simply generate random profiles, use the empirical distribution over $MMR(\mathbf{p}[k])$ as an estimate of the true distribution, and select the desired k based directly on properties of the empirical distribution (e.g., represented as histograms, as we illustrate in the next section).

Finally, we note that samples can be obtained in a variety of ways, e.g., drawn from a learned preference model, such as a Mallows model or Mallows mixture (e.g., using RIM), or simply obtained from historical problem instances. In multiround protocols, the GRIM model can be used to realize conditional sampling if needed. Our empirical methodology is especially attractive when k^* cannot easily be derived analytically (which may well be the case for Mallows, Plackett-Luce, and other common models).

5 Empirical Results

To explore the effectiveness of our methodology, we ran a suite of experiments, sampling voter preferences from Mallows models using a range of parameters, computing minimax regret for each sampled profile for various k, and estimating both the expected minimax regret and the MMR-distribution empirically. We also discuss experiments with two real-world data sets. Borda scoring is used in all experiments.

For the Mallows experiments, a preference profile is constructed by drawing n i.i.d. rankings, one per voter, from a fixed Mallows model. Each experiment varies the number of voters n, number of alternatives m, and dispersion ϕ, and uses 100 preference profiles. We simulate the elicitation of top-k preferences and measure both MMR and true regret (w.r.t. the true preferences and true winner) for $k = 1, \ldots, m-1$; results are "normalized" by reporting max regret and true regret *per voter*. Fig. 1 shows histograms reflecting the empirical distribution of both MMR and true regret for various k, ϕ, n, and m. That is, in each collection of histograms, as defined by particular (m, n, ϕ) parameter values, we generated 100 instances of random preference profiles. For each instance of a profile, and each k, we compute MMR of the partial votes when top-k preferences are revealed in the profile—this represents one data point along the horizontal axis, in the histogram corresponding to that particular k, and to parameters values (m, n, ϕ). Note that (normalized) MMR per voter can range from 0 to 9 since we use Borda scoring.

Clearly MMR is always zero when $k = m - 1 = 9$. For small ϕ (e.g., 0.1–0.4), preferences across voters are reasonably similar, and values of $k =$ 1–3 are usually sufficient to find the true winner, or one with small max regret. But even with $m = 10$, $n = 100$ and $\phi = 0.6$, $k = 4$ results in a very good approximate winner: $MMR \leq 0.6$ in $90/100$ instances. Even the most difficult case for partial elicitation—the uniform distribution with $\phi = 1$—gives reasonable MMR guarantees with high probability with less than full elicitation ($k =$ 5–7, depending on one's tolerance). The heuristic use of the empirical distribution in this fashion is likely to suffice in practice in a variety of settings; but we can apply the theoretical bounds above as well. Since we have a $t = 100$ (admittedly a small sample), by Eq. (4), we can set $\eta = 0.05$ and $\xi = 0.17$, and with $\delta = 0.9$, $\varepsilon = 0.5$, we obtain $\hat{k} = 4$. By Theorem 1, we are guaranteed with probability 0.95 that $\hat{k} \leq k^*$ and $q_{\hat{k}} > 0.56$. If we wanted $q_{\hat{k}}$ to be closer to 0.9, then requiring $t \geq 28842$ gives $\xi = 0.01$ and $q_{\hat{k}} > 0.88$.

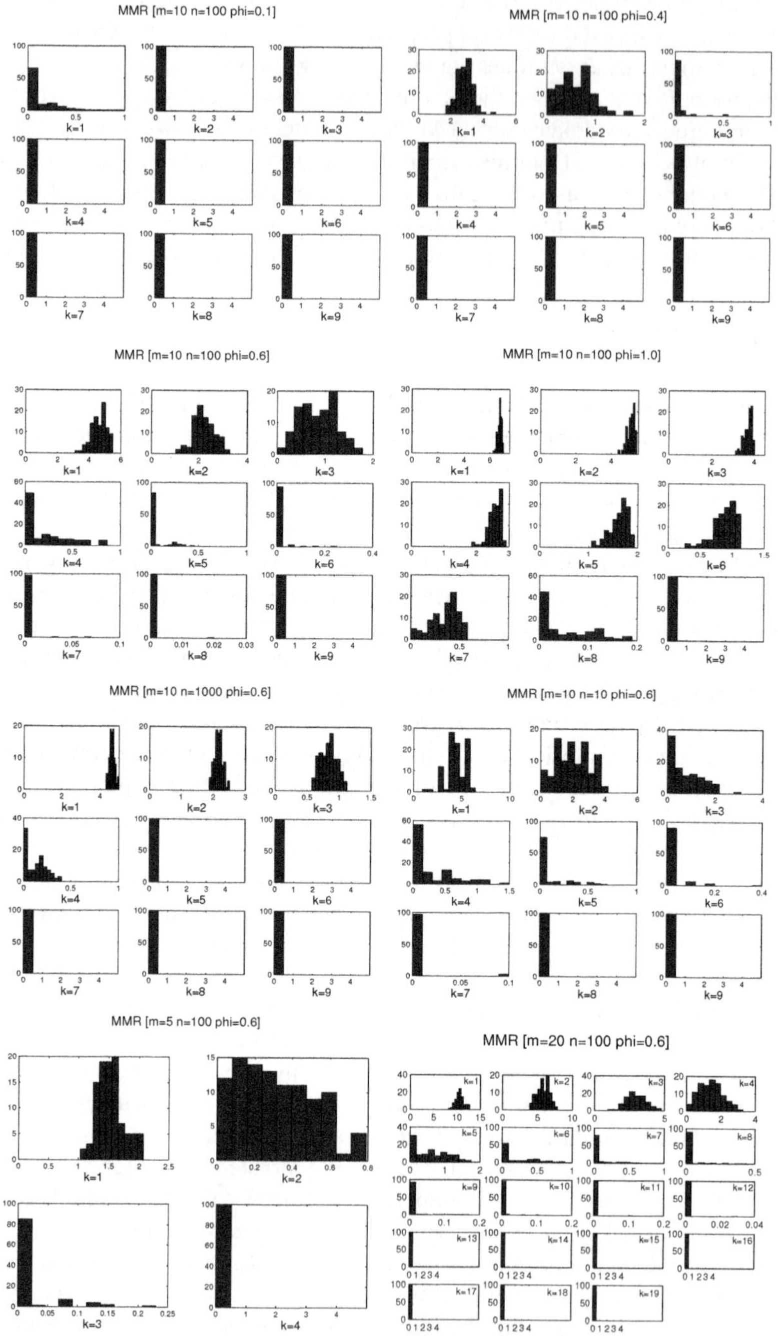

Fig. 1. MMR plots for various ϕ, n and m: for $m = 10$, $n = 100$ with $\phi \in \{0.1, 0.4, 0.6, 1.0\}$ and fixed $\phi = 0.6$ with $n \in \{10, 1000\}$; $m = 5$, $\phi = 0.6$; and $m = 20$, $\phi = 0.6$. Each histogram shows the distribution of MMR, normalized by n, after eliciting top-k.

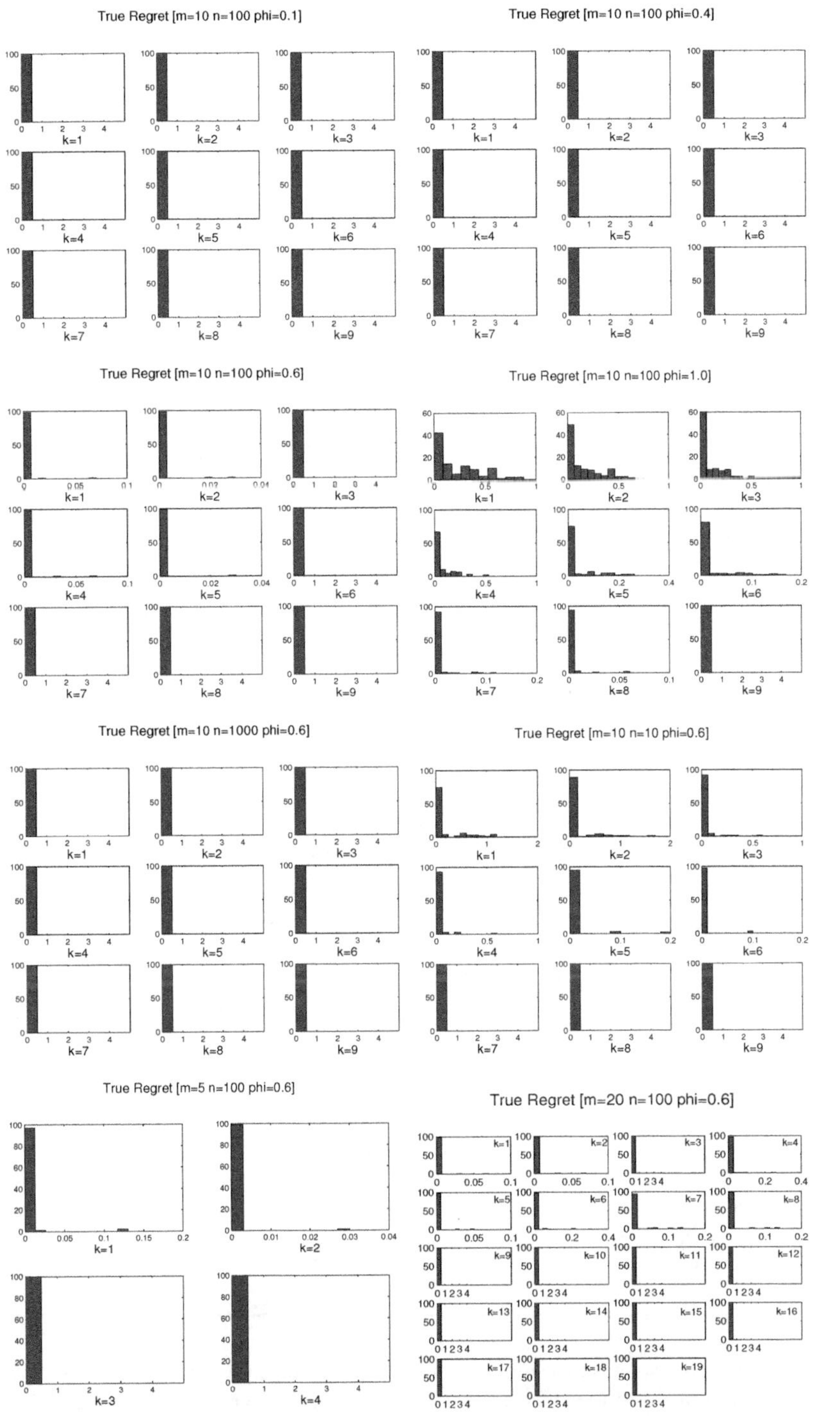

Fig. 2. The corresponding true regrets of experiments shown in Fig. 1

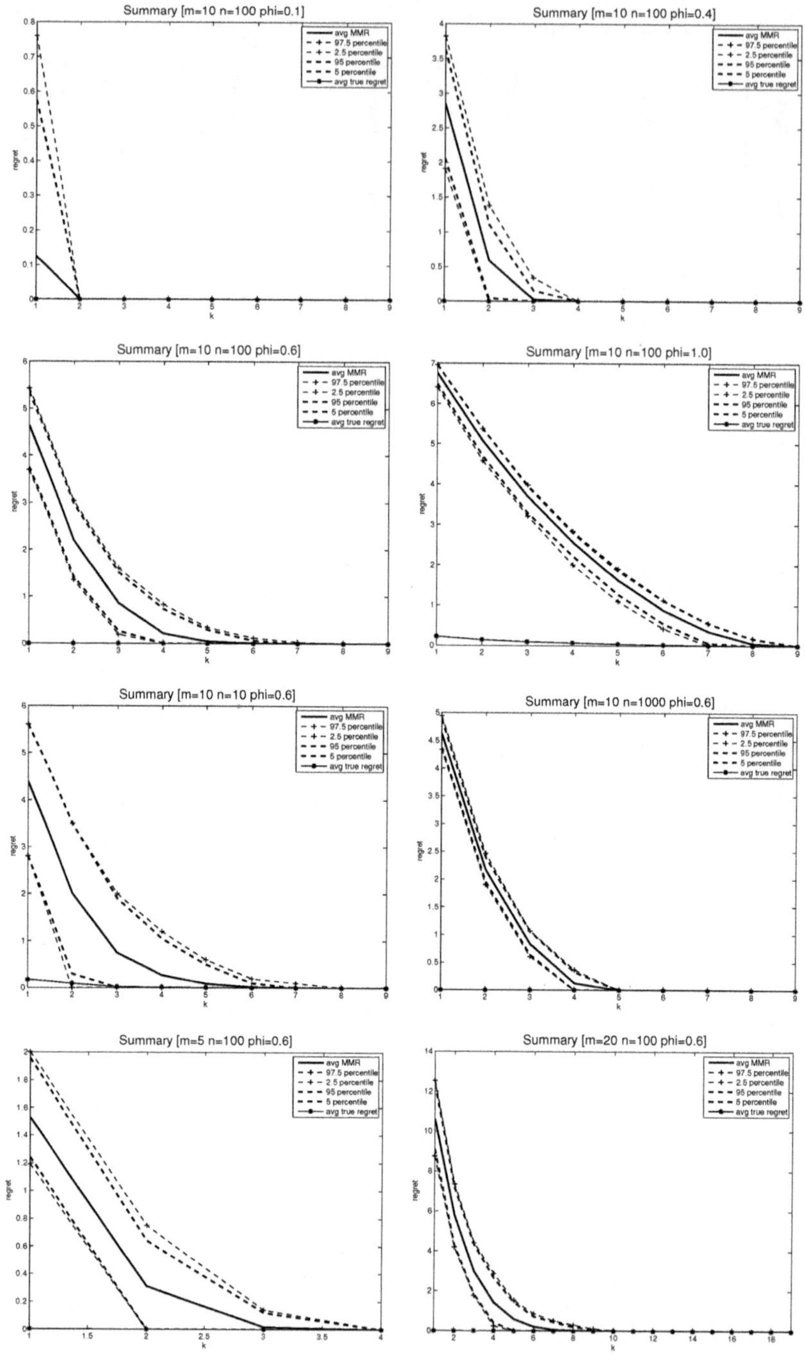

Fig. 3. Each plot corresponds to a summary of the experiments in Fig. 1, and shows the reduction in regret (avg. normalized (per voter) MMR and true regret over all instances) as k increases. Percentiles (.025, 0.05, 0.95, 0.975) for MMR are shown.

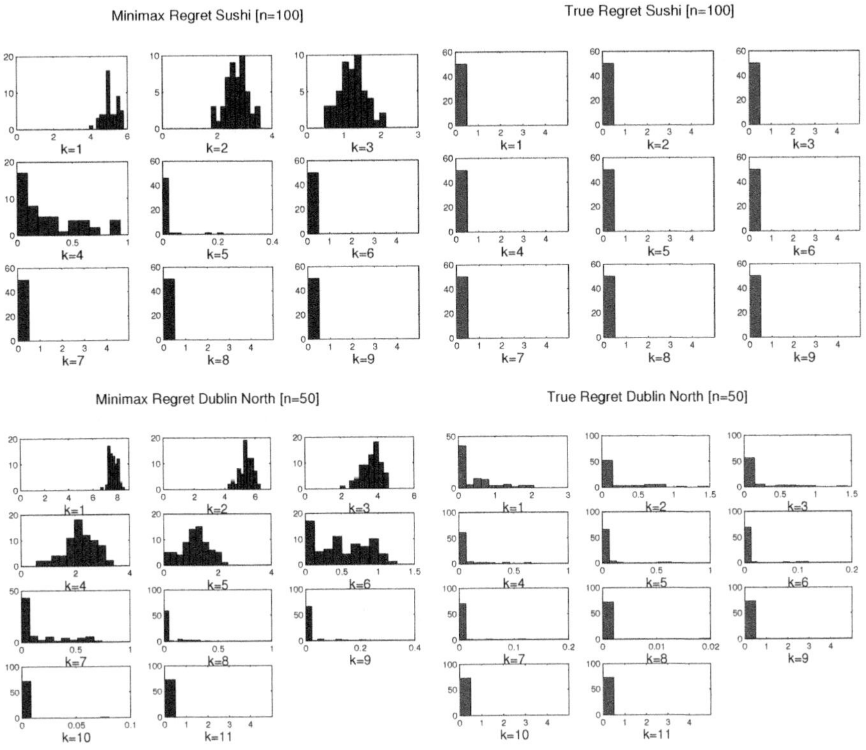

Fig. 4. Results on sushi rankings and Irish voting data

True regret (see Fig. 2) is even more illuminating: with $\phi = 0.6$, the MMR solution *after only top-*1 *queries to each voter* is nearly always the true winner; and true regret never exceeds 2. Even for the uniform distribution with $\phi = 1$, true regret is surprisingly small: after top-2 queries, regret is less than 0.5 in 97/100 cases. As we increase the number of voters n, the MMR distribution becomes more concentrated around the mean (e.g., $n = 1000$), and often resembles a Gaussian. Roughly, this is because with Borda scoring, (normalized) MMR can be expressed as the average of independent functions of p_i through pairwise max regret $PMR_i(a^*_{\mathbf{p}}, a') = \max_{v_i \in C(p_i)} B(v_i(a')) - B(v_i(a^*_{\mathbf{p}}))$, where a' is the adversarial witness (see Eq. (1)).

Fig. 3 provides a summary of the above experiments, showing average MMR as a function of k, along with average true regret and several percentile bounds. As above, we see that a smaller ϕ requires a smaller k to guarantee low MMR. It also illustrates the desirable anytime property of MMR: regret drops significantly with the "first few candidates" and levels off before reaching zero. For example, with $m = 10, n = 100, \phi = 0.6$, top-3 queries reduce MMR to 0.8 per voter from the MMR of 9 obtained with no queries; but an additional 3 candidates (i.e., top-6 queries) are needed to reduce regret from 0.8 per voter to 0. If we fix $\phi = 0.6$ and increase the number of candidates m, the k required for small MMR decreases in relation to m: we see that for $m = 5, 10, 20$

we need top-k queries with $k = 3, 6, 8$, respectively, to reach MMR of zero. This is, of course, specific to the Mallows model.

Fig. 4 show histograms on two real-world data sets: *Sushi* [7] (10 alternatives and 5000 rankings) and *Dublin*, voting data from the Dublin North constituency in 2002 (12 candidates and 3662 rankings).[4] With *Sushi*, we divided the 5000 rankings into 50 voting profile instances, each with $n = 100$ rankings, and plotted MMR histograms using the same protocol as in Fig. 1 and Fig. 2; similarly, *Dublin* was divided into 73 profiles each with $n = 50$. Sushi results suggest that with top-5 queries one can usually find a necessary winner; but top-4 queries are usually enough to obtain low MMR sufficient for such a low-stakes group decision (i.e., what sushi to order). True regret histograms show the minimax solution is almost always the true winner. With *Dublin*, top-5 queries virtually guarantee MMR of no more than 2 per voter; top-6, MMR of 1 per voter; and top-7, MMR of 0.5 per voter. True regret plots show minimax winner is either optimal or close to optimal in most profile instances.

6 Concluding Remarks

We have outlined a general framework for the design of multi-round elicitation protocols that are sensitive to tradeoffs between number of rounds of elicitation imposed on voters, the amount of information elicited per round, and the quality of the proposed winner. Our framework is probabilistic, allowing one to account for realistic distributions of voter preferences and profiles. We have formulated a probabilistic method for choosing the ideal threshold k for top-k elicitation in one-round protocols, and developed an empirical methodology that applies to *any voting rule* and *any preference distribution*. While the method can be used purely heuristically, our PAC-analysis provides our methodology with statistical guarantees. Experiments on random Mallows models, as well as real-world data sets (sushi preferences and Irish electoral data) demonstrate the practical viability and advantages of our empirical approach.

There are numerous opportunities for future research. We have dealt mainly with one-round elicitation of top-k candidates—developing algorithms for optimal multi-round instantiations of our framework is an important next step. Critically, we must deal with posterior distributions that are generally intractable, though GRIM-based techniques [9] may help. We are also interested in more flexible query classes such as batched pairwise comparisons. While the empirical framework is applicable to any preference distribution, we still wish to analyze the performance on additional distributions, including more flexible mixture models. On the theoretical side, we expect our PAC-analysis can be extended to different query classes and to multi-round protocols: we expect that probabilistic bounds on the amount of information required (e.g., k^* for top-k queries) will be significantly better than deterministic worst-case bounds [3] assuming, for example, a Mallows model. Bayesian approaches that assess candidate quality using expected regret rather than minimax regret are also of interest, especially in lower-stakes settings. We expect that combining expected regret and minimax regret might yield interesting solutions as well.

[4] There are $43,942$ ballots; 3662 are complete. See www.dublincountyreturningofficer.com

Acknowledgements. Thanks to Yann Chevaleyre, Jerôme Lang, and Nicolas Maudet for helpful discussions. This research was supported by NSERC.

References

1. Chevaleyre, Y., Endriss, U., Lang, J., Maudet, N.: A short introduction to computational social choice. In: van Leeuwen, J., Italiano, G.F., van der Hoek, W., Meinel, C., Sack, H., Plášil, F. (eds.) SOFSEM 2007. LNCS, vol. 4362, pp. 51–69. Springer, Heidelberg (2007)
2. Conitzer, V., Sandholm, T.: Vote elicitation: Complexity and strategy-proofness. In: Proceedings of the Eighteenth National Conference on Artificial Intelligence (AAAI 2002), Edmonton, pp. 392–397 (2002)
3. Conitzer, V., Sandholm, T.: Communication complexity of common voting rules. In: Proceedings of the Sixth ACM Conference on Electronic Commerce (EC 2005), Vancouver, pp. 78–87 (2005)
4. Doignon, J.-P., Pekec, A., Regenwetter, M.: The repeated insertion model for rankings: Missing link between two subset choice models. Psychometrika 69(1), 33–54 (2004)
5. Gaertner, W.: A Primer in Social Choice Theory. LSE Perspectives in Economic Analysis. Oxford University Press, USA (August 2006)
6. Kalech, M., Kraus, S., Kaminka, G.A., Goldman, C.V.: Practical voting rules with partial information. Journal of Autonomous Agents and Multi-Agent Systems 22(1), 151–182 (2011)
7. Kamishima, T., Kazawa, H., Akaho, S.: Supervised ordering: An empirical survey. In: IEEE International Conference on Data Mining, pp. 673–676 (2005)
8. Lang, J.: Vote and aggregation in combinatorial domains with structured preferences. In: Proceedings of the Twentieth International Joint Conference on Artificial Intelligence (IJCAI 2007), Hyderabad, India, pp. 1366–1371 (2007)
9. Lu, T., Boutilier, C.: Learning Mallows models with pairwise preferences. In: Proceedings of the Twenty-eighth International Conference on Machine Learning (ICML 2011), Bellevue, Washington (2011)
10. Lu, T., Boutilier, C.: Robust approximation and incremental elicitation in voting protocols. In: Proceedings of the Twenty-second International Joint Conference on Artificial Intelligence (IJCAI 2011), Barcelona (to appear, 2011)
11. Mallows, C.L.: Non-null ranking models. Biometrika 44, 114–130 (1957)
12. Marden, J.I.: Analyzing and modeling rank data. Chapman and Hall, Boca Raton (1995)
13. Murphy, T.B., Martin, D.: Mixtures of distance-based models for ranking data. Computational Statistics and Data Analysis 41, 645–655 (2003)
14. Regenwetter, M., Grofman, B., Marley, A.A.J., Tsetlin, I.: Behavioral Social Choice: Probabilistic Models, Statistical Inference, and Applications. Cambridge University Press, Cambridge (2006)
15. Valiant, L.G.: A theory of the learnable. Communications of the ACM 27(11), 1134–1142 (1984)
16. Xia, L., Conitzer, V.: Determining possible and necessary winners under common voting rules given partial orders. In: Proceedings of the Twenty-third AAAI Conference on Artificial Intelligence (AAAI 2008), Chicago, pp. 202–207 (2008)

Efficient Approximation Algorithms for Multi-objective Constraint Optimization

Radu Marinescu

IBM Research – Dublin
Mulhuddart, Dublin 15, Ireland
radu.marinescu@ie.ibm.com

Abstract. In this paper, we propose new depth-first heuristic search algorithms to approximate the set of Pareto optimal solutions in multi-objective constraint optimization. Our approach builds upon recent advances in multi-objective heuristic search over weighted AND/OR search spaces and uses an ϵ-dominance relation between cost vectors to significantly reduce the set of non-dominated solutions. Our empirical evaluation on various benchmarks demonstrates the power of our scheme which improves the resolution times dramatically over recent state-of-the-art competitive approaches.

Keywords: multi-objective constraint optimization, heuristic search, approximation, AND/OR search spaces.

1 Introduction

A *Constraint Optimization Problem* (COP) is the minimization (or maximization) of an objective function subject to a set of constraints (hard and soft) on the possible values of a set of independent decision variables [1]. Many real-world problems however involve multiple measures of performance or objectives that should be considered separately and optimized concurrently. *Multi-objective Constraint Optimization* (MO-COP) provides a general framework that can be used to model such problems involving multiple, conflicting and sometimes non-commensurate objectives that need to be optimized simultaneously [2,3,4,5]. In contrast with single function optimization, the solution space of these problems is typically only partially ordered and will, in general, contain several *non-inferior* or *non-dominated* solutions which must be considered equivalent in the absence of information concerning the relevance of each objective relative to the others. Therefore, solving a MO-COP is to find its *Pareto* or *efficient frontier*, namely the set of solutions with non-dominated costs.

In many practical situations the Pareto frontier may contain a very large (sometimes an exponentially large) number of solutions [6]. Producing the entire Pareto set in this case may induce prohibitive computation times and, it could possibly be useless to a decision maker. An alternative approach to overcome this difficulty, which gained attention in recent years, is to approximate the Pareto set while keeping a good representation of the various possible tradeoffs in the solution space. In this direction, several approximation methods based on either dynamic programming or best-first search and relying on the concept of ϵ-*dominance* between cost vectors as a relaxation of the Pareto

R.I. Brafman, F. Roberts, and A. Tsoukiàs (Eds.): ADT 2011, LNAI 6992, pp. 150–164, 2011.

dominance relation have been proposed to tackle various multi-objective optimization problems. In the context of multi-objective shortest paths problems, the methods due to Hansen [6] and Warburton [7] combine scaling and rounding techniques with pseudo-polynomial exact algorithms to decrease the size of the efficient frontier. The scheme proposed by Tsaggouris and Zaroliagis [8] is based on a generalized Bellman-Ford algorithm. The algorithm introduced by Papadimitriou and Yannakakis [9], although less specific to multi-objective shortest paths problems, maps the solution space onto a logarithmic grid in $(1+\epsilon)$ in order to generate an ϵ-covering of the Pareto frontier. For multi-objective knapsack problems, Erlebach *at al* [10] described a dynamic programming approach that partitions the profit space into intervals of exponentially increasing lengths, while Bazgan *et al* [11] proposed a dynamic programming algorithm that uses an extended family of the ϵ-dominance relation. In the context of multi-attribute theory, Dubus *et al* [12] presented a variable elimination algorithm that uses ϵ-dominance over generalized additive decomposable utility functions. The multi-objective A* search proposed recently by Perny and Spanjaard [13] for approximating the Pareto frontier of multi-objective shortest paths problems is a best-first search algorithm that uses the ϵ-dominance relation to trim the solution space. The latter method is limited to problems with a relatively small state space and requires an exponential amount of memory.

In contrast to the existing approaches, we propose in this paper a space efficient method for approximating the Pareto frontier. In particular, we introduce new depth-first Branch-and-Bound search algorithms to compute an ϵ-covering of the Pareto frontier in multi-objective constraint optimization. Our approach builds upon recent advances in multi-objective heuristic search over weighted AND/OR search spaces for MO-COPs. More specifically, we extend the depth-first multi-objective AND/OR Branch-and-Bound [5], a recent exact search algorithm that exploits the problem structure, to use the ϵ-dominance relation between cost vectors in order to significantly reduce the set of non-dominated solutions. The main virtue of an ϵ-covering is that its size can be significantly smaller than that of the corresponding Pareto frontier and, therefore, it can be used efficiently by the decision maker to determine interesting regions of the decision and objective space which can be explored in further optimization runs. In addition, the use of ϵ-dominance also makes the algorithms practical by allowing the decision maker to control the resolution of the Pareto set approximation by choosing an appropriate ϵ value. The proposed algorithms are guided by a general purpose heuristic evaluation function which is based on the multi-objective mini-bucket approximation scheme [3,5]. The mini-bucket heuristics can be either pre-compiled or generated dynamically at each node in the search tree. They are parameterized by a user controlled parameter called i-bound which allows for an adjustable tradeoff between the accuracy of the heuristic and its computational overhead. We evaluate empirically our approximation algorithms on two classes of problems: risk-conscious combinatorial auctions and multi-objective scheduling problems for smart buildings. Our results show that the new depth-first Branch-and-Bound search algorithms improve dramatically the resolution times over current state-of-the-art competitive approaches based on either multi-objective best-first search or dynamic programming [13,12].

Following background on MO-COPs and on weighted AND/OR search spaces for MO-COPs (Section 2), Section 3 introduces our depth-first AND/OR search approach for computing an ϵ-covering of the Pareto frontier. Section 4 is dedicated to our empirical evaluation while Section 5 concludes and outlines directions of future research.

2 Background

2.1 Multi-objective Constraint Optimization

Consider a finite set of objectives $\{1,\ldots,p\}$. A *bounded cost vector* $\boldsymbol{u} = (u_1,\ldots,u_p)$ is a vector of p components where each $u_j \in \mathbb{Z}_+$ represents the cost with respect to objective j and $0 \leq u_j \leq K$, respectively. We adopt the following notation. A cost vector which has all components equal to 0 is denoted by $\mathbf{0}$, while a cost vector having one or more components equal to K is denoted by $\boldsymbol{K}$.

A *Multi-objective Constraint Optimization Problem* (MO-COP) with $p > 1$ objectives is a tuple $\mathcal{M} = \langle \mathbf{X}, \mathbf{D}, \mathbf{F} \rangle$, where $\mathbf{X} = \{X_1, ..., X_n\}$ is a set of *variables*, $\mathbf{D} = \{D_1, ..., D_n\}$ is a set of finite *domains* and $\mathbf{F} = \{f_1, ..., f_r\}$ is a set of *multi-objective cost functions*. A multi-objective cost function $f_k(Y_k) \in \mathbf{F}$ is defined over a subset of variables $Y_k \subseteq \mathbf{X}$, called its *scope*, and associates a bounded cost vector $\boldsymbol{u} = (u_1, ..., u_p)$ to each assignment of its scope. The cost functions in $\mathbf{F}$ can be either *soft* or *hard* (constraints). Without loss of generality we assume that hard constraints are represented as multi-objective cost functions, where allowed and forbidden tuples have cost $\mathbf{0}$ and $\boldsymbol{K}$, respectively.

The sum of cost functions in $\mathbf{F}$ defines the objective function, namely $\mathcal{F}(\mathbf{X}) = \sum_{k=1}^{r} f_k(Y_k)$. A *solution* is a complete assignment of the variables $\bar{x} = (x_1, ..., x_n)$ and is characterized by a cost vector $\boldsymbol{u} = \mathcal{F}(\bar{x})$, where u_j is the value of $\bar{x}$ with respect to the jth objective. Hence, the comparison of solutions reduces to the comparison of their cost vectors. The set of all cost vectors attached to solutions is denoted by $\mathcal{S}$. We recall next some definitions related to Pareto dominance concepts.

Definition 1 (Pareto dominance). *Given two cost vectors $\boldsymbol{u}$ and $\boldsymbol{v} \in \mathbb{Z}_+^p$, we say that $\boldsymbol{u}$* dominates *$\boldsymbol{v}$, denoted by $\boldsymbol{u} \preccurlyeq \boldsymbol{v}$, if $\forall i\ u_i \leq v_i$. We say that $\boldsymbol{u}$* strictly dominates *$\boldsymbol{v}$, denoted by $\boldsymbol{u} \prec \boldsymbol{v}$, if $\boldsymbol{u} \preccurlyeq \boldsymbol{v}$ and $\boldsymbol{u} \neq \boldsymbol{v}$. Given two sets of cost vectors $\mathcal{U}$ and $\mathcal{V}$, we say that $\mathcal{U}$* dominates *$\mathcal{V}$, denoted by $\mathcal{U} \preccurlyeq \mathcal{V}$, if $\forall \boldsymbol{v} \in \mathcal{V}, \exists \boldsymbol{u} \in \mathcal{U}$ such that $\boldsymbol{u} \preccurlyeq \boldsymbol{v}$.*

Definition 2 (Pareto frontier). *Given a set of cost vectors $\mathcal{U}$, we define the* Pareto *or* efficient frontier *of $\mathcal{U}$, denoted by $\mathcal{ND}(\mathcal{U})$, to be the set consisting of the non-dominated cost vectors of $\mathcal{U}$, namely $\mathcal{ND}(\mathcal{U}) = \{\boldsymbol{u} \in \mathcal{U} \mid \nexists \boldsymbol{v} \in \mathcal{U}$ such that $\boldsymbol{v} \prec \boldsymbol{u}\}$. A cost vector $\boldsymbol{u} \in \mathcal{ND}(\mathcal{U})$ is called* Pareto optimal.

Solving a MO-COP is to minimize $\mathcal{F}$, namely to find the Pareto frontier of the set of solutions $\mathcal{S}$. Any MO-COP instance has an associated *primal graph*, which is computed as follows: nodes correspond to the variables and an edge connects any pair of nodes whose variables belong to the scope of the same multi-objective cost function.

Example 1. Figure 1(a) shows a simple MO-COP instance with 5 bi-valued variables and 3 bi-objective cost functions. Its corresponding primal graph is depicted in Figure 1(b). The solution space of the problem contains 32 cost vectors while the Pareto frontier has only 3 solutions: (00000), (00100) and (01100) with corresponding non-dominated cost vectors $(7,0)$, $(4,3)$ and $(3,9)$, respectively.

X_0	X_1	X_2	f_1
0	0	0	(5,0)
0	0	1	(2,3)
0	1	0	(3,2)
0	1	1	(2,5)
1	0	0	(2,1)
1	0	1	(3,4)
1	1	0	(0,3)
1	1	1	(2,6)

X_0	X_1	X_3	f_2
0	0	0	(1,0)
0	0	1	(4,4)
0	1	0	(0,2)
0	1	1	(2,6)
1	0	0	(6,1)
1	0	1	(5,5)
1	1	0	(6,3)
1	1	1	(5,7)

X_1	X_3	X_4	f_3
0	0	0	(1,0)
0	0	1	(3,5)
0	1	0	(5,4)
0	1	1	(4,9)
1	0	0	(1,2)
1	0	1	(3,7)
1	1	0	(5,6)
1	1	1	(4,11)

(a) bi-objective cost functions

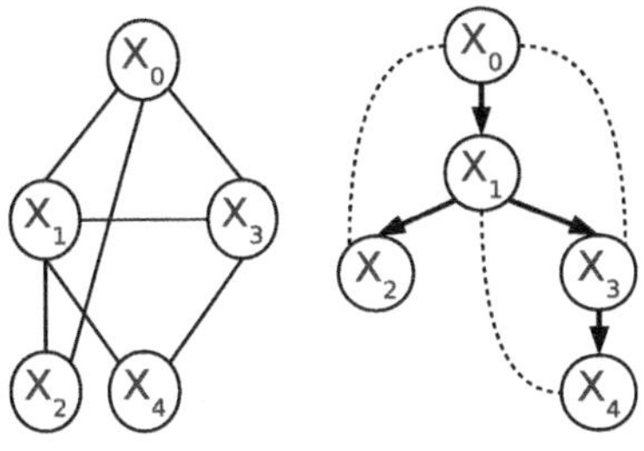

(b) primal graph (c) pseudo tree

Fig. 1. A simple MO-COP instance with 2 objectives

2.2 Approximation of the Pareto Frontier

The Pareto frontier may contain a very (sometimes exponentially) large number of solutions and, therefore, the determination of the entire Pareto set may be intractable in practice [6,7]. In order to overcome this difficulty, it is possible to relax the Pareto dominance relation and compute an approximation of the Pareto frontier by considering the notion of ϵ-*dominance* between cost vectors, defined as follows [9,14,13].

Definition 3 **(ϵ-dominance).** *Given two cost vectors $\boldsymbol{u}$ and $\boldsymbol{v} \in \mathbb{Z}_+^p$ and any $\epsilon > 0$, we say that $\boldsymbol{u}$* ϵ-dominates *$\boldsymbol{v}$, denoted by $\boldsymbol{u} \preccurlyeq_\epsilon \boldsymbol{v}$, if and only if $\boldsymbol{u} \preccurlyeq (1+\epsilon)\boldsymbol{v}$.*

The ϵ-dominance relation between cost vectors allows us to define an approximation of the Pareto frontier, called ϵ-*covering*, as follows:

Definition 4 **(ϵ-covering).** *For any $\epsilon > 0$ and any set of cost vectors $\mathcal{V}$, a subset $\mathcal{U} \subseteq \mathcal{V}$ is said to be an* ϵ-covering *of the Pareto frontier $\mathcal{ND}(\mathcal{V})$ of $\mathcal{V}$, if $\forall \boldsymbol{v} \in \mathcal{ND}(\mathcal{V})$, $\exists \boldsymbol{u} \in \mathcal{U}$ such that $\boldsymbol{u} \preccurlyeq_\epsilon \boldsymbol{v}$. We also say that $\mathcal{U}$ is an ϵ-covering of the entire set $\mathcal{V}$.*

In general, multiple ϵ-coverings of the Pareto frontier may exist, with different sizes, the most interesting being minimal with respect to set inclusion. Based on previous work by [9], it can be shown that given a MO-COP instance $\mathcal{M}$ with $p > 1$ objectives and cost vectors bounded by K, for any $\epsilon > 0$ there exists an ϵ-covering of the Pareto frontier that consists of at most $\lfloor \log K / \log(1+\epsilon) \rfloor^{p-1}$ solutions (or cost vectors).

This property can be explained by considering a logarithmic scaling function $\varphi : \mathbb{Z}_+^p \rightarrow \mathbb{Z}_+^p$ on the solution space $\mathcal{S}$ of the MO-COP instance $\mathcal{M}$ defined by: $\forall \boldsymbol{u} \in \mathcal{S}$, $\varphi(\boldsymbol{u}) = (\varphi(u_1), ..., \varphi(u_p))$ where $\forall i$, $\varphi(u_i) = \lfloor \log u_i / \log(1+\epsilon) \rfloor$. For every component u_i, the function returns an integer k such that $(1+\epsilon)^k \leq u_i \leq (1+\epsilon)^{k+1}$. Using φ we can define φ-dominance relation [13]:

Definition 5 **(φ-dominance).** *The φ-dominance relation on cost vectors in $\mathbb{Z}_+^p$ is defined by $\boldsymbol{u} \preccurlyeq_\varphi \boldsymbol{v}$ if and only if $\varphi(\boldsymbol{u}) \preccurlyeq \varphi(\boldsymbol{v})$.*

Proposition 1. *Let $\boldsymbol{u}, \boldsymbol{v}, \boldsymbol{w} \in \mathbb{Z}_+^p$. The following properties hold: (i) if $\boldsymbol{u} \preccurlyeq_\varphi \boldsymbol{v}$ and $\boldsymbol{v} \preccurlyeq_\varphi \boldsymbol{w}$ then $\boldsymbol{u} \preccurlyeq_\varphi \boldsymbol{w}$ (transitivity); (ii) if $\boldsymbol{u} \preccurlyeq_\varphi \boldsymbol{v}$ then $\boldsymbol{u} \preccurlyeq_\epsilon \boldsymbol{v}$.*

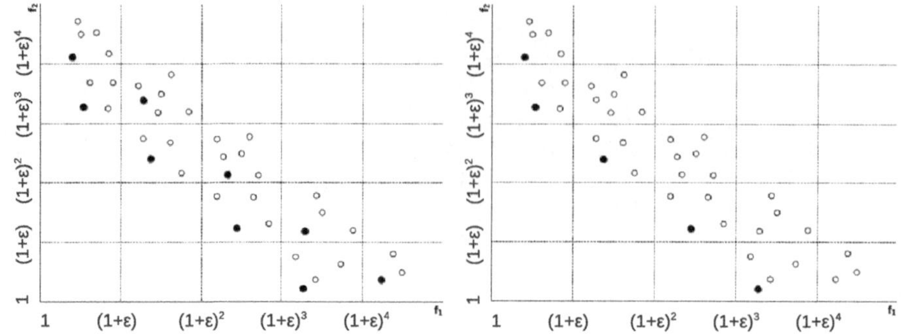

Fig. 2. Examples of ϵ-coverings

It is easy to see that function φ induces a logarithmic grid on the solution space $\mathcal{S}$, where any cell represents a different class of cost vectors having the same image through φ. Any vector belonging to a given grid cell ϵ-dominates any other vector of that cell. Hence, by choosing one representative in each cell of the grid we obtain an ϵ-covering of the entire set $\mathcal{S}$. The left part of Figure 2 illustrates this idea on a bi-objective MO-COP instance. The dotted lines form the logarithmic grid and an ϵ-covering of the Pareto frontier can be obtained by selecting one cost vector (black dots) from each of the non-empty cells of the grid. The resulting ϵ-covering can be refined further by keeping only the non-dominated vectors in the covering, as shown (in black) on Figure 2 right.

2.3 AND/OR Search Spaces for MO-COPs

The concept of AND/OR search spaces has recently been introduced as a unifying framework for advanced algorithmic schemes for graphical models to better capture the structure of the underlying graph [15]. Its main virtue consists in exploiting conditional independencies between variables, which can lead to exponential speedups. The search space was recently extended to multi-objective constraint optimization in [5] and is defined using a *pseudo tree* [16] which captures problem decomposition.

Definition 6 (pseudo tree). *Given an undirected graph $G = (V, E)$, a directed rooted tree $\mathcal{T} = (V, E')$ defined on all its nodes is called a* pseudo tree *if any edge of G that is not included in E' is a back-arc in $\mathcal{T}$, namely it connects a node to an ancestor in $\mathcal{T}$.*

Given a MO-COP instance $\mathcal{M} = \langle \mathbf{X}, \mathbf{D}, \mathbf{F} \rangle$, its primal graph G and a pseudo tree $\mathcal{T}$ of G, the AND/OR search tree associated with $\mathcal{M}$ and denoted by $S_{\mathcal{T}}(\mathcal{M})$ (or $S_{\mathcal{T}}$ for short) has alternating levels of OR and AND nodes. The OR nodes are labeled X_i and correspond to the variables. The AND nodes are labeled $\langle X_i, x_i \rangle$ (or just x_i) and correspond to value assignments of the variables. The structure of the AND/OR search tree is based on the underlying pseudo tree $\mathcal{T}$. The root of the AND/OR search tree is an OR node labeled with the root of $\mathcal{T}$. The children of an OR node X_i are AND nodes labeled with the values assignments in the domain of X_i. The children of an AND node $\langle X_i, x_i \rangle$ are OR nodes labeled with the children of variable X_i in $\mathcal{T}$. A *solution tree* T of an AND/OR search tree $S_{\mathcal{T}}$ is an AND/OR subtree such that: (1) it contains the root of $S_\mathcal{T}$, s; (2) if a non-terminal AND node $n \in S_\mathcal{T}$ is in T then all of its children are in T; (3) if a non-terminal OR node $n \in S_\mathcal{T}$ is in T then exactly one of its children is in T; (4) every tip node in T (*i.e.*, nodes with no children) is a terminal node. A *partial solution tree* T' is a subtree of an AND/OR search tree $S_\mathcal{T}$, whose definition is similar to that of a solution tree except that the tip nodes of T' are not necessarily terminal nodes of $S_\mathcal{T}$ (see also [15,5] for additional details).

The arcs from OR nodes X_i to AND nodes $\langle X_i, x_i\rangle$ in $S_\mathcal{T}$ are annotated by *weights* derived from the multi-objective cost functions in $\mathbf{F}$. Each node n in the weighted search tree is associated with a *value* $v(n)$ which stands for the answer to the optimization query restricted to the conditioned subproblem below n.

Definition 7 (arc weight). *The* weight $w(n, n')$ *of the arc from the OR node n labeled X_i to the AND node n' labeled x_i is a cost vector defined as the sum of all the multi-objective cost functions whose scope includes variable X_i and is fully assigned along the path from the root of the search tree to x_i, evaluated at the values along that path.*

Definition 8 (node value). *The* value $v(n)$ *of a node $n \in S_\mathcal{T}$ is defined recursively as follows (where $succ(n)$ are the children of n in the search tree):*

(1) $v(n) = \mathbf{0}$, *if $n = \langle X_i, x_i\rangle$ is a terminal AND node;*
(2) $v(n) = \sum_{n' \in succ(n)} v(n')$, *if $n = \langle X_i, x_i\rangle$ is a non-terminal AND node;*
(3) $v(n) = \mathcal{ND}(\{w(n,n') + v(n') \mid n' \in succ(n)\})$, *if $n = X_i$ is a non-terminal OR node.*

The sum of cost vectors in $\mathbb{Z}^p_+$ is the usual point-wise vector sum, namely $\boldsymbol{u} + \boldsymbol{v} = \boldsymbol{w}$ where $\forall 1 \leq i \leq p$, $w_i = u_i + v_i$. Given two sets of cost vectors $\mathcal{U}$ and $\mathcal{V}$, we define the sum $\mathcal{U} + \mathcal{V} = \{\boldsymbol{w} = \boldsymbol{u} + \boldsymbol{v} \mid \boldsymbol{u} \in \mathcal{U}, \boldsymbol{v} \in \mathcal{V}\}$.

It is easy to see that the value $v(n)$ of a node in $S_\mathcal{T}$ is the set of cost vectors representing the Pareto frontier of the subproblem rooted at n, conditioned on the variable assignment along the path from the root to n. If n is the root of $S_\mathcal{T}$, then $v(n)$ is the Pareto frontier of the initial problem.

Example 2. Figure 3 shows the weighted AND/OR search tree associated with the MO-COP instance from Figure 1, relative to the pseudo tree given in Figure 1(c). The cost vectors displayed on the OR-to-AND arcs are the weights corresponding to the input function values. A solution tree that represents the assignment $(X_0 = 0, X_1 = 1, X_2 = 1, X_3 = 0, X_4 = 0)$ with cost vector $(3, 9)$ is highlighted.

Based on previous work [16,15,5], it can be shown that given a MO-COP instance and a pseudo tree $\mathcal{T}$ of depth m, the size of the AND/OR search tree based on $\mathcal{T}$ is $O(n \cdot d^m)$, where d bounds the domains of variables. Moreover, a MO-COP instance having treewidth w^* has a pseudo tree of depth at most $w^* \log n$, and therefore it has an AND/OR search tree of size $O(n \cdot d^{w^* \log n})$ (see also [15] for more details).

3 Depth-First AND/OR Branch-and-Bound Search for Computing an ϵ-Covering of the Pareto Frontier

We present next a generic scheme for computing an ϵ-covering of the Pareto frontier based on depth-first search over weighted AND/OR search trees for MO-COPs.

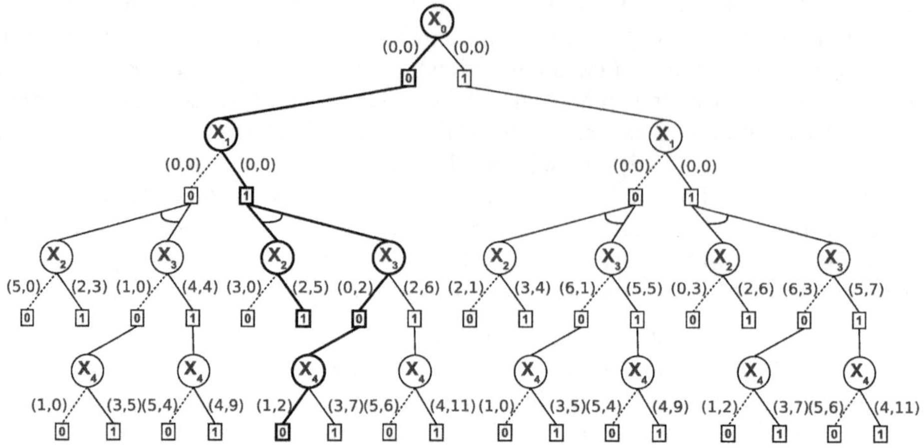

Fig. 3. Weighted AND/OR search tree for the MO-COP instance from Fig. 1

3.1 Multi-objective AND/OR Branch-and-Bound Search

One of the most effective heuristic search methods for computing Pareto frontiers in multi-objective constraint optimization is the multi-objective AND/OR Branch-and-Bound (MO-AOBB) introduced recently in [5]. We recall next the notion of heuristic evaluation function of a partial solution tree which is needed to describe the algorithm.

Definition 9 (heuristic evaluation function). *Given a partial solution tree T'_n rooted at node $n \in S_\mathcal{T}$ and an underestimate $h(n)$ of $v(n)$, the* heuristic evaluation function *$f(T'_n)$, is defined by: (1) if T'_n consists of a single node n, then $f(T'_n) = h(n)$; (2) if n is an OR node having the AND child m in T'_n, then $f(T'_n) = w(n,m) + f(T'_m)$; (3) if n is an AND node having OR children $m_1, ..., m_k$ in T'_n, then $f(T'_n) = \sum_{i=1}^{k} f(T'_{m_i})$.*

MO-AOBB is described by Algorithm 1. It performs a depth-first traversal of the weighted AND/OR search tree relative to a pseudo tree $\mathcal{T}$ by expanding alternating levels of OR and AND nodes (lines 3–13). The stack OPEN maintains the fringe of the search. Upon expansion, the node values are initialized as follows: $v(n)$ is set to **0** if n is an AND node, and is set to ∞ otherwise. At each step during search, an expanded node n having an empty set of successors propagates the value $v(n)$ to its parent p in the search tree which, in turn, updates the value $v(p)$ (lines 14–18). The OR nodes update their values by non-dominated closure with respect to Pareto dominance, while the AND nodes compute their values by summation (see Definition 8). The algorithm also discards any partial solution tree T' if the corresponding heuristic evaluation function $f(T')$ (see Definition 9) is dominated by the current upper bound $v(s)$ maintained by the root node s on the Pareto frontier (lines 9–13). For completeness, Algorithm 2 computes the non-dominated closure (Pareto frontier) of a set of cost vectors. When search terminates the value $v(s)$ of the root node s is the Pareto frontier.

Algorithm 1. MO-AOBB

Data: MO-COP $\mathcal{M} = \langle \mathbf{X}, \mathbf{D}, \mathbf{F} \rangle$, pseudo tree $\mathcal{T}$, heuristic function h.
Result: Pareto frontier of $\mathcal{M}$.

create an OR node s labeled by the root of $\mathcal{T}$
$OPEN \leftarrow \{s\}$; $CLOSED \leftarrow \emptyset$; set $v(s) = \infty$
while $OPEN \neq \emptyset$ **do**
    move top node n from OPEN to CLOSED
    expand n by creating its successors $succ(n)$
    foreach $n' \in succ(n)$ **do**
        evaluate $h(n')$ and add n' on top of OPEN
        set $v(n') = \mathbf{0}$ if n' is an AND node, and $v(n') = \infty$ otherwise
        if n' *is AND* **then**
            let T' be the current partial solution tree with n' as tip node
            let $f(T') \leftarrow \texttt{evaluate}(T')$
            if $v(s) \preccurlyeq f(T')$ **then**
                remove n' from OPEN and $succ(n)$
    while $\exists n \in CLOSED$ *s.t.* $succ(n) = \emptyset$ **do**
        remove n from CLOSED and let p be n's parent
        if p *is AND* **then** $v(p) \leftarrow v(p) + v(n)$
        else $v(p) \leftarrow \mathcal{ND}(v(p) \cup \{w(p,n) + v(n)\})$
        remove n from $succ(p)$
return $v(s)$

Theorem 1 ([5]). *Given a MO-COP instance with $p > 1$ objectives MO-AOBB is sound and complete. It uses $O(n \cdot K^p)$ space and $O(K^{2p} \cdot n \cdot d^m)$ time, where n is the number of variables, d bounds their domains and m is the depth of the pseudo tree.*

3.2 Logarithmic Scaling Based Approximation

Computing an ϵ-covering using depth-first AND/OR search is similar to computing the Pareto frontier. However, it is not possible to update the values of the OR nodes during search using the non-dominated closure with respect to the ϵ-dominance (or φ-dominance) relation because we might exceed the desired error threshold $(1+\epsilon)$ due to error propagation as we will see next.

Let $S = \{\boldsymbol{x}, \boldsymbol{y}, \boldsymbol{z}, \boldsymbol{w}\}$ be a solution space consisting of four non-dominated cost vectors such that $\boldsymbol{x} \preccurlyeq_\epsilon \boldsymbol{y}$, $\boldsymbol{z} \preccurlyeq_\epsilon \boldsymbol{w}$ and $\boldsymbol{z} \preccurlyeq_\epsilon \boldsymbol{x}$, for some $\epsilon > 0$. Since $\boldsymbol{x} \preccurlyeq_\epsilon \boldsymbol{y}$, assume that $\boldsymbol{y}$ is discarded. We can also discard $\boldsymbol{w}$ because $\boldsymbol{z} \preccurlyeq_\epsilon \boldsymbol{w}$. Finally, since $\boldsymbol{z} \preccurlyeq_\epsilon \boldsymbol{x}$, it is easy to see that in this case the non-dominated closure with respect to ϵ-dominance contains a single vector, namely $\boldsymbol{z}$. However, it is clear that the set $\{\boldsymbol{z}\}$ is not a valid ϵ-covering of S because the cost vector $\boldsymbol{y}$ is not ϵ-covered by $\boldsymbol{z}$. In fact, we only have $\boldsymbol{z} \preccurlyeq (1+\epsilon)\boldsymbol{x} \preccurlyeq (1+\epsilon)^2\boldsymbol{y}$. This example suggests that we could have replaced $(1+\epsilon)$ with $(1+\epsilon)^{\frac{1}{2}}$ (also referred to as $\epsilon/2$-dominance) to ensure a valid ϵ-covering of the solution space. We will use a finer dominance relation defined as follows [9,14,13,12].

Algorithm 2. $\mathcal{ND}(\mathcal{U})$

$\mathcal{V} \leftarrow \emptyset$;
foreach $\boldsymbol{u} \in \mathcal{U}$ **do**
  if $\nexists \boldsymbol{v} \in \mathcal{V}$ *such that* $\boldsymbol{v} \preccurlyeq \boldsymbol{u}$ **then**
    remove from $\mathcal{V}$ all $\boldsymbol{v}$ such that $\boldsymbol{u} \preccurlyeq \boldsymbol{v}$;
    $\mathcal{V} \leftarrow \mathcal{V} \cup \{\boldsymbol{u}\}$;
return $\mathcal{V}$;

Algorithm 3. $\mathcal{ND}_{(\epsilon,1/m)}(\mathcal{U})$

$\mathcal{G} \leftarrow \emptyset$; $\mathcal{V} \leftarrow \emptyset$;
foreach $\boldsymbol{u} \in \mathcal{U}$ **do**
  if $\varphi(\boldsymbol{u}) \notin \mathcal{G}$ *and* $\nexists \boldsymbol{v} \in \mathcal{V}$ *such that* $\boldsymbol{v} \preccurlyeq_{\epsilon}^{1/m} \boldsymbol{u}$ **then**
    remove from $\mathcal{V}$ all $\boldsymbol{v}$ such that $\boldsymbol{u} \preccurlyeq \boldsymbol{v}$;
    $\mathcal{V} \leftarrow \mathcal{V} \cup \{\boldsymbol{u}\}$; $\mathcal{G} \leftarrow \mathcal{G} \cup \{\varphi(\boldsymbol{u})\}$;
return $\mathcal{V}$;

Definition 10. *Let* $\boldsymbol{u}, \boldsymbol{v} \in \mathbb{Z}_{+}^{p}$ *be two positive cost vectors and let* $\lambda > 0$. *We say that* $\boldsymbol{u}$ (ϵ, λ)-dominates $\boldsymbol{v}$, *denoted by* $\boldsymbol{u} \preccurlyeq_{\epsilon}^{\lambda} \boldsymbol{v}$, *iff* $\boldsymbol{u} \preccurlyeq (1+\epsilon)^{\lambda}\boldsymbol{v}$. *A set of* (ϵ, λ)*-non-dominated positive cost vectors is called an* (ϵ, λ)-covering.

Proposition 2. *Let* $\boldsymbol{u}, \boldsymbol{v}, \boldsymbol{w} \in \mathbb{Z}_{+}^{p}$ *and* $\lambda, \lambda' > 0$. *The following properties hold: (i) if* $\boldsymbol{u} \preccurlyeq_{\epsilon}^{\lambda} \boldsymbol{v}$ *then* $\boldsymbol{u} + \boldsymbol{w} \preccurlyeq_{\epsilon}^{\lambda} \boldsymbol{v} + \boldsymbol{w}$, *and (ii) if* $\boldsymbol{u} \preccurlyeq_{\epsilon}^{\lambda} \boldsymbol{v}$ *and* $\boldsymbol{v} \preccurlyeq_{\epsilon}^{\lambda'} \boldsymbol{w}$ *then* $\boldsymbol{u} \preccurlyeq_{\epsilon}^{\lambda+\lambda'} \boldsymbol{w}$.

Consider a MO-COP instance $\mathcal{M}$ and a pseudo tree $\mathcal{T}$ of its primal graph. Clearly, if the depth of $\mathcal{T}$ is m, then the corresponding weighted AND/OR search tree $S_{\mathcal{T}}$ has m levels of OR nodes. Let $\pi_{s,t}$ be a path in $S_{\mathcal{T}}$ from the root node s to a terminal AND node t. The bottom-up revision of the OR nodes values along $\pi_{s,t}$ requires chaining at most m (ϵ, λ_i)-dominance tests, $i = 1, \ldots, m$. Therefore, a sufficient condition to obtain a valid ϵ-covering is to choose the λ_i's such that they sum to 1, namely $\lambda_i = 1/m$. Given a set of cost vectors $\mathcal{U}$, Algorithm 3 describes the procedure for computing an $(\epsilon, 1/m)$-covering of $\mathcal{U}$. Consequently, we can redefine the value $v(n)$ of an OR node $n \in S_{\mathcal{T}}$ as $v(n) = \mathcal{ND}_{(\epsilon,1/m)}(\{w(n,n') + v(n') \mid n' \in succ(n)\})$.

The first approximation algorithm, called MO-AOBB-C$_{\epsilon}$, is obtained from Algorithm 1 by two simple modifications. First, the revision of the OR node values in line 17 is replaced by $v(p) \leftarrow \mathcal{ND}_{(\epsilon,1/m)}(v(p) \cup \{w(p,n) + v(n)\})$. Second, a partial solution tree T' is safely discarded in line 12 if $f(T')$ is $(\epsilon, 1/m)$-dominated by the current value $v(s)$ of the root node. We can show the following properties.

Proposition 3. *Let* n *be an OR node labeled* X_i *in the AND/OR search tree* $S_{\mathcal{T}}$ *such that the subtree of* $\mathcal{T}$ *rooted at* X_i *has depth* k, *where* m *is the depth of* $\mathcal{T}$ *and* $1 \leq k \leq m$. *Then,* $v(n)$ *is an* $(\epsilon, k/m)$*-covering of the conditioned subproblem below* n.

Proposition 4. *Given a MO-COP instance with* $p > 1$ *objectives, for any finite* $\epsilon > 0$ *algorithm MO-AOBB-C*$_{\epsilon}$ *computes an* ϵ*-covering of the Pareto frontier.*

Proposition 5. *The time and space complexities of algorithm MO-AOBB-C$_\epsilon$are bounded by $O((m \log K/\epsilon)^{2p} \cdot n \cdot d^m)$ and $O(n \cdot (m \log K/\epsilon)^p)$, respectively, where m is the depth of the guiding pseudo tree and K bounds the cost vectors.*

3.3 A More Aggressive Approximation Algorithm

Rather than requiring an upper bound on the size of the solution trees, it is possible to compute an ϵ-covering of the Pareto frontier by considering a more aggressive pruning rule that uses the ϵ-dominance relation only, thus allowing for an early termination of the unpromising partial solution trees. Consequently, the second approximation algorithm, called MO-AOBB-A$_\epsilon$ extends Algorithm 1 by discarding the partial solution tree T' in line 12 if its corresponding heuristic evaluation function $f(T')$ is ϵ-dominated by the current value $v(s)$ of the root node. During search, the values of the OR nodes in the search tree are updated using the regular (Pareto) non-dominated closure.

We can see that with this pruning rule the root node of the search tree maintains an ϵ-covering of the solution space. Specifically, if T' is the current partial solution tree and n is the current search node, then for all $\boldsymbol{v} \in \mathcal{U}^*$ there exists $\boldsymbol{u} \in f(T')$ such that $\boldsymbol{u} \preccurlyeq \boldsymbol{v}$, where $\mathcal{U}^*$ is the Pareto frontier obtained by solving the problem conditioned on T'. Hence, $v(s) \preccurlyeq_\epsilon f(T')$ implies $v(s) \preccurlyeq_\epsilon \mathcal{U}^*$, meaning that the current upper bound already ϵ-covers the current conditioned problem. Unlike the previous method, this approach does not provide any guarantees regarding the size of the ϵ-covering generated, and therefore, the time complexity of MO-AOBB-A$_\epsilon$ is bounded in the worst case by $O(K^{2p} \cdot n \cdot d^m)$, the size of the AND/OR search tree.

4 Experiments

We evaluated the performance of our depth-first Branch-and-Bound search approximation algorithms on two classes of MO-COP benchmarks: risk conscious combinatorial auctions and multi-objective scheduling problems for smart buildings. All experiments were carried out on a 2.4GHz quad-core processor with 8GB of RAM.

For our purpose, the algorithms MO-AOBB-C$_\epsilon$ and MO-AOBB-A$_\epsilon$ were guided by the multi-objective mini-bucket heuristics presented in [5]. The algorithms using *static mini-bucket heuristics* (SMB) are denoted by MO-AOBB-C$_\epsilon$+SMB(i) and MO-AOBB-A$_\epsilon$+SMB(i), while those using *dynamic mini-bucket heuristics* (DMB) are denoted by MO-AOBB-C$_\epsilon$+DMB(i) and MO-AOBB-A$_\epsilon$+DMB(i), respectively, where i is the mini-bucket i-bound and controls the accuracy of the corresponding heuristic. The static mini-bucket heuristics are pre-compiled, have a reduced computational overhead during search but are typically less accurate. Alternatively, the dynamic mini-bucket heuristics are computed dynamically at each node in the search tree, are far more accurate than the pre-compiled ones for the same i-bound value but have a much higher computational overhead.

We compared our algorithms against two recent state-of-the-art approaches for computing an ϵ-covering of the Pareto frontier, as follows:

- BE$_\epsilon$ – a multi-objective variable elimination algorithm proposed recently by [12]
- MOA$^*_\epsilon$ – a multi-objective A* search introduced in [13] which we extended here to use the mini-bucket based heuristics as well.

We note that algorithms BE_ϵ and MOA^*_ϵ require time and space exponential in the treewidth and, respectively, the number of variables of the problem instance. For reference, we also ran two exact search algorithms for computing Pareto frontiers: the multi-objective Russian Doll Search algorithm (MO-RDS) from [17] and the baseline AND/OR Branch-and-Bound with mini-bucket heuristics (MO-AOBB) from [5].

In all experiments we report the average CPU time in seconds and the number of nodes visited for solving the problems. We also record the size of the Pareto frontier as well as the size of the corresponding ϵ-covering generated for different ϵ values. We also specify the problem parameters such as the treewidth (w^*) and depth of the pseudo tree (h). The pseudo trees were computed using the classic *minfill* heuristic [15]. The data points shown in each plot represent an average over 10 random instances generated for the respective problem size.

4.1 Risk Conscious Combinatorial Auctions

In combinatorial auctions, an auctioneer has a set of goods to sell and the buyers submit a set of bids on indivisible subsets of goods. In *risk conscious* auctions, the auctioneer wants also to control the risk of not being paid after a bid has been accepted, because it may cause large losses in revenue. Let $M = \{1, ..., n\}$ be the set of goods to be auctioned and let $\mathcal{B} = \{B_1, ..., B_m\}$ be the set of bids. A bid B_j is defined by a triple (S_j, p_j, r_j), where $S_j \subseteq M$, p_j is the bid price and r_j is the *probability of failure*, respectively. The auctioneer must decide which bids to accept under the constraint that each good is allocated to at most one bid. The first objective is to maximize the auctioneer profit. The second objective is to minimize risk of not being paid. Assuming independence and after a logarithmic transformation of probabilities, this objective can also be expressed as an additive function [4,5].

We generated combinatorial auctions from the `paths` distribution of the CATS suite (*http://cats.stanford.edu/*) and randomly added failure probabilities to the bids in the range 0 to 0.3. These problems simulate the auction of paths in space with real-world applications such as bidding for truck routes, natural gas pipelines, network bandwidth allocation, as well as bidding for the right to use railway tracks. Figure 4 displays the results obtained on auctions with 30 goods and increasing number of bids, for $\epsilon \in \{0.01, 0.1, 0.3, 0.5\}$. Due to space reasons, we report only on algorithms using static mini-bucket heuristics with $i = 12$. As can be observed, the depth-first AND/OR search algorithms MO-AOBB-C_ϵ+SMB(12) and MO-AOBB-A_ϵ+SMB(12) clearly outperformed their competitors MOA^*_ϵ+SMB(12) and BE_ϵ, in many cases by several orders of magnitude of improved resolution time. The poor performance of MOA^*_ϵ+SMB(12) and BE_ϵ can be explained by their exponential space requirements. More specifically, MO-AOBB-A_ϵ+SMB(12) was the fastest algorithm on this domain, across all ϵ values. At the smallest reported ϵ value ($\epsilon = 0.01$), the algorithm is only slightly faster than the baseline MO-AOBB+SMB(12) because the ϵ-dominance based pruning rule is almost identical to the Pareto dominance based one used by the latter (ie, $1 + \epsilon \approx 1$), and therefore its performance is dominated by the size of the search space explored which is slightly smaller. As ϵ increases, the running time of MO-AOBB-A_ϵ+SMB(12) improves considerably because it prunes the search space more aggressively, which translates into additional time savings. We also see that the performance of MO-AOBB-C_ϵ+SMB(12)

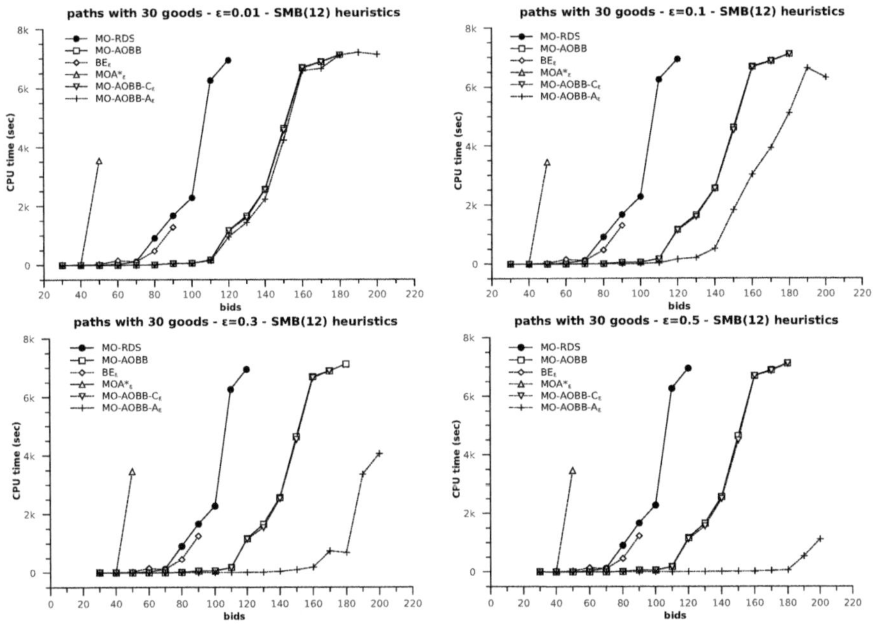

Fig. 4. CPU time (in seconds) obtained for risk conscious combinatorial auctions with 30 goods and increasing number of bids ($w^* \in [8, 80]$, $h \in [16, 119]$). Time limit 2 hours.

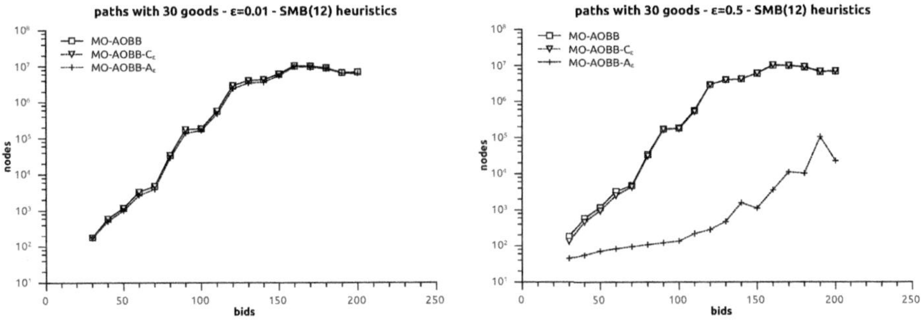

Fig. 5. Number of nodes visited for risk conscious combinatorial auctions with 30 goods and increasing number of bids ($w^* \in [8, 80]$, $h \in [16, 119]$). Time limit 2 hours.

is almost identical to that of MO-AOBB+SMB(12), across all reported ϵ values. This demonstrates that the pruning strategy with respect to the finer $(\epsilon, 1/m)$-dominance relation is rather conservative and does not prune the search space significantly. In particular, the pruning rule based on $(\epsilon, 1/m)$-dominance is almost identical to the one based on regular Pareto dominance (because $(1 + \epsilon)^{1/m} \approx 1$ for relatively large m), for all ϵ, and therefore both algorithms explore almost the same search space. Figure 5 displays the size of the search space explored by MO-AOBB+SMB(12), MO-AOBB-C_ϵ+SMB(12) and MO-AOBB-A_ϵ+SMB(12), for $\epsilon = 0.01$ and $\epsilon = 0.5$, respectively.

On this domain, the Pareto frontier contained on average 7 solutions while the size of the ϵ-coverings computed by both MO-AOBB-C$_\epsilon$+SMB(12) and MO-AOBB-A$_\epsilon$+SMB (12) varied between 3 ($\epsilon = 0.01$) and 1 ($\epsilon = 0.5$), respectively. MO-RDS performs poorly in this case solving relatively small problems only.

4.2 Scheduling Maintenance Tasks

Consider an office building where a set $\{1, ..., n\}$ of maintenance tasks must be scheduled daily during one of the following four dayparts: *morning*, *afternoon*, *evening* or *overnight*, subject to m binary hard constraints that forbid pairs of tasks to be scheduled during the same daypart. Each task i is defined by a tuple (w_i, p_i, o_i), where w_i is the electrical energy consumed during each daypart, p_i represents the financial costs incurred for each daypart and o_i is the overtime associated if the task is scheduled overnight. The goal is to assign each task to a daypart such that the number of hard constraints satisfied is maximized and three additional objectives are minimized: *energy waste* ($\sum_i^n w_i$), *financial penalty* ($\sum_i^n p_i$) and *overtime* ($\sum_i^n o_i$).

We generated a class of random problems with medium connectivity having n tasks and $2n$ binary hard constraints. For each task, the values w_i, p_i and o_i were generated uniformly randomly from the intervals $[0, 10]$, $[0, 40]$ and $[0, 20]$, respectively. Figure 6 summarizes the results obtained on problems with increasing number of tasks. We report only on algorithms with dynamic mini-bucket heuristics with $i = 2$, due computational issues associated with larger i-bounds. We observe again that MO-AOBB-A$_\epsilon$+DMB(2) offers the best performance, especially for larger ϵ values, while its

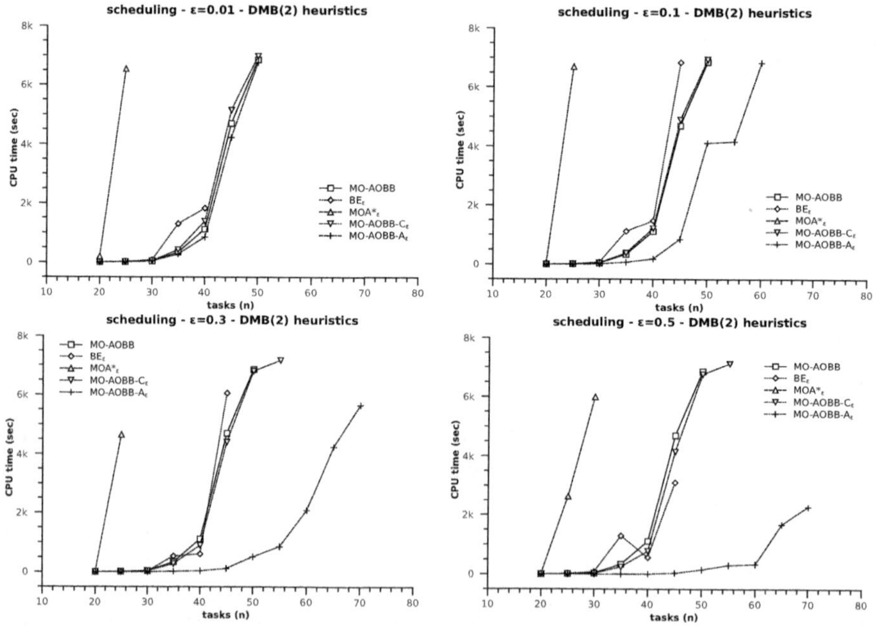

Fig. 6. CPU time (in seconds) for multi-objective scheduling problems with increasing number of tasks ($w^* \in [6, 15]$, $h \in [11, 26]$). Time limit 2 hours.

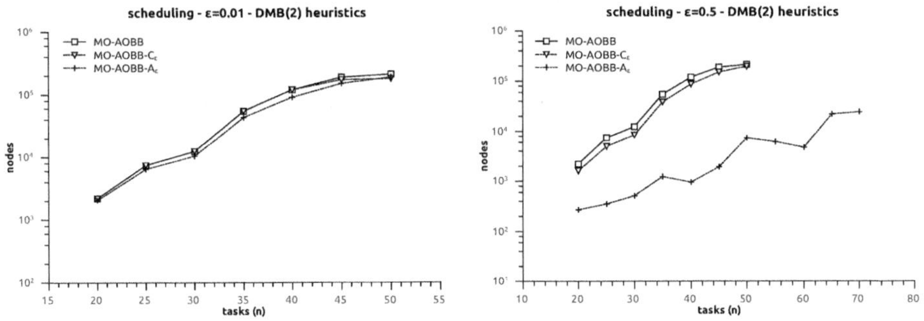

Fig. 7. Number of nodes visited for multi-objective scheduling problems with increasing number of tasks ($w^* \in [6, 15]$, $h \in [11, 26]$). Time limit 2 hours.

competitors MOA$^*_\epsilon$+DMB(2) and BE$_\epsilon$ could solve only relatively small problems due to their prohibitive memory requirements. MO-AOBB-C$_\epsilon$+DMB(2) is only slightly faster than MO-AOBB+DMB(2), across all ϵ values, showing that in this case as well the conservative pruning rule is not cost effective and outweighs the savings caused by manipulating smaller frontiers. In this case, MO-RDS could not solve any instance. Figure 7 displays the number of nodes visited for $\epsilon = 0.01$ and $\epsilon = 0.5$, respectively.

We noticed a significant reduction in the size of the ϵ-coverings generated on this domain, especially for larger ϵ-values. For instance, on problems with 50 tasks, the Pareto frontier contained on average 557 solutions, while the average size of the ϵ-coverings generated by MO-AOBB-C$_\epsilon$+DMB(2) and MO-AOBB-A$_\epsilon$+DMB(2) with $\epsilon = 0.5$ was 120 and 68, respectively.

In our experimental evaluation, we also investigated the impact of the mini-bucket i-bound on the performance of the proposed algorithms. For relatively small i-bounds, the algorithms using dynamic mini-buckets are typically faster than the ones guided by static mini-buckets, because the dynamic heuristics are more accurate than the pre-compiled ones. The picture is reversed for larger i-bounds because the computational overhead of the dynamic heuristics outweighs their pruning power. We also experimented with sparse and densely connected multi-objective scheduling problems. The results displayed a similar pattern to those presented here and therefore were omitted.

5 Conclusion

The paper rests on two contributions. First, we proposed two depth-first Branch-and-Bound search algorithms that traverse a weighted AND/OR search tree and use an ϵ-relaxation of the Pareto dominance relation between cost vectors to reduce the set of non-dominated solutions for multi-objective constraint optimization problems. The algorithms are guided by a general purpose heuristic evaluation function which was based on the multi-objective mini-bucket approximation scheme. Second, we carried out an empirical evaluation on MO-COPs simulating real-world applications that demonstrated the power of this new approach which improves dramatically the resolution times over state-of-the-art competitive algorithms based on either multi-objective best-first search or dynamic programming, in many cases by several orders of magnitude.

Future work includes extending the approximation scheme to explore an AND/OR search *graph* rather than a tree, via caching, as well as investigating alternative search regimes such as a linear space AND/OR best-first search strategy.

References

1. Dechter, R.: Constraint Processing. Morgan Kaufmann Publishers, San Francisco (2003)
2. Junker, U.: Preference-based inconsistency proving: when the failure of the best is sufficient. In: European Conference on Artificial Intelligence (ECAI), pp. 118–122 (2006)
3. Rollon, E., Larrosa, J.: Bucket elimination for multi-objective optimization problems. Journal of Heuristics 12, 307–328 (2006)
4. Rollon, E., Larrosa, J.: Multi-objective propagation in constraint programming. In: European Conference on Artificial Intelligence (ECAI), pp. 128–132 (2006)
5. Marinescu, R.: Exploiting problem decomposition in multi-objective constraint optimization. In: Gent, I.P. (ed.) CP 2009. LNCS, vol. 5732, pp. 592–607. Springer, Heidelberg (2009)
6. Hansen, P.: Bicriterion path problems. In: Multicriteria Decision Making (1980)
7. Warburton, A.: Approximation of Pareto optima in multiple-objective shortest problems. Operations Research 35(1), 70–79 (1987)
8. Tsaggouris, G., Zaroliagis, C.: Multiobjective optimization: improved fptas for shortest paths and non-linear objectives with applications. Theory of Comp. Sys. 45(1), 162–186 (2009)
9. Papadimitriou, C., Yannakakis, M.: On the approximability of trade-offs and optimal access to web sources. In: FOCS, pp. 86–92 (2000)
10. Erlebach, T., Kellerer, H., Pferschy, U.: Approximating multiobjective knapsack problems. Management Sciences 48(12), 1603–1612 (2002)
11. Bazgan, C., Hugot, H., Vanderpooten, D.: A practical efficient fptas for the 0-1 multi-objective knapsack problem. In: Arge, L., Hoffmann, M., Welzl, E. (eds.) ESA 2007. LNCS, vol. 4698, pp. 717–728. Springer, Heidelberg (2007)
12. Dubus, J.-P., Gonzales, C., Perny, P.: Multiobjective optimization using GAI models. In: International Conference on Artificial Intelligence (IJCAI), pp. 1902–1907 (2009)
13. Perny, P., Spanjaard, O.: Near admissible algorithms for multiobjective search. In: European Conference on Artificial Intelligence (ECAI), pp. 490–494 (2008)
14. Laumanns, M., Thiele, L., Deb, K., Zitzler, E.: Combining convergence and diversity in evolutionary multiobjective optimization. Evolutionary Computation 3(10), 263–282 (2002)
15. Dechter, R., Mateescu, R.: AND/OR search spaces for graphical models. Artificial Intelligence 171(2-3), 73–106 (2007)
16. Freuder, E.C., Quinn, M.J.: Taking advantage of stable sets of variables in constraint satisfaction problems. In: IJCAI, pp. 1076–1078 (1985)
17. Rollon, E., Larrosa, J.: Multi-objective Russian doll search. In: AAAI Conference on Artificial Intelligence, pp. 249–254 (2007)

Empirical Evaluation of Voting Rules with Strictly Ordered Preference Data

Nicholas Mattei

University of Kentucky
Department of Computer Science
Lexington, KY 40506, USA
nick.mattei@uky.edu

Abstract. The study of voting systems often takes place in the theoretical domain due to a lack of large samples of sincere, strictly ordered voting data. We derive several million elections (more than all the existing studies combined) from a publicly available data, the Netflix Prize dataset. The Netflix data is derived from millions of Netflix users, who have an incentive to report sincere preferences, unlike random survey takers. We evaluate each of these elections under the Plurality, Borda, k-Approval, and Repeated Alternative Vote (RAV) voting rules. We examine the Condorcet Efficiency of each of the rules and the probability of occurrence of Condorcet's Paradox. We compare our votes to existing theories of domain restriction (e.g., single-peakedness) and statistical models used to generate election data for testing (e.g., Impartial Culture). We find a high consensus among the different voting rules; almost no instances of Condorcet's Paradox; almost no support for restricted preference profiles, and very little support for many of the statistical models currently used to generate election data for testing.

1 Introduction

Voting rules and social choice methods have been used for centuries in order to make group decisions. Increasingly, in computer science, data collection and reasoning systems are moving towards distributed and multi-agent design paradigms [17]. With this design shift comes the need to aggregate these (possibly disjoint) observations and preferences into a total, group ordering in order to synthesize knowledge and data.

One of the most common methods of preference aggregation and group decision making in human systems is voting. Many societies, both throughout history and across the planet, use voting to arrive at group decisions on a range of topics from deciding what to have for dinner to declaring war. Unfortunately, results in the field of social choice prove that there is no perfect voting system and, in fact, voting systems can succumb to a host of problems. Arrow's Theorem demonstrates that any preference aggregation scheme for three or more alternatives will fail to meet a set of simple fairness conditions [2]. Each voting method violates one or more properties that most would consider important for a voting rule (such as non-dictatorship) [12]. Questions about voting and preference aggregation have circulated in the math and social choice communities for centuries [1, 8, 18].

R.I. Brafman, F. Roberts, and A. Tsoukiàs (Eds.): ADT 2011, LNAI 6992, pp. 165–177, 2011.

Many scholars wish to empirically study how often and under what conditions individual voting rules fall victim to various voting irregularities [7, 12]. Due to a lack of large, accurate datasets, many computer scientists and political scientists are turning towards statistical distributions to generate election scenarios in order to verify and test voting rules and other decision procedures [21, 24]. These statistical models may or may not be grounded in reality and it is an open problem in both the political science and social choice fields as to what, exactly, election data looks like [23].

A fundamental problem in research into properties of voting rules is the lack of large data sets to run empirical experiments [19, 23]. There have been studies of some datasets but these are limited in both number of elections analyzed [7] and size of individual elections within the datasets analyzed [12, 23]. While there is little agreement about the frequency that voting paradoxes occur or the consensus between voting methods, all the studies so far have found little evidence of *Condorcet's Voting Paradox* [13] (a cyclical majority ordering) or *preference domain restrictions* such as *single peakedness* [5] (where one candidate out of a set of three is never ranked last). Additionally, most of the studies find a strong consensus between most voting rules except Plurality [7, 12, 19].

As the computational social choice community continues to grow there is increasing attention on empirical results (see, e.g., [24]). The empirical data will support and justify the theoretical concerns [10, 11]. Walsh explicitly called for the establishment of a repository of voting data in his COMSOC 2010 talk [25]. We begin to respond to this call through the identification, analysis, and posting of a new repository of voting data.

We evaluate a large number of distinct 3 and 4 candidate elections derived from a novel data set, under the voting rules: Plurality, Copeland, Borda, Repeated Alternative Vote, and k-Approval. Our research question is manifold: Do different voting rules often produce the same winner? How often does Condorcet's Voting Paradox occur? Do basic statistical models of voting accurately describe our domain? Do any of the votes we analyze show single-peaked preferences [5] or other domain restrictions [22]?

2 Related Work

The literature on the empirical analysis of large voting datasets is somewhat sparse and many studies use the same datasets [12, 23]. These problems can be attributed to the lack of large amounts of data from real elections [19]. Chamberlin et al. [7] provide empirical analysis of five elections of the American Psychological Association (APA). These elections range in size from 11,000 to 15,000 ballots (some of the largest elections studied). Within these elections there are no cyclical majority orderings and, of the six voting rules under study, only Plurality fails to coincide with the others on a regular basis. Similarly, Regenwetter et al. analyse APA data from later years [20] and observe the same phenomena: a high degree of stability between elections rules. Felsenthal et al. [12] analyze a dataset of 36 unique voting instances from unions and other professional organizations in Europe. Under a variety of voting rules Felsenthal et al. also find a high degree of consensus between voting rules (with the notable exception of Plurality).

All of the empirical studies surveyed [7, 12, 16, 19, 20, 23] come to a similar conclusion: that there is scant evidence for occurrences of Condorcet's Paradox [18]. Many of these studies find no occurrence of majority cycles (and those that find cycles find them in rates of less than 1% of elections). Additionally, each of these (with the exception of Niemi and his study of university elections, which he observes is a highly homogenous population [16]) find almost no occurrences of either single-peaked preferences [5] or the more general value restricted preferences [22].

Given this lack of data and the somewhat surprising results regarding voting irregularities, some authors have taken a more statistical approach. Over the years multiple statistical models have been proposed to generate election pseudo-data to analyze (e.g., [19, 23]). Gehrlein [13] provides an analysis of the probability of occurrence of Condorcet's Paradox in a variety of election cultures. Gehrlein exactly quantifies these probabilities and concludes that Condorcet's Paradox probably will only occur with very small electorates. Gehrlein states that some of the statistical cultures used to generate election pseudo-data, specifically the Impartial Culture, may actually represent a worst-case scenario when analyzing voting rules for single-peaked preferences and the likelihood of observing Condorcet's Paradox [13]

Tideman and Plassmann have undertaken the task of verifying the statistical cultures used to generate pseudo-election data [23]. Using one of the largest datasets available Tideman and Plassmann find little evidence supporting the models currently in use to generate election data. Regenwetter et al. undertake a similar exercise and also find small support for the existing models of election generation [19]. The studies by both Regenwetter et al. and Tideman and Plassmann propose new statistical models with which to generate election pseudo-data that are better fits for their respective datasets.

3 The Data

We have mined strict preference orders from the Netflix Prize Dataset [3]. The Netflix dataset offers a vast amount of preference data; compiled and publically released by Netflix for its Netflix Prize [3]. There are 100,480,507 distinct ratings in the database. These ratings cover a total of 17,770 movies and 480,189 distinct users. Each user provides a numerical ranking between 1 and 5 (inclusive) of some subset of the movies. While all movies have at least one ranking it is not that case that all users have rated all movies. The dataset contains every movie rating received by Netflix, from its users, between when Netflix started tracking the data (early 2004) up to when the competition was announced (late 2005). This data has been perturbed to protect privacy and is conveniently coded for use by researchers.

The Netflix data is rare in preference studies: it is more sincere than most other preference data sets. Since users of the Netflix service will receive better recommendations from Netflix if they respond truthfully to the rating prompt, there is an incentive for each user to express sincere preference. This is in contrast to many other datasets which are compiled through surveys or other methods where the individuals questioned about their preferences have no stake in providing truthful responses.

We define an election as $E(m,n)$, where m is a set of candidates, $\{c_1,\ldots,c_m\}$, and n is a set of votes. A vote is a strict preference ordering over all the candidates $c_1 > c_2 > \cdots > c_m$. For convenience and ease of exposition we will often speak in the terms of a three candidate election and label the candidates as A,B,C and preference profiles as $A > B > C$. All results and discussion can be extended to the case of more than three candidates. A voting rule takes, as input, a set of candidates and a set of votes and returns a set of winners which may be empty or contain one or more candidates. In our discussion, elections return a complete ordering over all the candidates in the election with no ties between candidates (after a tiebreaking rule has been applied). The candidates in our data set correspond to movies from the Netflix dataset and the votes correspond to strict preference orderings over these movies. We break ties according to the lowest numbered movie identifier in the Netflix set; this is a random, sequential number assigned to every movie.

We construct vote instances from this dataset by looking at combinations of three movies. If we find a user with a strict preference ordering over the three moves, we tally that as a vote. For example, given movies A,B, and C: if a user rates movie $A = 1$, $B = 3$, and $C = 5$, then the user has a strict preference profile over the three movies we are considering and hence a vote. If we can find 350 or more votes for a particular movie triple then we regard that movie triple as an election and we record it. We use 350 as a cutoff for an election as it is the number of votes used by Tideman and Plassmann [23] in their study of voting data. While this is a somewhat arbitrary cutoff, Tideman and Plassmann claim it is a sufficient number to eliminate random noise in the elections [23] and we use it to generate comparable results.

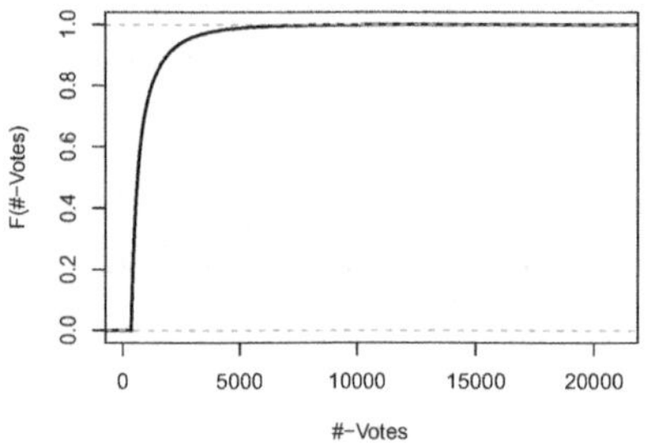

Fig. 1. Empirical CDF of Set 3A

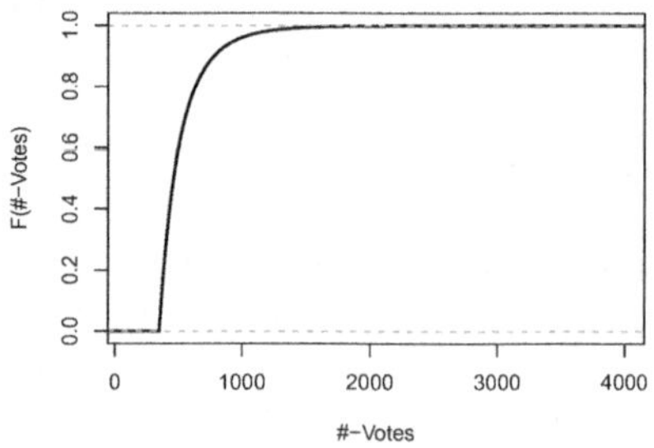

Fig. 2. Empirical CDF of Set 4A

The dataset is too large to use completely ($\binom{17770}{3} \approx 1 \times 10^{12}$). Therefore, we have drawn 3 independent (non-overlapping with respect to movies) samples of 2000 movies randomly from the set of all movies. We then, for each sample, search all the $\binom{2000}{3} \approx 1.33 \times 10^9$ possible elections for those with more than 350 votes. This search generated 1,553,611, 1,331,549, and 2,049,732 distinct movie triples within each of the respective samples. Not all users have rated all movies so the actual number of elections for each set is not consistent. The maximum election size found in the dataset is 22,079 votes; metrics of central tendency are presented in Table 1. Figures 1 and 2 show the empirical cumulative distribution functions (ECFD) for Set3A and 4A respectively. All of the datasets show similar ECDF's to those pictured.

Using the notion of item-item extension [14] we attempted to extend every triple found in the initial search. Item-item extension allows us to trim our search space by only searching for 4 movie combinations which contain a combination of 3 movies

Table 1. Summary Statistics for the election data

	3 Candidate Sets			4 Candidate Sets		
	Set 3A	Set 3B	Set 3C	Set4A	Set 4B	Set 4C
Min.	350.0	350.0	350.0	350.0	350.0	350.0
1st Qu.	444.0	433.0	435.0	394.0	393.0	384.0
Median	617.0	579.0	581.0	461.0	461.0	438.0
Mean	963.8	881.8	813.4	530.9	530.5	494.6
3rd Qu.	1,041.0	931.0	901.0	588.0	591.0	539.0
Max.	22,079.0	18,041.0	20,678.0	3830.0	3396.0	3639.0
Elements	1,553,611.0	1,331,549.0	2,049,732.0	2,721,235.0	1,222,009.0	1,243,749.0

which was a valid voting instance. For each set we only searched for extensions within the same draw of 2000 movies, making sure to remove any duplicate 4-item extensions. The results of this search are also summarized in Table 1. We found no 5-item extensions with more than 350 votes in the >30 billion possible extensions. Our constructed dataset contains more than 5 orders of magnitude more distinct elections than all the previous studies *combined* and the largest single election contains slightly more votes than the largest previously studied distinct election.

The data mining and experiments were performed on a pair of dedicated machines with dual-core Athlon 64x2 5000+ processors and 4 gigabytes of RAM. All the programs for searching the dataset and performing the experiments were written in C++. All of the statistical analysis was performed in R using RStudio. The initial search of three movie combinations took approximately 24 hours (parallelized over the two cores) for each of the three independently drawn sets. The four movie extension searches took approximately 168 hours per dataset while the five movie extensions took about 240 hours per dataset. Computing the results of the various voting rules, checking for domain restrictions, and checking for cycles took approximately 20 hours per dataset. Calibrating and verifying the statistical distributions took approximately 15 hours per dataset. All the computations for this project are straightforward, the benefit of modern computational power allows our parallelized code to more quickly search the billions of possible movie combinations.

4 Analysis and Discussion

We have found a large correlation between each of the voting rules under study with the exception of Plurality (when $m = 3,4$) and 2-Approval (when $m = 3$). A *Condorcet Winner* is a candidate who is preferred by a majority of the voters to each of the other candidates in an election [12]. The voting rules under study, with the exception of Copeland, are not *Condorcet Consistent*: they do not necessarily select a Condorcet Winner if one exists [18]. Therefore we also analyze the voting rules in terms of their *Condorcet Efficiency*, the rate at which the rule selects a Condorcet Winner if one exists [15]. The results in Section 4.1 show extremely small evidence for cases of single peaked preferences and very low rates of occurrence of preference cycles. In Section 4.2 we see that the voting rules exhibit a high degree of Condorcet Efficiency in our dataset. Finally, the experiments in Section 4.3 indicate that several statistical models currently in use for testing new voting rules [21] do not reflect the reality of our dataset. All of these results are in keeping with the analysis of other, distinct, datasets [7, 12, 16, 19, 20, 23] and provide support for their conclusions.

4.1 Domain Restrictions and Preference Cycles

Condorcet's Paradox of Voting is the observation that rational group preferences can be aggregated, through a voting rule, into an irrational total preference [18]. It is an important theoretical and practical concern to evaluate how often the scenario arises in empirical data. In addition to analyzing instances of *total cycles* (Condorcet's Paradox) involving all candidates in an election, we check for two other types of cyclic preferences. We also search our results for both *partial cycles*, a cyclic ordering that does not include the top candidate (Condorcet Winner), and *partial top cycles*, a cycle that includes the top candidate but excludes one or more other candidates [12].

Table 2. Number of elections demonstrating various types of voting cycles

		Partial Cycle	Partial Top	Total
$m = 3$	Set 3A	635 (0.041%)	635 (0.041%)	635 (0.041%)
	Set 3B	591 (0.044%)	591 (0.044%)	591 (0.044%)
	Set 3C	1,143 (0.056%)	1,143 (0.056%)	1,143 (0.056%)
$m = 4$	Set 4A	3,837 (0.141%)	2,882 (0.106%)	731 (0.027%)
	Set 4B	1,864 (0.153%)	1,393 (0.114%)	462 (0.035%)
	Set 4C	3,233 (0.258%)	2,367 (0.189%)	573 (0.046%)

Table 2 is a summary of the rates of occurrence of the different types of voting cycles found in our data set. The cycle counts for $m = 3$ are all equivalent due to the fact that there is only one type of possible cycle when $m = 3$. There is an extremely low instance of total cycles for all our data ($< 0.06\%$ of all elections). This corresponds to findings in the empirical literature that support the conclusion that Condorcet's Paradox has a low incidence of occurrence. Likewise, cycles of any type occur in rates $< 0.2\%$ and therefore seem of little practical importance in our dataset as well. Our results for cycles that do not include the winner mirror those of Felsenthal et al. [12]: many cycles occur in the lower ranks of voters' preference orders in the election due to the voters' inability to distinguish between, or indifference towards, candidates the voter has a low ranking for or considers irrelevant.

Black first introduced the notion of single-peaked preferences [5]; a domain restriction that states that the candidates can be ordered along one axis of preference and there is a single peak to the graph of all votes by all voters if the candidates are ordered along this axis. Informally, it is the idea that some candidate, in a three candidate election, is never ranked last. The notion of restricted preference profiles was extended by Sen [22] to include the idea of candidates who are never ranked first (single-bottom) and candidates who are always ranked in the middle (single-mid). Domain restrictions can be expanded to the case where elections contain more than three candidates [1]. Preference restrictions have important theoretical applications and are widely studied in the area of election manipulation. Many election rules become trivially easy to manipulate when electorates preferences are single-peaked [6].

Table 3. Number of elections demonstrating various value restricted preferences

		Single-Peak	Single-Mid	Single-Bottom
$m = 3$	Set 3A	342 (0.022%)	0 (0.0%)	198 (0.013%)
	Set 3B	227 (0.017%)	0 (0.0%)	232 (0.017%)
	Set 3C	93 (0.005%)	0 (0.0%)	100 (0.005%)
$m = 4$	Set 4A	1 (0.022%)	0 (0.000%)	1 (0.013%)
	Set 4B	0 (0.000%)	0 (0.000%)	0 (0.000%)
	Set 4C	0 (0.000%)	0 (0.000%)	0 (0.000s%)

Table 3 summarizes our results for the analysis of different restricted preference profiles. There is (nearly) a complete lack of preference profile restrictions when $m = 4$ and near lack ($< 0.03\%$) when $m = 3$. It is important to remember that the underlying objects in this dataset are movies, and individuals, most likely, evaluate movies for many different reasons. Therefore, as the results of our analysis confirm, there are very few items that users rate with respect to a single dimension.[1]

4.2 Voting Rules

The variety of voting rules and election models that have been implemented or "improved" over time is astounding. For a comprehensive history and survey of voting rules see Nurmi [18]. Arrow shows that any preference aggregation scheme for three or more alternatives cannot meet some simple fairness conditions [2]. This leads most scholars to question "which voting rule is the best?" We analyze our dataset under the voting rules Plurality, Borda, 2-Approval, and Repeated Alternative Vote (RAV). We briefly describe the voting rules under analysis. A more complete treatment of voting rules and their properties can be found in Nurmi [18] and in Arrow, Sen, and Suzumura [1].

Plurality: Plurality is the most widely used voting rule [18] (and, to many Americans, synonymous with the term voting). The Plurality score of a candidate is the sum of all the first place votes for that candidate. No other candidates in the vote are considered besides the first place vote. The winner is the candidate with the highest score.

k-Approval: Under k-Approval voting, when a voter casts a vote, the first k candidates each receive the same number of points. In a 2-Approval scheme, the first 2 candidates

[1] Set 3B contains the movies *Star Wars: Return of the Jedi* and *The Shawshank Redemption.* Both are widely considered to be "good" movies; all but 15 of the 227 elections exhibiting single-peaked preferences share one of these two movies.

of every voter's preference order would receive the same number of points. The winner of a k-Approval election is the candidate with the highest total score.

Copeland: In a Copeland election each pairwise contest between candidates is considered. If candidate a defeats candidate b in a head-to-head comparison of first place votes then candidate a receives 1 point; a loss is -1 and a tie is worth 0 points. After all head-to-head comparisons are considered, the candidate with the highest total score is the winner of the election.

Borda: Borda's System of Marks involves assigning a numerical score to each position. In most implementations [18] the first place candidate receives $c-1$ points, with each candidate later in the ranking receiving 1 less point down to 0 points for the last ranked candidate. The winner is the candidate with the highest total score.

Repeated Alternative Vote: Repeated Alternative Vote (RAV) is an extension of the Alternative Vote (AV) into a rule which returns a complete order over all the candidates [12]. For the selection of a single candidate there is no difference between RAV and AV. Scores are computed for each candidate as in Plurality. If no candidate has a strict majority of the votes the candidate receiving the fewest first place votes is dropped from all ballots and the votes are re-counted. If any candidate now has a strict majority, they are the winner. This process is repeated up to $c-1$ times [12]. In RAV this procedure is repeated, removing the winning candidate from all votes in the election after they have won, until no candidates remain. The order in which the winning candidates were removed is the total ordering of all the candidates.

We follow the analysis outlined by Felsenthal et al. [12]. We establish the Copeland order as "ground truth" in each election; Copeland always selects the Condorcet Winner if one exists and many feel the ordering generated by the Copeland rule is the "most fair" when no Condorcet Winner exists [12, 18]. After determining the results of each election, for each voting rule, we compare the order produced by each rule to the Copeland order and compute the Spearman's Rank Order Correlation Coefficient (Spearman's ρ) to measure similarity [12]. This procedure has the disadvantage of demonstrating if voting rules fail to correspond closely to the results from Copeland. Another method, not used in this paper, would be to consider each of the voting rules as a maximum likelihood estimator of some "ground truth." We leave this track for future work [9].

Table 4 lists the mean and standard deviation for Spearman's Rho between the various voting rules and Copeland. All sets had a median value of 1.0. Our analysis supports other empirical studies in the field that find a high consensus between the various voting rules [7, 12, 20]. Plurality performs the worst as compared to Copeland across all the datasets. 2-Approval does fairly poorly when $m=3$ but does surprisingly well when $m=4$. We suspect this discrepancy is due to the fact that when $m=3$, individual voters are able to select a full 2/3 of the available candidates. Unfortunately, our data is not split into enough independent samples to accurately perform any statistical hypothesis testing. Computing a paired t-test with all $> 10^6$ elections within a sample set would provide trivially significant results due to the extremely large sample size.

There are many considerations one must make when selecting a voting rule for use within a given system. Merrill suggests that one of the most powerful metrics is

Table 4. Voting results (Spearman's ρ) for Sets A,B, and C

		Plurality	2-Approval	Borda	RAV
Set 3A	Mean	0.9300	0.9149	0.9787	0.9985
	SD	0.1999	0.2150	0.1029	0.0336
Set 3B	Mean	0.9324	0.9215	0.9802	0.9985
	SD	0.1924	0.2061	0.0995	0.0341
Set 3C	Mean	0.9238	0.9177	0.9791	0.9980
	SD	0.208	0.2130	0.1024	0.0394
Set 4A	Mean	0.9053	0.9578	0.9787	0.9978
	SD	0.1691	0.0956	0.0673	0.0273
Set 4B	Mean	0.9033	0.9581	0.9798	0.9980
	SD	0.1627	0.0935	0.0651	0.0263
Set 4C	Mean	0.8708	0.9516	0.9767	0.9956
	SD	0.2060	0.1029	0.0706	0.0404

Condorcet Efficiency [15]. Table 5 shows the proportion of Condorcet Winners selected by the various voting rules under study. We eliminated all elections that did not have a Condorcet Winner in this analysis. All voting rules select the Condorcet Winner a surprising majority of the time. 2-Approval, when $m = 3$, results in the lowest rate of Condorcet Winner selection in our dataset.

Table 5. Condorcet Efficiency of the various voting rules

		Condorcet Winners	Plurality	2-Approval	Borda	RAV
$m = 3$	Set 3A	1,548,553	0.9665	0.8714	0.9768	0.9977
	Set 3B	1,326,902	0.9705	0.8842	0.9801	0.9980
	Set 3C	2,041,756	0.9643	0.8814	0.9795	0.9971
$m = 4$	Set 4A	2,701,464	0.9591	0.9213	0.9630	0.9966
	Set 4B	1,212,370	0.9626	0.9290	0.9693	0.9971
	Set 4C	1,241,762	0.9550	0.9253	0.9674	0.9940

Overall, we find a consensus between the various voting rules in our tests. This supports the findings of other empirical studies in the field [7, 12, 20]. Merrill finds much different rates for Condorcet Efficiency than we do in our study [15]. However, Merrill uses statistical models to generate elections rather than empirical data to compute his numbers and this is likely the cause of the discrepancy [13].

4.3 Statistical Models of Elections

We evaluate our dataset to see how it matches up to different probabilistic distributions found in the literature. We briefly detail several probability distributions (or "cultures") here that we test. Tideman and Plassmann provide a more complete discussion of the variety of statistical cultures in the literature [23]. There are other election generating cultures that we do not analyze because we found no support for restricted preference profiles (either single-peaked or single-bottomed). These cultures, such as weighted Independent Anonymous Culture, generate preference profiles that are skewed towards single-peakedness or single-bottomness (a further discussion and additional election generating statistical models can be found in [23]). We follow the general outline in Tideman and Plassmann to guide us in this study. For ease of discussion we divide the models into two groups: probability models (IC, DC, UC, UUP) and generative models (IAC, Urn, IAC-Fit). Probability models define a probability vector over each of the $m!$ possible strict preference rankings. We note these probabilities as $pr(ABC)$, which is the probability of observing a vote $A > B > C$ for each of the possible orderings. In order to compare how the statistical models describe the empirical data, we compute the mean Euclidean distance between the empirical probability distribution and the one predicted by the model.

Impartial Culture (IC): An even distribution over every vote exists. That is, for the $m!$ possible votes, each vote has probability $1/m!$

Dual Culture (DC): The dual culture assumes that the probability of opposite preference orders is equal. So, $pr(ABC) = pr(CAB)$, $pr(ACB) = pr(BCA)$ etc. This culture is based on the idea that some groups are polarized over certain issues.

Uniform Culture (UC): The uniform culture assumes that the probability of distinct pairs of lexicographically neighboring orders are equal. For example, $pr(ABC) = pr(ACB)$ and $pr(BAC) - pr(BCA)$ but not $pr(ACB) = pr(CAB)$ (as, for three candidates, we pair them by the same winner). This culture corresponds to situations where voters have strong preferences over the top candidates but may be indifferent over candidates lower in the list.

Unequal Unique Probabilities (UUP): The unequal unique probabilities culture defines the voting probabilities as the maximum likelihood estimator over the entire dataset. We determine, for each of the data sets, the UUP distribution as described below.

For DC and UC each election generates its own statistical model according to the definition of the given culture. For UUP we need to calibrate the parameters over the entire dataset. We follow the method described in Tideman and Plassmann [23]: first re-label each empirical election in the dataset such that the order with the most votes becomes the labeling for all the other votes. This requires reshuffling the vector so that the most likely vote is always $A > B > C$. Then, over all the reordered vectors, we maximize the log-likelihood of

$$f(N_1,\ldots,N_6;N,p_1,\ldots,p_6) = \frac{N!}{\prod_{r=1}^{6} N_r!} \prod_{r=1}^{6} p_r^{N_r} \tag{1}$$

where $N_1,\ldots,N_6$ is the number of votes received by a vote vector and $p_1,\ldots,p_6$ are the probabilities of observing a particular order over all votes (we expand this equation to 24 vectors for the $m = 4$ case). To compute the error between the culture's distribution and the empirical observations, we re-label the culture distribution so that preference order with the most votes in the empirical distribution matches the culture distribution and compute the error as the mean Euclidean distance between the discrete probability distributions.

Urn Model: The Polya Eggenberger urn model is a method designed to introduce some correlation between votes and does not assume a complete uniform random distribution [4]. We use a setup as described by Walsh [24]; we start with a jar containing one of each possible vote. We draw a vote at random and place it back into the jar with a additional votes of the same kind. We repeat this procedure until we have created a sufficient number of votes.

Impartial Anonymous Culture (IAC): Every distribution over orders is equally likely. For each generated election we first randomly draw a distribution over all the $m!$ possible voting vectors and then use this model to generate votes in an election.

IAC-Fit: For this model we first determine the vote vector that maximizes the log-likelihood of Equation 1 without the reordering described for UUP. Using the probability vector obtained for $m = 3$ and $m = 4$ we randomly generate elections. This method generates a probability distribution or culture that represents our entire dataset.

For the generative models we must generate data in order to compare them to the culture distributions. To do this we average the total elections found for $m = 3$ and $m = 4$ and generate 1,639,070 and 1,718,532 elections, respectively. We then draw the individual election sizes randomly from the distribution represented in our dataset. After we generate these random elections we compare them to the probability distributions predicted by the various cultures.

Table 6. Mean Euclidean distance between the empirical data set and different statistical cultures (standard error in parentheses)

		IC	DC	UC	UUP
	Set 3A	0.3304 (0.0159)	0.2934 (0.0126)	0.1763 (0.0101)	0.3025 (0.0372)
$m = 3$	Set 3B	0.3192 (0.0153)	0.2853 (0.0121)	0.1685 (0.0095)	0.2959 (0.0355)
	Set 3C	0.3041 (0.0151)	0.2709 (0.0121)	0.1650 (0.0093)	0.2767 (0.0295)
	Urn	0.6226 (0.0249)	0.4744 (0.0225)	0.4743 (0.0225)	0.4909 (0.1054)
$m = 3$	IAC	0.2265 (0.0056)	0.1690 (0.0056)	0.1689 (0.0056)	0.2146 (0.0063)
	IAC-Fit	0.0372 (0.0002)	0.0291 (0.0002)	0.0260 (0.0002)	0.0356 (0.0002)
	Set 4A	0.2815 (0.0070)	0.2282 (0.0042)	0.1141 (0.0034)	0.3048 (0.0189)
$m = 4$	Set 4B	0.2596 (0.0068)	0.2120 (0.0041)	0.1011 (0.0026)	0.2820 (0.0164)
	Set 4C	0.2683 (0.0080)	0.2149 (0.0049)	0.1068 (0.0034)	0.2811 (0.0166)
	Urn	0.6597 (0.0201)	0.4743 (0.0126)	0.4743 (0.0126)	0.6560 (0.1020)
$m = 4$	IAC	0.1257 (0.0003)	0.0899 (0.0003)	0.0899 (0.0003)	0.1273 (0.0004)
	IAC-Fit	0.0528 (0.0001)	0.0415 (0.0001)	0.3176 (0.0001)	0.0521 (0.0001)

Table 6 summarizes our results for the analysis of different statistical models used to generate elections. In general, none of the probability models captures our empirical data. UC has the lowest error in predicting the distributions found in our empirical data. The data generated by our IAC-Fit model fits very closely to the various statistical models. This is most likely due to the fact that the distributions generated by the IAC-Fit procedure closely resemble an IC. We, like Tideman and Plassmann, find little support for the static cultures' ability to model real data [23]

5 Conclusion

We have identified and thoroughly evaluated a novel dataset as a source of sincere election data. We find overwhelming support for many of the existing conclusions in the empirical literature. Namely, we find a high consensus among a variety of voting methods; low occurrences of Condorcet's Paradox and other voting cycles; low occurrences of preference domain restrictions such as single-peakedness; and a lack of support for existing statistical models which are used to generate election pseudo-data. Our study is significant as it adds more results to the current discussion of what is an election and how often do voting irregularities occur? Voting is a common method by which agents make decisions both in computers and as a society. Understanding the unique statistical and mathematical properties of voting rules, as verified by empirical evidence across multiple domains, is an important step. We provide a new look at this question with a novel dataset that is several orders of magnitude larger than the sum of the data in previous studies.

The collection and public dissemination of the datasets is a central point our work. We plan to establish a repository of election data so that theoretical researchers can validate with empirical data. A clearing house for data was discussed at COMSOC 2010 by Toby Walsh and others in attendance [25]. We plan to identify several other free, public datasets that can be viewed as "real world" voting data. The results reported in our study imply that our data is reusable as real world voting data. Therefore, it seems that the Netflix dataset, and its $> 10^{12}$ possible elections, can be used as a source of election data for future empirical validation of theoretical voting studies.

There are many directions for future work that we would like to explore. We plan to evaluate how many of the elections in our data set are manipulable and evaluate the frequency of occurrence of easily manipulated elections. We would like to, instead of comparing how voting rules correspond to one another, evaluate their power as maximum likelihood estimators [9]. Additionally, we would like to expand our evaluation of statistical models to include several new models proposed by Tideman and Plassmann, and others [23].

Acknowledgements. Thanks to Dr. Florenz Plassmann for his helpful discussions on this paper and guidance on calibrating statistical models. Also thanks to Dr. Judy Goldsmith and Elizabeth Mattei for their helpful discussion and comments on preliminary drafts of this paper. We gratefully acknowledge the support of NSF EAGER grant CCF-1049360.

References

1. Arrow, K., Sen, A., Suzumura, K. (eds.): Handbook of Social Choice and Welfare, vol. 1. North-Holland, Amsterdam (2002)
2. Arrow, K.: Social choice and individual values. Yale Univ. Press, New Haven (1963)

3. Bennett, J., Lanning, S.: The Netflix Prize. In: Proceedings of KDD Cup and Workshop (2007), `www.netflixprize.com`
4. Berg, S.: Paradox of voting under an urn model: The effect of homogeneity. Public Choice 47(2), 377–387 (1985)
5. Black, D.: On the rationale of group decision-making. The Journal of Political Economy 56(1) (1948)
6. Brandt, F., Brill, M., Hemaspaandra, E., Hemaspaandra, L.A.: Bypassing combinatorial protections: Polynomial-time algorithms for single-peaked electorates. In: Proc. of the 24th AAAI Conf. on Artificial Intelligence, pp. 715–722 (2010)
7. Chamberlin, J.R., Cohen, J.L., Coombs, C.H.: Social choice observed: Five presidential elections of the American Psychological Association. The Journal of Politics 46(2), 479–502 (1984)
8. Condorcet, M.: Essay sur l'application de l'analyse de la probabilit des decisions: Redues et pluralit des voix, Paris (1785)
9. Conitzer, V., Sandholm, T.: Common voting rules as maximum likelihood estimators. In: Proc. of the 21st Annual Conf. on Uncertainty in AI (UAI), pp. 145–152 (2005)
10. Conitzer, V., Sandholm, T., Lang, J.: When are elections with few candidates hard to manipulate? Journal of the ACM 54(3), 1–33 (2007)
11. Faliszewski, P., Hemaspaandra, E., Hemaspaandra, L.A., Rothe, J.: A richer understanding of the complexity of election systems. In: Ravi, S., Shukla, S. (eds.) Fundamental Problems in Computing: Essays in Honor of Professor D.J. Rosenkrantz, pp. 375–406. Springer, Heidelberg (2009)
12. Felsenthal, D.S., Maoz, Z., Rapoport, A.: An empirical evaluation of six voting procedures: Do they really make any difference? British Journal of Political Science 23, 1–27 (1993)
13. Gehrlein, W.V.: Condorcet's paradox and the likelihood of its occurance: Different perspectives on balanced preferences. Theory and Decisions 52(2), 171–199 (2002)
14. Han, J., Kamber, M. (eds.): Data Mining. Morgan Kaufmann, San Francisco (2006)
15. Merrill III, S.: A comparison of efficiency of multicandidate electoral systems. American Journal of Politial Science 28(1), 23–48 (1984)
16. Niemi, R.G.: The occurrence of the paradox of voting in university elections. Public Choice 8(1), 91–100 (1970)
17. Nisan, N., Roughgarden, T., Tardos, E., Vazirani, V. (eds.): Algorithmic Game Theory. Cambridge Univ. Press, Cambridge (2007)
18. Nurmi, H.: Voting procedures: A summary analysis. British Journal of Political Science 13, 181–208 (1983)
19. Regenwetter, M., Grogman, B., Marley, A.A.J., Testlin, I.M.: Behavioral Social Choice: Probabilistic Models, Statistical Inference, and Applications. Cambridge Univ. Press, Cambridge (2006)
20. Regenwetter, M., Kim, A., Kantor, A., Ho, M.R.: The unexpected empirical consensus among consensus methods. Psychological Science 18(7), 629–635 (2007)
21. Rivest, R.L., Shen, E.: An optimal single-winner preferential voting system based on game theory. In: Conitzer, V., Rothe, J. (eds.) Proc. of the 3rd Intl. Workshop on Computational Social Choice (COMSOC), pp. 399–410 (2010)
22. Sen, A.K.: A possibility theorem on majority decisions. Econometrica 34(2), 491–499 (1966)
23. Tideman, N., Plassmann, F.: Modeling the outcomes of vote-casting in actual elections. To appear in Springer publshed book, `http://bingweb.binghamton.edu/~fplass/papers/Voting_Springer.pdf`
24. Walsh, T.: An empirical study of the manipulability of single transferable voting. In: Proc. of the 19th European Conf. on AI (ECAI 2010), pp. 257–262. IOS Press, Amsterdam (2010)
25. Walsh, T.: Where are the hard manipulation problems? In: Conitzer, V., Rothe, J. (eds.) Proc. of the 3rd Intl. Workshop on Computational Social Choice (COMSOC), pp. 9–11 (2010)

A Reduction of the Complexity of Inconsistencies Test in the MACBETH 2-Additive Methodology

Brice Mayag[1], Michel Grabisch[2], and Christophe Labreuche[3]

[1] Laboratoire Génie industriel, École Centrale Paris, Grande Voie des Vignes, F-92295 Châtenay-Malabry Cedex, France
brice.mayag@ecp.fr

[2] University of Paris 1, 106-112 Boulevard de l'Hôpital, 75013 Paris, France
michel.grabisch@univ-paris1.fr

[3] T.R.T France, 1 avenue Augustin Fresnel, 91767 Palaiseau Cedex, France
christophe.labreuche@thalesgroup.com

Abstract. MACBETH 2-additive is the generalization of the Choquet integral to the MACBETH approach, a MultiCriteria Decision Aid method. In the elicitation of a 2-additive capacity step, the inconsistencies of the preferential information, given by the Decision Maker on the set of binary alternatives, is tested by using the MOPI conditions. Since a 2-additive capacity is related to all binary alternatives, this inconsistencies checking can be more complex if the set of alternatives is very large. In this paper, we show that it is possible to limited the test of MOPI conditions to the only alternatives used in the preferential information.

Keywords: MCDA, Preference modeling, MOPI conditions, Choquet integral, MACBETH.

1 Introduction

Multiple Criteria Decision Aid (MCDA) aims at helping a decision maker (DM) in the representation of his preferences over a set of alternatives, on the basis of several criteria which are often contradictory. One possible model is the transitive decomposable one where an overall utility is determined for each option. In this category, we have the model based on Choquet integral, especially the 2-additive Choquet integral (Choquet integral w.r.t. a 2-additive) [6,8,14]. The 2-additive Choquet integral is defined w.r.t. a capacity (or nonadditive monotonic measure, or fuzzy measure), and can be viewed as a generalization of the arithmetic mean. Any interaction between two criteria can be represented and interpreted by a Choquet integral w.r.t. a 2-additive capacity, but not more complex interaction.

Usually the DM is supposed to be able to express his preference over the set of all alternatives X. Because this is not feasible in most of practical situations (the cardinality of X may be very large), the DM is asked to give, using pairwise comparisons, an ordinal information (a preferential information containing only

R.I. Brafman, F. Roberts, and A. Tsoukiàs (Eds.): ADT 2011, LNAI 6992, pp. 178–189, 2011.

a strict preference and an indifference relations) on a subset $X' \subseteq X$, called reference set. The set X' we use in this paper is the set of binary alternatives or binary actions denoted by $\mathcal{B}$. A binary action is an (fictitious) alternative representing a prototypical situation where on a given subset of at most two criteria, the attributes reach a satisfactory level **1**, while on the remaining ones, they are at a neutral level (neither satisfactory nor unsatisfactory) **0**. The characterization theorem of the representation of an ordinal information by a 2-additive Choquet integral [13] is based on the MOPI property. The inconsistencies test of this condition is done on every subsets of three criteria.

We are interested in the following problem: *how to reduce the complexity of this test of inconsistencies when the number of criteria is large?* We propose here a simplification of the MOPI property based only on the binary alternatives related to the ordinal information.

After some basic notions given in the next section, we present in Section 3 our main result.

2 Basic Concepts

Let us denote by $N = \{1, \ldots, n\}$ a finite set of n criteria and $X = X_1 \times \cdots \times X_n$ the set of actions (also called alternatives or options), where $X_1, \ldots, X_n$ represent the point of view or attributes. For all $i \in N$, the function $u_i : X_i \to \mathbb{R}$ is called a utility function. Given an element $x = (x_1, \ldots, x_n)$, we set $U(x) = (u_1(x_1), \ldots, u_n(x_n))$. For a subset A of N and actions x and y, the notation $z = (x_A, y_{N-A})$ means that z is defined by $z_i = x_i$ if $i \in A$, and $z_i = y_i$ otherwise.

2.1 Choquet Integral w.r.t. a 2-Additive Capacity

The Choquet integral w.r.t. a 2-additive capacity [6], called for short a *2-additive Choquet integral*, is a particular case of the Choquet integral [8,9,14]. This integral generalizes the arithmetic mean and takes into account interactions between criteria. A 2-additive Choquet integral is based on a 2-additive capacity [4,8] defined below and its Möbius transform [3,7]:

Definition 1

1. *A capacity on N is a set function $\mu : 2^N \to [0,1]$ such that:*
   (a) $\mu(\emptyset) = 0$
   (b) $\mu(N) = 1$
   (c) $\forall A, B \in 2^N,\ [A \subseteq B \Rightarrow \mu(A) \leq \mu(B)]$ (monotonicity).
2. *The Möbius transform [3] of a capacity μ on N is a function $m : 2^N \to \mathbb{R}$ defined by:*
$$m(T) := \sum_{K \subseteq T} (-1)^{|T \setminus K|} \mu(K), \forall T \in 2^N. \quad (1)$$

When m is given, it is possible to recover the original μ by the following expression:

$$\mu(T) := \sum_{K \subseteq T} m(K), \forall T \in 2^N. \tag{2}$$

For a capacity μ and its Möbius transform m, we use the following shorthand: $\mu_i := \mu(\{i\})$, $\mu_{ij} := \mu(\{i,j\})$, $m_i := m(\{i\})$, $m_{ij} := m(\{i,j\})$, for all $i, j \in N$, $i \neq j$. Whenever we use i and j together, it always means that they are different.

Definition 2. *A capacity μ on N is said to be 2-additive if*

- *For all subsets T of N such that $|T| > 2$, $m(T) = 0$;*
- *There exists a subset B of N such that $|B| = 2$ and $m(B) \neq 0$.*

The following important Lemma shows that a 2-additive capacity is entirely determined by the value of the capacity on the singletons $\{i\}$ and pairs $\{i, j\}$ of 2^N:

Lemma 1

1. *Let μ be a 2-additive capacity on N. We have for all $K \subseteq N$, $|K| \geq 2$,*

$$\mu(K) = \sum_{\{i,j\} \subseteq K} \mu_{ij} - (|K| - 2) \sum_{i \in K} \mu_i. \tag{3}$$

2. *If the coefficients μ_i and μ_{ij} are given for all $i, j \in N$, then the necessary and sufficient conditions that μ is a 2-additive capacity are:*

$$\sum_{\{i,j\} \subseteq N} \mu_{ij} - (n-2) \sum_{i \in N} \mu_i = 1 \tag{4}$$

$$\mu_i \geq 0, \ \forall i \in N \tag{5}$$

For all $A \subseteq N, |A| \geq 2, \forall k \in A$,

$$\sum_{i \in A \setminus \{k\}} (\mu_{ik} - \mu_i) \geq (|A| - 2)\mu_k. \tag{6}$$

Proof. See [6].

For an alternative $x := (x_1, ..., x_n) \in X$, the expression of the Choquet integral w.r.t. a capacity μ is given by:

$$C_\mu(U(x)) := \sum_{i=1}^{n} (u_{\tau(i)}(x_{\tau(i)}) - u_{\tau(i-1)}(x_{\tau(i-1)})) \ \mu(\{\tau(i), \ldots, \tau(n)\})$$

where τ is a permutation on N such that $u_{\tau(1)}(x_{\tau(1)}) \leq u_{\tau(2)}(x_{\tau(2)}) \leq \cdots \leq u_{\tau(n-1)}(x_{\tau(n-1)}) \leq u_{\tau(n)}(x_{\tau(n)})$, and $u_{\tau(0)}(x_{\tau(0)}) := 0$.

The 2-additive Choquet integral can be written also as follows [9]:

$$C_\mu(U(x)) = \sum_{i=1}^{n} v_i u_i(x_i) - \frac{1}{2} \sum_{\{i,j\} \subseteq N} I_{ij} |u_i(x_i) - u_j(x_j)| \tag{7}$$

where $v_i = \sum_{K \subseteq N \setminus i} \frac{(n - |K| - 1)!|K|!}{n!} (\mu(K \cup i) - \mu(K))$ is the importance of criterion i corresponding to the Shapley value of μ [17] and $I_{ij} = \mu_{ij} - \mu_i - \mu_j$ is the interaction index between the two criteria i and j [6,15].

2.2 Binary Actions and Relations

MCDA methods based on multiattribute utility theory, e.g, UTA [19], robust methods [1,5,11], require in practice a preferential information of the DM on a subset X_R of X because of the cardinality of X which can be very large. The set X_R is called reference subset and it is generally chosen by the DM. His choice may be guided by his knowledge about the problem addressed, his experience or his sensitivity to one or more particular alternatives, etc. This task is often difficult for the DM, especially when the alternatives are not known in advance, and sometimes his preferences on X_R are not sufficient to specify all the parameters of the model as interaction between criteria. For instance, in the problem of the design of a complex system for the protection of a strategic site [16], it is not easy for the DM to choose X_R himself because these systems are not known a priori. For these reasons, we suggest him to use as a reference subset a set of fictitious alternatives called binary actions defined below. We assume that the DM is able to identify for each criterion i two reference levels:

1. A reference level $\mathbf{1}_i$ in X_i which he considers as good and completely satisfying if he could obtain it on criterion i, even if more attractive elements could exist. This special element corresponds to the *satisficing level* in the theory of bounded rationality of Simon [18].
2. A reference level $\mathbf{0}_i$ in X_i which he considers neutral on i. The neutral level is the absence of attractiveness and repulsiveness. The existence of this neutral level has roots in psychology [20], and is used in bipolar models [21].

We set for convenience $u_i(\mathbf{1}_i) = 1$ and $u_i(\mathbf{0}_i) = 0$. Because the use of Choquet integral requires to ensure the commensurateness between criteria, the previous reference levels can be used in order to define the same scale on each criterion [10,12]. More details about these reference levels can be found in [8,9].

We call a *binary action or binary alternative*, an element of the set

$$\mathcal{B} = \{\mathbf{0}_N,\ (\mathbf{1}_i, \mathbf{0}_{N-i}),\ (\mathbf{1}_{ij}, \mathbf{0}_{N-ij}),\ i, j \in N, i \neq j\} \subseteq X$$

where

- $\mathbf{0}_N = (\mathbf{1}_\emptyset, \mathbf{0}_{N-\emptyset}) =: a_0$ is an action considered neutral on all criteria.
- $(\mathbf{1}_i, \mathbf{0}_{N-i}) =: a_i$ is an action considered satisfactory on criterion i and neutral on the other criteria.
- $(\mathbf{1}_{ij}, \mathbf{0}_{N-ij}) =: a_{ij}$ is an action considered satisfactory on criteria i and j and neutral on the other criteria.

Using the Choquet integral, we get the following consequences:

1. For any capacity μ,

$$C_\mu(U((\mathbf{1}_A, \mathbf{0}_{N-A}))) = \mu(A), \ \forall A \subseteq N. \tag{8}$$

2. Using Equation (2), we have for any 2-additive capacity μ:

$$C_\mu(U(a_0)) = 0 \tag{9}$$

$$C_\mu(U(a_i)) = \mu_i = v_i - \frac{1}{2} \sum_{k \in N, \ k \neq i} I_{ik} \tag{10}$$

$$C_\mu(U(a_{ij})) = \mu_{ij} = v_i + v_j - \frac{1}{2} \sum_{k \in N, \ k \notin \{i,j\}} (I_{ik} + I_{jk}) \tag{11}$$

With the arithmetic mean, we are able to compute the weights by using the reference subset $X_R = \{a_0, a_i, \forall i \in N\}$ (see MACBETH methodology [2]). For the 2-additive Choquet integral model, these alternatives are not sufficient to compute interaction between criteria, hence the elaboration of $\mathcal{B}$ by adding the alternatives a_{ij}. The Equations (10) and (11) show that the binary actions are directly related to the parameters of the 2-additive Choquet integral model. Therefore a preferential information on $\mathcal{B}$ given by the DM permits to determine entirely all the parameters of the model.

As shown by the previous equations (9),(10), (11) and Lemma 1, it should be sufficient to get some preferential information from the DM only on binary actions. To entirely determine the 2-additive capacity this information is expressed by the following relations:

- $P = \{(x, y) \in \mathcal{B} \times \mathcal{B} :$ DM strictly prefers x to $y\}$,
- $I = \{(x, y) \in \mathcal{B} \times \mathcal{B} :$ DM is indifferent between x and $y\}$.

The relation P is irreflexive and asymmetric while I is reflexive and symmetric. Here P does not contradict the classic dominance relation.

Definition 3. *The ordinal information on $\mathcal{B}$ is the structure $\{P, \ I\}$.*

These two relations are completed by adding the relation M which models the natural relations of monotonicity between binary actions coming from the monotonicity conditions $\mu(\{i\}) \geq 0$ and $\mu(\{i, j\}) \geq \mu(\{i\})$ for a capacity μ. For $(x, y) \in \{(a_i, a_0), i \in N\} \cup \{(a_{ij}, a_i), i, j \in N, i \neq j\}$,

$$x \ M \ y \text{ if } \mathrm{not}(x \ (P \cup I) \ y).$$

Example 1. *Mary wants to buy a digital camera for her next trip. To do this, she consults a website where she finds six propositions based on three criteria:*

resolution of the camera (expressed in million of pixels), price (expressed in euros) and zoom (expressed by a real number)

Cameras	1 : *Resolution*	2 : *Price*	3 : *Zoom*
a : *Nikon*	6	150	5
b : *Sony*	7	180	5
c : *Panasonic*	10	155	4
d : *Casio*	12	175	5
e : *Olympus*	10	160	3
f : *Kodak*	8	165	4

The criteria 1 and 3 have to be maximize while criterion 2 have to minimize.

Using our notations, we have $N = \{1, 2, 3\}$, $X_1 = [6, 12]$, $X_2 = [150, 180]$, $X_3 = [3, 5]$ *and* $X = X_1 \times X_2 \times X_3$.

Mary chooses for each criterion the following reference levels with some understanding of meaning in her mind.

	1 : *Resolution*	2 : *Price*	3 : *Zoom*
Satisfactory level	12	150	4
Neutral level	9	160	3.5

Based on these reference levels, the set of binary actions is $\mathcal{B} = \{a_0, a_1, a_2, a_3, a_{12}, a_{13}, a_{23}\}$, *where for instance the alternative* a_{12} *refers to a camera for which Mary is satisfied on resolution and price, but neutral on zoom. In order to make her choice, Mary gives also the following ordinal information:* $I = \{(a_{12}, a_3)\}$, $P = \{(a_{13}, a_1), (a_2, a_0)\}$. *Hence we have* $M = \{(a_1, a_0), (a_3, a_0), (a_{12}, a_1), (a_{12}, a_2), (a_{13}, a_3), (a_{23}, a_2), (a_{23}, a_3)\}$.

2.3 The Representation of Ordinal Information by the Choquet Integral

An ordinal information $\{P, I\}$ is said to be *representable by a 2-additive Choquet integral* if there exists a 2-additive capacity μ such that:

1. $\forall x, y \in \mathcal{B},\ x\ P\ y \Rightarrow C_\mu(U(x)) > C_\mu(U(y))$
2. $\forall x, y \in \mathcal{B},\ x\ I\ y \Rightarrow C_\mu(U(x)) = C_\mu(U(y))$.

A characterization of an ordinal information is given by Mayag et al. [13]. This result, presented below, is based on the following property called MOPI:

Definition 4. *[MOPI property]*

1. *For a binary relation* $\mathcal{R}$ *on* $\mathcal{B}$ *and* x, y *elements of* $\mathcal{B}$, $\{x_1, x_2, \cdots, x_p\} \subseteq \mathcal{B}$ *is a* path *of* $\mathcal{R}$ *from* x *to* y *if* $x = x_1\ \mathcal{R}\ x_2\ \mathcal{R} \cdots \mathcal{R}\ x_{p-1}\ \mathcal{R}\ x_p = y$. *A path of* $\mathcal{R}$ *from* x *to* x *is called a* cycle *of* $\mathcal{R}$.

- *We denote* $x\ TC\ y$ *if there exists a path of* $(P \cup I \cup M)$ *from* x *to* y.
- *A path* $\{x_1, x_2, ..., x_p\}$ *of* $(P \cup I \cup M)$ *is said to be a* strict path *from* x *to* y *if there exists* i *in* $\{1, ..., p-1\}$ *such that* $x_i\ P\ x_{i+1}$. *In this case, we will write* $x\ TC_P\ y$.
- *We write* $x \sim y$ *if there exists a nonstrict cycle of* $(P \cup I \cup M)$ *(hence a cycle of* $(I \cup M)$*) containing* x *and* y.

2. *Let* $i, j, k \in N$. *We call* Monotonicity of Preferential Information in $\{i, j, k\}$ w.r.t. i *the following property (denoted by* **(**$\{i,j,k\}$**,**i**)-MOPI***):*

$$\begin{cases} a_{ij} \sim a_i \\ a_{ik} \sim a_k \end{cases} \Rightarrow not(a_j\ TC_P\ a_0)$$
and
$$\begin{cases} a_{ij} \sim a_j \\ a_{ik} \sim a_k \end{cases} \Rightarrow not(a_i\ TC_P\ a_0)$$
and
$$\begin{cases} a_{ij} \sim a_j \\ a_{ik} \sim a_i \end{cases} \Rightarrow not(a_k\ TC_P\ a_0).$$

3. *We say that, the set* $\{i, j, k\}$ *satisfies the property of* MOnotonicity of Preferential Information *(MOPI) if* $\forall l \in \{i, j, k\}$, $(\{i, j, k\}, l)$*-MOPI is satisfied.*

Theorem 1. *An ordinal information* $\{P, I\}$ *is representable by a 2-additive Choquet integral on* $\mathcal{B}$ *if and only if the following two conditions are satisfied:*

1. $(P \cup I \cup M)$ *contains no strict cycle;*
2. *Any subset* K *of* N *such that* $|K| = 3$ *satisfies the MOPI property.*

Proof. See [13].

Using this characterization theorem, we deal with inconsistencies in the ordinal information [14]. But, the inconsistencies test of MOPI conditions requires to test them on all subsets of three criteria. Therefore, all the binary alternatives are used in the MOPI conditions test. If the number of elements of $\mathcal{B}$ is large ($n > 2$), it can be impossible to show to the DM a graph, where vertices are binary actions, for the explanation of inconsistencies. To solve this problem, we give an equivalent characterization of an ordinal information which concerns only the binary actions related the preferences $\{P, I\}$. This is done by extending the relation M to some couples (a_{ij}, a_0). Therefore, this new characterization theorem can be viewed as a reduction of complexity of inconsistencies test.

3 Reduction of the Complexity in the Inconsistencies Test of Ordinal Information

Let us consider the following sets:

$$\mathcal{B}' = \{a_0\} \cup \{x \in \mathcal{B} \mid \exists y \in \mathcal{B} \text{ such that } (x, y) \in (P \cup I) \text{ or } (y, x) \in (P \cup I)\}$$

$$M' = M \cup \{(a_{ij}, a_0) \mid a_{ij} \in \mathcal{B}', a_i \notin \mathcal{B}' \text{ et } a_j \notin \mathcal{B}'\}$$

$$(P \cup I \cup M')_{|\mathcal{B}'} = \{(x, y) \in \mathcal{B}' \times \mathcal{B}' \mid (x, y) \in (P \cup I \cup M')\}$$

The set $\mathcal{B}'$ is the set of all binary actions related to the preferential information of the DM. The relation on M' on $\mathcal{B}$ is an extension of the monotonicity relation on $\mathcal{B}$. The restriction of the relation $(P \cup I \cup M')$ to the set $\mathcal{B}'$ corresponds to $(P \cup I \cup M')_{|\mathcal{B}'}$.

The following result shows that, when it is possible to extend the monotonicity relation M to the set $\mathcal{B}'$, then the test of inconsistencies for the representation of ordinal information can be only limited to the elements of $\mathcal{B}'$.

Proposition 1. *Let be $\{P, I\}$ an ordinal information on $\mathcal{B}$.*

The ordinal information $\{P, I\}$ is representable by a 2-additive Choquet integral if and only if the following two conditions are satisfied:

1. *$(P \cup I \cup M')_{|\mathcal{B}'}$ contains no strict cycle;*
2. *Every subset K of N such that $|K| = 3$ satisfies the MOPI conditions restricted to $\mathcal{B}'$ (Only the elements of $\mathcal{B}'$ are concerned in this condition and paths considered in these conditions are paths of $(P \cup I \cup M')_{|\mathcal{B}'}$).*

Proof. See Section 3.1.

Example 2. $N = \{1, 2, 3, 4, 5, 6\}$, $P = \{(a_5, a_{12})\}$, $I = \{(a_3, a_5)\}$, $\mathcal{B} = \{a_0, a_1, a_2, a_3, a_4, a_5, a_6, a_{12}, a_{13}, a_{14}, a_{15}, a_{16}, a_{23}, a_{24}, a_{25}, a_{26}, a_{34}, a_{35}, a_{36}, a_{45}, a_{46}, a_{56}\}$.

According to our notations, we will have

$$\mathcal{B}' = \{a_0, a_{12}, a_3, a_5\},$$

$$M' = M \cup \{(a_{12}, a_0)\},$$

$$(P \cup I \cup M')_{|\mathcal{B}'} = \{(a_5, a_{12}), (a_3, a_5), (a_5, a_3), (a_3, a_0), (a_5, a_0), (a_{12}, a_0)\}$$

Hence, Proposition 1 shows that the inconsistencies test of the ordinal information $\{P, I\}$ will be limited on $\mathcal{B}'$ by satisfying the following conditions:

- *$(P \cup I \cup M')_{|\mathcal{B}'}$ contains no strict cycle;*
- *MOPI conditions written only by using elements of $\mathcal{B}'$ and paths of $(P \cup I \cup M')_{|\mathcal{B}'}$.*

3.1 Proof of Proposition 1

Let be $\{P, I\}$ an ordinal information on $\mathcal{B}$. In this section, for all elements $x, y \in \mathcal{B}$, we denote by:

1. $x\ TC'\ y$ a path of $(P \cup I \cup M')$ from x to y.
2. $x\ TC'_{|\mathcal{B}'}\ y$ a path of $(P \cup I \cup M')_{|\mathcal{B}'}$ from x to y i.e. a path from x to y containing only the elements of $\mathcal{B}'$.
3. $x \sim' y$ if one of the two following conditions happens:
   (a) $x = y$;
   (b) there is a non strict cycle $(P \cup I \cup M')$ containing x and y.
4. $x \sim'_{|\mathcal{B}'} y$ if one of the two following conditions happens:
   (a) $x = y$;
   (b) there is a non strict cycle of $(P \cup I \cup M')_{|\mathcal{B}'}$ containing x and y.

We will use the following lemmas in the proof of the result:

Lemma 2. *If $(x_1, x_2, \ldots, x_p)$ is a cycle of $(P \cup I \cup M)$, then every elements of $\mathcal{B}'$ of this cycle are contained in a cycle of $(P \cup I \cup M')_{|\mathcal{B}'}$.*

Proof. For all x_l, elements of the cycle $(x_1, x_2, \ldots, x_p)$ which are not in $\mathcal{B}'$, there exists necessarily $i, j \in N$ such that $a_{ij}\ M\ a_i\ M\ a_0$ (see Figure 1) where $x_{l-1} = a_{ij}$, $x_l = a_i$ and $x_{l+1} = a_0$ ($x_0 = x_p$ and $x_{p+1} = x_1$). Therefore, We can cancel the element a_i of the cycle because the elements a_{ij} and a_0 can be related as follows:

- if $a_j \notin \mathcal{B}'$, we will have $a_{ij}\ M'\ a_0$;
- if $a_j \in \mathcal{B}'$, we will have $a_{ij}\ (P \cup I \cup M)\ a_j\ (P \cup I \cup M)\ a_0$. This element a_j, which is not necessarily an element of the cycle $(x_1, x_2, \ldots, x_p)$, will be an element of the new cycle of $(P \cup I \cup M')_{|\mathcal{B}'}$.

The cycle of $(P \cup I \cup M')_{|\mathcal{B}'}$ obtained is then constituted by the elements of $(x_1, x_2, \ldots, x_p)$ belonging in $\mathcal{B}'$ and eventually the elements a_j coming from the cancelation of the elements a_i of $(x_1, x_2, \ldots, x_p)$ which are not in $\mathcal{B}'$.

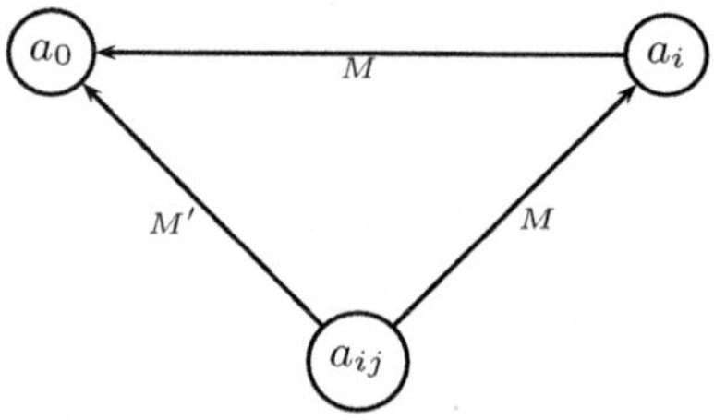

Fig. 1. Relation M' between a_{ij}, a_i and a_0

Lemma 3

1. *Let $i, j \in N$ such that $a_{ij} \sim a_i$. We have the following results:*
   (a) $a_{ij} \in \mathcal{B}'$;
   (b) If $a_i \notin \mathcal{B}'$ then $a_{ij} \sim'_{|\mathcal{B}'} a_0$;
   (c) If $a_i \in \mathcal{B}'$ then $a_{ij} \sim'_{|\mathcal{B}'} a_i$.
2. *Let $i, j \in N$ such that $a_{ij} \sim a_j$. We have the following results:*
   (a) $a_{ij} \in \mathcal{B}'$;
   (b) If $a_j \notin \mathcal{B}'$ then $a_{ij} \sim'_{|\mathcal{B}'} a_0$;
   (c) If $a_j \in \mathcal{B}'$ then $a_{ij} \sim'_{|\mathcal{B}'} a_j$.

Proof

1. If $a_{ij} \sim a_i$ then there exists $x \in \mathcal{B}$ such that $x\ (P \cup I \cup M)\ a_{ij}$. Using the definition of M, one may not have $x\ M\ a_{ij}$. Hence $a_{ij} \in \mathcal{B}'$ by the definition of $\mathcal{B}'$.
2. $a_{ij} \sim a_i \Rightarrow a_{ij}\ M\ a_i\ M\ a_0\ TC\ a_{ij}$ because $a_i \notin \mathcal{B}'$. Using Lemma 2, a_{ij} and a_0 are contained in a cycle of $(P \cup I \cup M')_{|\mathcal{B}'}$ i.e. $a_{ij} \sim'_{|\mathcal{B}'} a_0$.
3. Since a_{ij} and a_i are in $\mathcal{B}'$, then using Lemma 2, they are contained in a cycle of $(P \cup I \cup M')_{|\mathcal{B}'}$ i.e. $a_{ij} \sim'_{|\mathcal{B}'} a_i$.

The proof of the second point of the Lemma is similar to the previous one by replacing a_i by a_j.

Lemma 4. *If* $(P \cup I \cup M')_{|\mathcal{B}'}$ *contains no strict cycle then* $(P \cup I \cup M)$ *contains no strict cycle.*

Proof. Let $(x_1, x_2, \ldots, x_p)$ a strict cycle of $(P \cup I \cup M)$. Using Lemma 2, all the elements of $(x_1, x_2, \ldots, x_p)$ belonging to $\mathcal{B}'$ are contained in a cycle $\mathcal{C}$ de $(P \cup I \cup M')_{|\mathcal{B}'}$. Since $(x_1, x_2, \ldots, x_p)$ is a strict cycle of $(P \cup I \cup M)$, there exists $x_{i_o}, x_{i_o+1} \in \{x_1, x_2, \ldots, x_p\}$ such that $x_{i_o}\ P\ x_{i_o+1}$. Therefore $\mathcal{C}$ is a strict cycle of $(P \cup I \cup M')_{|\mathcal{B}'}$ because $x_{i_o}, x_{i_o+1} \in \mathcal{B}'$, a contradiction with the hypothesis.

Lemma 5. *Let* $x \in \mathcal{B}$. *If* $x\ TC_P\ a_0$ *then* $x \in \mathcal{B}'$ *and for each strict path* $(P \cup I \cup M)$ *from* x *to* a_0, *there exists a strict path of* $(P \cup I \cup M')_{|\mathcal{B}'}$ *from* x *to* a_0.

Proof. If $x \notin \mathcal{B}'$ then we can only have $x\ M\ a_0$. Therefore we will not have $x\ TC_P\ a_0$, a contradiction. Hence we have $x \in \mathcal{B}'$.

Let $x\ (P \cup I \cup M)\ x_1\ (P \cup I \cup M) \ldots x_p\ (P \cup I \cup M)\ a_0$ a strict path of $(P \cup I \cup M)$ from x to a_0. If there exists an element $y \notin \mathcal{B}'$ belonging to this path, then there necessarily exists $i, j \in N$ such that $y = a_i$ and $x\ TC_P\ a_{ij}\ M\ a_i\ M\ a_0$. So we can suppress the element y and have the path $x\ TC_P\ a_{ij}\ M'\ a_0$ if $a_j \notin \mathcal{B}'$ or the path $x\ TC_P\ a_{ij}\ (P \cup I \cup M)\ a_j\ (P \cup I \cup M)\ a_0$ if $a_j \in \mathcal{B}'$. If we suppress all the elements of $\mathcal{B} \setminus \mathcal{B}'$ like this, then we obtain a strict path of $(P \cup I \cup M')_{|\mathcal{B}'}$ containing only elements of $\mathcal{B}'$.

Lemma 6. *Let us suppose that* $(P \cup I \cup M')_{|\mathcal{B}'}$ *contains no strict cycle.*

1. *If we have* $\begin{cases} a_{ij} \sim a_i \\ a_{ik} \sim a_k \end{cases}$ *and* $(a_j\ TC_P\ a_0)$ *then* a_i, a_k *and* a_j *are the elements of* $\mathcal{B}'$.
2. *If we have* $\begin{cases} a_{ij} \sim a_j \\ a_{ik} \sim a_i \end{cases}$ *and* $(a_k\ TC_P\ a_0)$ *then* a_i, a_j *and* a_k *are the elements* $\mathcal{B}'$.
3. *If we have* $\begin{cases} a_{ij} \sim a_j \\ a_{ik} \sim a_k \end{cases}$ *and* $(a_i\ TC_P\ a_0)$ *then* a_j, a_k *and* a_i *are the elements of* $\mathcal{B}'$.

Proof

1. a_j is an element of $\mathcal{B}'$ using Lemma 5.
   - If $a_i \notin \mathcal{B}'$ then using Lemma 3 we have $a_{ij}\ \sim'_{|\mathcal{B}'}\ a_0$. Since $a_j\ TC_P\ a_0$, then using Lemma 5, we have $a_j\ TC'_{P|\mathcal{B}'}\ a_0$ a strict path from a_j to a_0. Hence, we will have $a_0\ \sim'_{|\mathcal{B}'}\ a_{ij}\ (P \cup I \cup M)\ a_j\ TC'_{P|\mathcal{B}'}\ a_0$. Therefore we obtain un strict cycle of $(P \cup I \cup M')_{|\mathcal{B}'}$, which is a contradiction with the hypothesis. Hence $a_i \in \mathcal{B}'$
   - If $a_k \notin \mathcal{B}'$ then using Lemma 3, $a_{ik}\ \sim'_{|\mathcal{B}'}\ a_0$. Therefore, since $a_i \in \mathcal{B}'$ (using the previous point), we will have the following cycle $(P \cup I \cup M')_{|\mathcal{B}'}$ of

$a_0 \sim'_{|\mathcal{B}'} a_{ik}\ M\ a_i\ TC'_{|\mathcal{B}'}\ a_{ij}\ (P \cup I \cup M)\ a_j\ TC'_{P|_{\mathcal{B}'}}\ a_0$.
This cycle is strict because $a_j\ TC'_{P|_{\mathcal{B}'}}\ a_0$ is a strict path from a_j to a_0 using Lemma 5, a contradiction. Hence $a_k \in \mathcal{B}'$.

2. The proof of the two last points is similar to the first point.

Proof of the Proposition 1: Il is obvious that if $\{P, I\}$ is representable by a 2-additive Choquet integral then the two following conditions are satisfied:

- $(P \cup I \cup M')_{|\mathcal{B}'}$ contains no strict cycle;
- Every subset K of N such that $|K| = 3$ satisfies the MOPI conditions reduced to $\mathcal{B}'$ (Only the elements of $\mathcal{B}'$ are concerned in this condition).

The converse of the proposition is a consequence of Lemmas 4 and 6.

References

1. Angilella, S., Greco, S., Matarazzo, B.: Non-additive robust ordinal regression: A multiple criteria decision model based on the Choquet integral. European Journal of Operational Research 41(1), 277–288 (2009)
2. Bana e Costa, C.A., De Corte, J.-M., Vansnick, J.-C.: On the mathematical foundations of MACBETH. In: Figueira, J., Greco, S., Ehrgott, M. (eds.) Multiple Criteria Decision Analysis: State of the Art Surveys, pp. 409–437. Springer, Heidelberg (2005)
3. Chateauneuf, A., Jaffray, J.Y.: Some characterizations of lower probabilities and other monotone capacities through the use of Möbius inversion. Mathematical Social Sciences 17, 263–283 (1989)
4. Clivillé, V., Berrah, L., Mauris, G.: Quantitative expression and aggregation of performance measurements based on the MACBETH multi-criteria method. International Journal of Production economics 105, 171–189 (2007)
5. Figueira, J.R., Greco, S., Slowinski, R.: Building a set of additive value functions representing a reference preorder and intensities of preference: Grip method. European Journal of Operational Research 195(2), 460–486 (2009)
6. Grabisch, M.: k-order additive discrete fuzzy measures and their representation. Fuzzy Sets and Systems 92, 167–189 (1997)
7. Grabisch, M.: The Möbius transform on symmetric ordered structures and its application to capacities on finite sets. Discrete Mathematics 287(1-3), 17–34 (2004)
8. Grabisch, M., Labreuche, C.: Fuzzy measures and integrals in MCDA. In: Figueira, J., Greco, S., Ehrgott, M. (eds.) Multiple Criteria Decision Analysis: State of the Art Surveys, pp. 565–608. Springer, Heidelberg (2005)
9. Grabisch, M., Labreuche, C.: A decade of application of the Choquet and Sugeno integrals in multi-criteria decision aid. 4OR 6, 1–44 (2008)
10. Grabisch, M., Labreuche, C., Vansnick, J.-C.: On the extension of pseudo-Boolean functions for the aggregation of interacting bipolar criteria. Eur. J. of Operational Research 148, 28–47 (2003)
11. Greco, S., Mousseau, V., Slowinski, R.: Ordinal regression revisited: Multiple criteria ranking using a set of additive value functions. European Journal of Operational Research 51(2), 416–436 (2008)
12. Labreuche, C., Grabisch, M.: The Choquet integral for the aggregation of interval scales in multicriteria decision making. Fuzzy Sets and Systems 137, 11–26 (2003)

13. Mayag, B., Grabisch, M., Labreuche, C.: A representation of preferences by the Choquet integral with respect to a 2-additive capacity. Theory and Decision, forthcoming, `http://www.springerlink.com/content/313t22t08v722h82/`, doi:10.1007/s11238-010-9198-3
14. Mayag, B.: Elaboration d'une démarche constructive prenant en compte les interactions entre critères en aide multicritère à la décision. PhD thesis, University of Paris 1 Panthéon-Sorbonne, Paris (2010), `http://sites.google.com/site/bricemayag/about-my-phd`
15. Murofushi, T., Soneda, S.: Techniques for reading fuzzy measures (III): interaction index. In: 9th Fuzzy System Symposium, Japan, pp. 693–696 (May 1993) (in Japanese)
16. Pignon, J.P., Labreuche, C.: A methodological approach for operational and technical experimentation based evaluation of systems of systems architectures. In: Int. Conference on Software & Systems Engineering and their Applications (ICSSEA), Paris, France (December 4-6, 2007)
17. Shapley, L.S.: A value for n-person games. In: Kuhn, H.W., Tucker, A.W. (eds.) Contributions to the Theory of Games. Annals of Mathematics Studies, vol. II(28), pp. 307–317. Princeton University Press, Princeton (1953)
18. Simon, H.: Rational choice and the structure of the environment. Psychological Review 63(2), 129–138 (1956)
19. Siskos, Y., Grigoroudis, E., Matsatsinis, N.F.: Uta methods. In: Figueira, J., Greco, S., Ehrgott, M. (eds.) Multiple Criteria Decision Analysis: State of the Art Surveys, pp. 297–343. Springer, Heidelberg (2005)
20. Slovic, P., Finucane, M., Peters, E., MacGregor, D.G.: The affect heuristic. In: Gilovitch, T., Griffin, D., Kahneman, D. (eds.) Heuristics and Biases: The Psychology of Intuitive Judgment, pp. 397–420. Cambridge University Press, Cambridge (2002)
21. Tversky, A., Kahneman, D.: Advances in prospect theory: cumulative representation of uncertainty. J. of Risk and Uncertainty 5, 297–323 (1992)

On Minimizing Ordered Weighted Regrets in Multiobjective Markov Decision Processes

Wlodzimierz Ogryczak[1], Patrice Perny[2], and Paul Weng[2]

[1] ICCE, Warsaw University of Technology, Warsaw, Poland
wogrycza@elka.pw.edu.pl
[2] LIP6 - UPMC, Paris, France
{patrice.perny,paul.weng}@lip6.fr

Abstract. In this paper, we propose an exact solution method to generate fair policies in Multiobjective Markov Decision Processes (MMDPs). MMDPs consider n immediate reward functions, representing either individual payoffs in a multiagent problem or rewards with respect to different objectives. In this context, we focus on the determination of a policy that fairly shares regrets among agents or objectives, the regret being defined on each dimension as the opportunity loss with respect to optimal expected rewards. To this end, we propose to minimize the ordered weighted average of regrets (OWR). The OWR criterion indeed extends the minimax regret, relaxing egalitarianism for a milder notion of fairness. After showing that OWR-optimality is state-dependent and that the Bellman principle does not hold for OWR-optimal policies, we propose a linear programming reformulation of the problem. We also provide experimental results showing the efficiency of our approach.

Keywords: Ordered Weighted Regret, Fair Optimization, Multiobjective MDP.

1 Introduction

Markov Decision Process (MDP) is a standard model for planning problems under uncertainty [15,10]. This model admits various extensions developed to address different questions that emerge in applications of Operations Research and Artificial Intelligence, depending on the structure of state space, the definition of actions, the representation of uncertainty, and the definition of preferences over policies. We consider here the latter point. In the standard model, preferences over actions are represented by immediate rewards represented by scalar numbers. The value of a sequence of actions is defined as the sum of these rewards and the value of a policy as the expected discounted reward. However, there are various contexts in which the value of a sequence of actions is defined using several reward functions. It is the case in multiagent planning problems [2,7] where every agent has its own value system and its own reward function. It is also the case of multiobjective problems [1,13,3], for example path-planning problems under uncertainty when one wishes to minimize length, time, energy consumption

R.I. Brafman, F. Roberts, and A. Tsoukiàs (Eds.): ADT 2011, LNAI 6992, pp. 190–204, 2011.

and risk simultaneously. In all these problems, n distinct reward functions need to be considered. In general, they cannot be reduced to a single reward function even if each of them is additive over sequences of actions, and even if the value of a policy can be synthesized into a scalar overall utility through an aggregation function (except for linear aggregation). This is why we need to develop specific approaches to determine compromise solutions in Multiobjective or Multiagent MDPs.

Many studies on Multiobjective MDPs (MMDP) concentrate on the determination of the entire set of Pareto-optimal solutions, i.e., policies having a reward vector that cannot be improved on a component without being downgraded on another one. However, the size of the Pareto set is often very large due to the combinatorial nature of the set of deterministic policies, its determination induces prohibitive response times and requires very important memory space as the number of states and/or criteria increases. Fortunately, there is generally no need to determine the entire set of Pareto-optimal policies, but only specific compromise policies achieving a well-balanced tradeoff between criteria or equivalently, in a multiagent context, policies that fairly shares expected rewards among agents. Motivated by such examples, we study in this paper the determination of fair policies in MMDPs. To this end, we propose to minimize the ordered weighted average of regrets (OWR). The OWR criterion indeed extends the minimax regret, relaxing egalitarianism on regrets for a milder notion of fairness.

The paper is organized as follows: In Section 2, we recall the basic notions related to Markov decision processes and their multiobjective extension. In Section 3, we discuss the choice of a scalarizing function to generate fair solutions. This leads us to adopt the ordered weighted regret criterion (OWR) as a proper scalarizing function to be minimized. Section 4 is devoted to the search of OWR-optimal policies. Finally, Section 5 presents some experimental results showing the effectiveness of our approach for finding fair policies.

2 Background

A *Markov Decision Process* (MDP) [15] is described as a tuple (S, A, T, R) where S is a finite set of states, A is a finite set of actions, transition function $T(s, a, s')$ gives the probability of reaching state s' by executing action a in state s, reward function $R(s, a) \in \mathbb{R}$ gives the immediate reward obtained for executing action a in state s.

In this context, a *decision rule* δ is a procedure that determines which action to choose in each state. A decision rule can be *deterministic*, i.e., defined as $\delta : S \to A$, or more generally, *randomized*, i.e., defined as $\delta : S \to \mathbf{Pr}(A)$ where $\mathbf{Pr}(A)$ is the set of probability distributions over A.

A *policy* π is a sequence of decision rules $(\delta_0, \delta_1, \ldots, \delta_t, \ldots)$ that indicates which decision rule to apply at each step. It is said to be *deterministic* if each decision rule is deterministic and *randomized* otherwise. If the same decision rule δ is applied at each step, the policy is said *stationary* and is denoted δ^∞.

The value of a policy π is defined by a function $v^\pi : S \to \mathbb{R}$, called *value function*, which gives the expected discounted total reward yielded by applying π from each initial state. For $\pi = (\delta_0, \delta_1, \ldots, \delta_t, \ldots)$, they are given $\forall h > 0$ by:

$$\begin{aligned} v_0^\pi(s) &= 0 && \forall s \in S \\ v_t^\pi(s) &= R(s, \delta_{h-t}(s)) + \gamma \sum_{s' \in S} T(s, \delta_{h-t}(s), s') v_{t-1}^\pi(s') && \forall s \in S, \forall t = 1, \ldots, h \end{aligned}$$

where $\gamma \in [0, 1[$ is the discount factor. This sequence converges to the value function of π.

In this framework, there exists an optimal stationary policy that yields the best expected discounted total reward in each state. Solving an MDP amounts to finding one of those policies and its associated value function. The optimal value function $v^* : S \to \mathbb{R}$ can be determined by solving the *Bellman equations*:

$$\forall s \in S, \quad v^*(s) = \max_{a \in A} R(s, a) + \gamma \sum_{s' \in S} T(s, a, s') v^*(s')$$

There are three main approaches for solving MDPs. Two are based on dynamic programming: value iteration and policy iteration. The third is based on linear programming. We recall the last approach as it is needed for the exposition of our results. The linear program $(\mathcal{P})$ for solving MDPs can be written as follows:

$$(\mathcal{P}) \begin{cases} \min \sum_{s \in S} \mu(s) v(s) \\ \text{s.t. } v(s) - \gamma \sum_{s' \in S} T(s, a, s') v(s') \geq R(s, a) \quad \forall s \in S, \forall a \in A \end{cases}$$

where weights μ could be interpreted as the probability of starting in a given state. Any positive μ can in fact be chosen to determine the optimal value function. Program $\mathcal{P}$ is based on the idea that the Bellman equations imply that functions satisfying the constraints of $\mathcal{P}$ are upper bounds of the optimal value function. Writing the dual $(\mathcal{D})$ of this program is interesting as it uncovers the dynamic of the system:

$$(\mathcal{D}) \begin{cases} \max \sum_{s \in S} \sum_{a \in A} R(s, a) \, x_{sa} \\ \left.\begin{aligned} &\text{s.t. } \sum_{a \in A} x_{sa} - \gamma \sum_{s' \in S} \sum_{a \in A} T(s', a, s) \, x_{s'a} = \mu(s) \quad \forall s \in S \\ &x_{sa} \geq 0 \quad \forall s \in S, \forall a \in A \end{aligned}\right\} (\mathcal{C}) \end{cases}$$

To interpret variables x_{sa}, we recall the following two propositions relating feasible solutions of $\mathcal{D}$ to stationary randomized policies in the MDP [15].

Proposition 1. *For a policy π, if x^π is defined as $x^\pi(s, a) = \sum_{t=0}^{\infty} \gamma^t p_t^\pi(s, a)$, $\forall s \in S, \forall a \in A$ where $p_t^\pi(s, a)$ is the probability of reaching state s and choosing a at step t, then x^π is a feasible solution of $\mathcal{D}$.*

Proposition 2. *If x_{sa} is a solution of $\mathcal{D}$, then the stationary randomized policy δ^∞, defined by $\delta(s,a) = x_{sa}/\sum_{a' \in A} x_{sa'}, \forall s \in S, \forall a \in A$ defines $x^{\delta^\infty}(s,a)$ as in Proposition 1, that are equal to x_{sa}.*

Thus, the set of randomized policies is completely characterized by constraints $(\mathcal{C})$. Besides, the basic solutions of $\mathcal{D}$ correspond to deterministic policies. Moreover, the basic solutions of $\mathcal{P}$ correspond to the value functions of deterministic policies. Those of randomized policies are in the convex hull of those basic solutions. Note that in an MDP, any feasible value function can be obtained with a randomized policy.

Multiobjective MDP. MDPs have been extended to take into account multiple dimensions or criteria. A *multiobjective MDP* (MMDP) is an MDP where the reward function is redefined as: $R: S \times A \to \mathbb{R}^n$ where n is the number of objectives, $R(s,a) = (R_1(s,a), \ldots, R_n(s,a))$ and $R_i(s,a)$ is the immediate reward for objective $i \subset O = \{1, \ldots, n\}$.

Now, a policy π is valued by a value function $V^\pi : S \to \mathbb{R}^n$, which gives the expected discounted total reward vector in each state. To compare the value of policies in a given state s, the basic model adopted in most previous studies [5,17,18] is *Pareto dominance* defined as follows:

$$\forall x, y \in \mathbb{R}^n, x \succ_P y \text{ iff } [x \neq y \text{ and } \forall i \in O, x_i \geq y_i] \tag{1}$$

Hence, for any two policies π, π', π is preferred to π' in a state s if and only if $V^\pi(s) \succ_P V^{\pi'}(s)$. For a set $X \subset \mathbb{R}^n$, a vector $x \in X$ is said to be *Pareto-optimal* in X if there is no $y \in X$ such that $y \succ_P x$. Due to the incompleteness of Pareto dominance, there may exist several Pareto-optimal vectors in a given state.

Standard methods for MDPs can be extended to solve MMDPs [18,17]. As shown by Viswanathan et al. [17], the dual linear program $(\mathcal{D})$ can be extended to a multiobjective linear program for finding Pareto-optimal solutions in a MMDP since the dynamics of a MDP and that of a MMDP are identical. Thus, we obtain the following multiobjective linear program $v\mathcal{D}$:

$$(v\mathcal{D}) \begin{cases} \max f_i(x) = \sum_{s \in S} \sum_{a \in A} R_i(s,a)\, x_{sa} & \forall i = 1, \ldots, n \\ \text{s.t. } (\mathcal{C}) & \end{cases}$$

Looking for all Pareto-optimal solutions can be difficult and time-consuming as there are instances of problems where the number of Pareto-optimal value functions of deterministic policies is exponential in the number of states [8].

Besides, in practice, one is generally only interested in specific compromise solutions among Pareto-optimal solutions achieving interesting tradeoffs between objectives. To this end, one could try to optimize one of the objectives subject to constraints over the other objectives (see for instance [1]). However, this approach reveals to be cumbersome to reach well-balanced tradeoffs, as the number of objectives grows. A more natural approach for that could be to use a *scalarizing function* $\psi : \mathbb{R}^n \to \mathbb{R}$, monotonic with respect to Pareto dominance, that defines the value v^π of a policy π in a state s by: $v^\pi(s) = \psi(V_1^\pi(s), \ldots, V_n^\pi(s))$. The problem can then be reformulated as the search for a policy π optimizing $v^\pi(s)$ in an initial state s. We discuss now about a proper choice of ψ in order to achieve a fair satisfaction of objectives.

3 Fair Regret Optimization

Weighted Sum. The most straightforward choice for ψ seems to be *weighted sum* (WS), i.e., $\forall y \in \mathbb{R}^n$, $\psi(y) = \lambda \cdot y$ where $\lambda \in \mathbb{R}^n_+$. By linearity of WS and that of mathematical expectation, optimizing v is equivalent to solving the standard MDP obtained from the MMDP where the reward function is defined as: $r(s, a) = \lambda \cdot R(s, a), \forall s, a$. In that case, an optimal stationary deterministic policy exists and standard solution methods can then be applied. However, using WS is not a good procedure for reaching balanced solutions as weighted sum is a fully compensatory operator. For example, with WS, $(5, 5)$ would never be stricly preferred to $(10, 0)$ and $(0, 10)$ simultaneously, whatever the weights.

MaxMin. In opposition to the previous utilitarist approach, we could adopt egalitarianism that consists in maximizing the value of the least satisfied objective ($\psi = \min$). This approach obviously includes an idea of fairness as for example, here, $(5, 5)$ is strictly preferred to both $(10, 0)$ and $(0, 10)$. However, it has two significant drawbacks: (i) min does not take into account the potentialities of each objective with respect to the maximum values that each objective can achieve. For instance, if objective 1 can reach a maximum of 10 while objective 2 can reach a maximum of 6, a solution leading to $(6, 6)$ might be seemed less fair than another valued by $(8, 4)$ since the second better distributes the opportunity losses; (ii) reducing a vector to its worst component is too pessimistic and creates drowning effects, i.e., $(1, 0)$ is seen as equivalent to $(10, 0)$, whereas the latter Pareto-dominates the former.

Minmax Regret. A standard answer to (i) is to consider *Minmax regret* (MMR), which is defined as follows. Let Y be a set of valuation vectors in $\mathbb{R}^n$ and $I \in \mathbb{R}^n$ denote the ideal point defined by $I_i = \sup_{y \in Y} y_i$ for all $i \in O$. The regret of choosing $y \in Y$ according to objective i is defined by $\eta_i = I_i - y_i$. Then, MMR is defined for all $y \in Y$ by $\psi(y) = \max_{i \in O}(\eta_i)$. However, MMR does not address issue (ii). In order to guarantee the Pareto monotonicity, MMR may be further generalized to take into account all the regret values according to the Ordered Weighted Average (OWA) aggregation [19], thus using the following scalarizing function [20]:

$$\rho_w(y) = \sum_{i \in O} w_i \eta_{\langle i \rangle} \tag{2}$$

where $(\eta_{\langle 1 \rangle}, \eta_{\langle 2 \rangle}, \ldots, \eta_{\langle n \rangle})$ denotes the vector obtained from the regret vector η by rearranging its components in the non-increasing order (i.e., $\eta_{\langle 1 \rangle} \geq \eta_{\langle 2 \rangle} \geq \ldots \geq \eta_{\langle n \rangle}$ and there exists a permutation τ of set O such that $\eta_{\langle i \rangle} = \eta_{\tau(i)}$ for $i \in O$) and weights w_i are non-negative and normalized to meet $\sum_{i \in O} w_i = 1$.

Example 1. *We illustrate how* ρ_w *is computed (see Table 1) with ideal point* $I = (9, 7, 6)$ *and weights* $w = (1/2, 1/3, 1/6)$*. One first computes the regrets* η*, then reorders them. Finally,* ρ_w *can be computed, inducing the preference order* $x \succ z \succ y$*.*

Table 1. Example of computation of ρ_w

	1 2 3	η_1	η_2	η_3	$\eta_{\langle 1\rangle}$	$\eta_{\langle 2\rangle}$	$\eta_{\langle 3\rangle}$	ρ_w
x	8 4 5	1	3	1	3	1	1	12/6
y	9 2 6	0	5	0	5	0	0	15/6
z	6 7 4	3	0	2	3	2	0	13/6

Note that ρ_w is a symmetric function of regrets. Indeed, weights w_i's are assigned to the specific positions within the ordered regret vector rather than to the individual regrets themselves. These rank-dependent weights allow to control the importance attached to small or large regrets. For example, if $w_1 = 1$ and $w_2 = \ldots = w_n = 0$, one can recognize the standard MMR, which focuses on the worst regret.

Augmented Tchebycheff norm. This criterion, classically used in multiobjective optimization [16], is defined by $\psi(y) = \max_{i \in O} \eta_i + \epsilon \sum_{i \in O} \eta_i$ where ϵ is a small positive real. It addresses issues (i) and (ii). However, it has some drawbacks as soon as $n \geq 3$. Indeed, when several vectors have the same max regret, then they are discriminated with a weighted sum, which does not provide any control on fairness.

Ordered Weighted Regret. In order to convey an idea of fairness, we now consider the subclass of scalarizing functions defined by Equation (2) with the additional constraints: $w_1 > \ldots > w_n > 0$. Any function in this subclass is named *Ordered Weighted Regret* (OWR) in the sequel. This additional constraint on weights can easily be explained by the following two propositions:

Proposition 3. $[\forall y, z \in \mathbb{R}^n, y \succ_P z \Rightarrow \rho_w(y) < \rho_w(z)] \Leftrightarrow \forall i \in O, w_i > 0$

Proposition 4. $\big[\forall y \in \mathbb{R}^n, \forall i, k \in O, \forall \varepsilon,$ *s.t.* $0 < \varepsilon < \eta_k - \eta_i,$
$\rho_w(y_1, \ldots, y_i - \varepsilon, \ldots, y_k + \varepsilon, \ldots, y_n) < \rho_w(y_1, y_2, \ldots, y_n)\big] \Leftrightarrow w_1 > \ldots > w_n > 0.$

Proposition 3 states that OWR is Pareto-monotonic. It follows from monotonicity of the OWA aggregation [11]. Consequently, OWR-optimal solutions are Pareto-optimal. Proposition 4 is the Schur-convexity of ρ_w, a key property in inequality measurement [12], and it follows from the Schur-convexity of the OWA aggregation with monotonic weights [9]. In MMDPs, it says that a reward transfer reducing regret inequality, i.e., a transfer of any small reward from an objective to any other objective whose regret is greater, results in a preferred valuation vector (a smaller OWR value). For example, if $w = (3/5, 2/5)$ and $I = (10, 10)$, $\rho_w(5, 5) = 5$ whereas $\rho_w(10, 0) = \rho_w(0, 10) = 6$, which means that $(5, 5)$ is preferred to the two others. Due to Proposition 4, if x is an OWR-optimal solution, x cannot be improved by any reward transfer reducing regret inequality, thus ensuring the fairness of OWR-optimal solutions.

Due to Propositions 3 and 4, minimizing OWR leads to a Pareto-optimal solution that fairly distributes regrets over the objectives (see the left part of Figure 1). Moreover, whenever the objectives (criteria or agents) do not have the same importance, it is possible to break the symmetry of OWR by introducing scaling factors $\lambda_i > 0, \forall i \in O$ in Equation (2) so as to deliberately deliver biased (Pareto-optimal) compromise solutions (see the right part of Figure 1). To this end, we generalize OWR by considering:

$$\rho_w^\lambda(y) = \sum_{i \in O} w_i \eta_{\langle i \rangle}^\lambda \quad \text{with} \quad \eta_i^\lambda = \lambda_i (I_i - y_i) \quad \forall\, i \in O \tag{3}$$

where $\lambda = (\lambda_1, \ldots, \lambda_n)$ and $(\eta_{\langle 1 \rangle}^\lambda, \eta_{\langle 2 \rangle}^\lambda, \ldots, \eta_{\langle n \rangle}^\lambda)$ denotes the vector obtained from the scaled regret vector η^λ by rearranging its components in the non-increasing order. For the sake of simplicity, ρ_w^λ is also called an OWR.

Fig. 1. *Fair* (left) and *biased* (right) compromises

Using OWR, a policy π is weakly preferred to a policy π' in a state s (denoted $\pi \succsim_s \pi'$) iff $\rho_w^\lambda(V^\pi(s)) \le \rho_w^\lambda(V^{\pi'}(s))$. Hence, an optimal policy π^* in s can be found by solving:

$$v^{\pi^*}(s) = \min_\pi \rho_w^\lambda(V^\pi(s)). \tag{4}$$

As a side note, ρ_w^λ can be used to explore interactively the set of Pareto solutions by solving problem (4) for various scaling factors λ_i and a proper choice of OWR weights w_i. Indeed, we have:

Proposition 5. *For any polyhedral compact feasible set $F \subset \mathbb{R}^n$, for any feasible Pareto-optimal vector $\bar{y} \in F$ such that $\bar{y}_i < I_i, \forall i \in O$, there exist weights $w_1 > \ldots > w_n > 0$, and scaling factors $\lambda_i > 0, \forall i \in O$ such that $\bar{y}$ is a ρ_w^λ-optimal solution.*

Proof. Let $\bar{y} \in F$ be a feasible Pareto-optimal vector such that $\bar{y}_i < I_i, \forall i \in O$. Since, F is a polyhedral compact feasible set, there exists $\Delta > 0$ such that for any feasible vector $y \in F$ the implication

$$y_i > \bar{y}_i \text{ and } y_k < \bar{y}_k \Rightarrow (y_i - \bar{y}_i)/(\bar{y}_k - y_k) \le \Delta \tag{5}$$

is valid for any $i, k \in O$ [6].

Let us set the scaling factors $\lambda_i = 1/(I_i - \bar{y}_i), \forall i \in O$ and define weights $w_1 > \ldots > w_n > 0$ such that $w_1 \geq L\Delta\sum_{i=1}^{n} w_i$, where $L \geq \lambda_i/\lambda_k$ for any $i, k \in O$. We will show that $\bar{y}$ is a ρ_w^λ-optimal solution.

Suppose there exists a feasible vector $y \in F$ with better OWR value $\rho_w^\lambda(\bar{y}) = \sum_{i\in O} w_i \bar{\eta}_{\langle i\rangle}^\lambda < \sum_{i\in O} w_i \eta_{\langle i\rangle}^\lambda = \rho_w^\lambda(y)$. Note that $\bar{\eta}_i^\lambda = \lambda_i(I_i - \bar{y}_i) = 1$ for all $i \in O$. Hence, $\eta_{\langle i\rangle}^\lambda - \bar{\eta}_{\langle i\rangle}^\lambda = \eta_{\tau(i)}^\lambda - \bar{\eta}_{\tau(i)}^\lambda$ for all $i \in O$ where τ is the ordering permutation for the regret vector η^λ with $\eta_i^\lambda = \lambda_i(I_i - y_i) = 1$ for $i \in O$. Moreover, $\bar{\eta}_{\tau(i)}^\lambda - \eta_{\tau(i)}^\lambda = \lambda_{\tau(i)}(y_{\tau(i)} - \bar{y}_{\tau(i)})$ and, due to Pareto-optimality of $\bar{y}$, $0 > \bar{\eta}_{\tau(1)}^\lambda - \eta_{\tau(1)}^\lambda = \lambda_{\tau(1)}(y_{\tau(1)} - \bar{y}_{\tau(1)})$. Thus, taking advantages of inequalities (5) for $k = \tau(1)$ one gets

$$\sum_{i=2}^{m} w_i \lambda_{\tau(i)}(y_{\tau(i)} - \bar{y}_{\tau(i)}) \leq -\sum_{i=2}^{m} w_i L\Delta\lambda_{\tau(1)}(y_{\tau(1)} - \bar{y}_{\tau(1)}) \leq -w_1\lambda_{\tau 1}(y_{\tau(1)} - \bar{y}_{\tau(1)})$$

which contradicts to the inequality $\sum_{i\in O} w_i \bar{\eta}_{\langle i\rangle}^\lambda < \sum_{i\subset O} w_i \eta_{\langle i\rangle}^\lambda$ and thereby it confirms ρ_w^λ-optimality of $\bar{y}$. ■

Note that the condition $\bar{y}_i < I_i, \forall i \in O$ is not restrictive in practice: one can replace I_i by $I_i + \epsilon$ for any arbitrary small positive ϵ to extend the result to any $\bar{y}$ in F.

4 Solution Method

We now address the problem of solving problem (4). First, remark that, for all scalarizing functions considered in the previous section (apart from WS), finding an optimal policy in an MMDP cannot be achieved by aggregating first the immediate vectorial rewards and solving the resulting MDP. Optimizing OWR implies some subtleties that we present now.

Randomized Policies. When optimizing OWR, searching for a solution among the set of stationary deterministic policies may be suboptimal. Let us illustrate this point on an example where $n = 2$. Assume that points on Figure 2 represent the value of deterministic policies in a given state. The Pareto-optimal solutions are then a, b, c and d. If we were searching for a fair policy, we could consider c as a good candidate solution. However, by considering also randomized policies, we could obtain an even better solution. Indeed, the valuation vectors of randomized policies are in the convex hull of the valuation vectors of deterministic policies, represented by the light-greyed zone (Figure 3). The dotted lines linking points a, b and d represent all Pareto-optimal valuation vectors. The dark greyed zone represents all feasible valuation vectors that are preferred to point c. Those vectors that are Pareto-optimal seem to be good candidate solutions. Therefore, we will not restrict ourselves to deterministic policies and we will consider any feasible randomized policy.

OWR-Optimality is State-Dependent. Contrary to standard MDPs where optimal policies are optimal in every initial state, the optimality notion based on

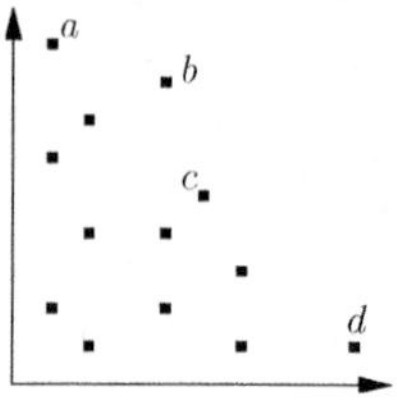

Fig. 2. Valuation vectors

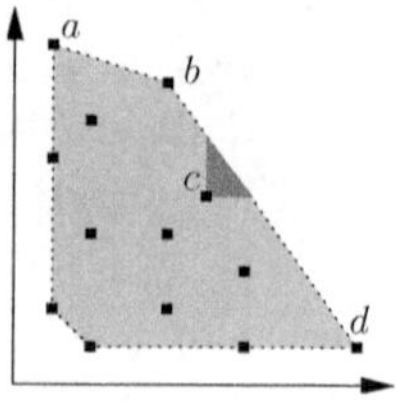

Fig. 3. Better solutions

OWR depends on the initial state, i.e., an OWR-optimal policy in a given initial state may not be an OWR-optimal solution in another state.

Example 2. *Consider the deterministic MMDP represented on Figure 4 with two states* ($S = \{1, 2\}$) *and two actions* ($A = \{a, b\}$)*. The vectorial rewards can be read on Figure 4.*

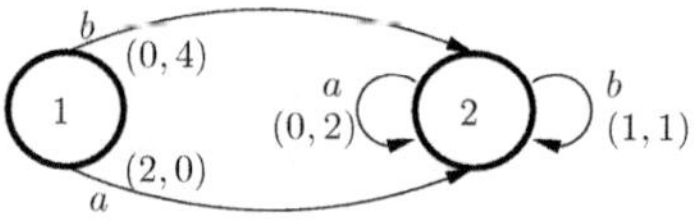

Fig. 4. Representation of the MMDP

Set $\gamma = 0.5$, $w = (0.9, 0.1)$ *and* $\lambda = (1, 1)$*. The ideal point from state* 1 *is* $I_1 = (3, 6)$*. Reward* 3 *is obtained by first choosing* a *in state* 1 *and then repeatedly* b *in state* 2 *while reward* 6 *is obtained by first choosing* b *in state* 1 *and then repeatedly* a *in state* 2*. By similar computations, the ideal point from state* 2 *is* $I_2 = (2, 4)$*. There are four stationary deterministic policies, denoted* δ_{xy}*, which consists in choosing action* x *in state* 1 *and action* y *in state* 2*.*

The OWR-optimal policies in state 2 *are* δ^∞_{aa} *and* δ^∞_{ba} *with the same value in state* 2*:* $V^{\delta^\infty_{aa}}(2) = V^{\delta^\infty_{ba}}(2) = (0, 4)$ *(OWR of* 1.8 *with* I_2*). One can indeed check that no randomized policy can improve this score. However, none of these policies are optimal in state* 1 *as they are beaten by* δ^∞_{bb}*. Indeed,* $V^{\delta^\infty_{bb}}(1) = (1, 5)$ *(OWR of* 1.9 *with* I_1*) whereas* $V^{\delta^\infty_{aa}}(1) = (2, 2)$ *(OWR of* 3.7 *with* I_1*) and* $V^{\delta^\infty_{ba}}(1) = (0, 6)$ *(OWR of* 2.7 *with* I_1*). This shows that a policy that is optimal when viewed from one state is not necessarily optimal when viewed from another.*

Therefore the OWR-optimality is state-dependent.

Violation of the Bellman Optimality Principle. The Bellman Optimality Principle, which says that any subpolicy of any optimal policy is optimal is not guaranteed to be valid anymore when optimizing OWR as it is not a linear scalarizing function. We illustrate this point on Example 2.

Example 2 (continued). *We have* $V^{\delta^\infty_{aa}}(1) = (2, 2)$ *(OWR of* 3.7*) and* $V^{\delta^\infty_{ab}}(1) = (3, 1)$ *(OWR of* 4.5*). Thus,* $\delta^\infty_{aa} \succ_1 \delta^\infty_{ab}$ *(seen from state* 1*). Now, if we consider policy* $(\delta_{bb}, \delta^\infty_{aa})$ *and policy* $(\delta_{bb}, \delta^\infty_{ab})$ *that consist in applying* δ_{bb} *first, then policy*

δ^∞_{aa} or policy δ^∞_{ab} respectively, we get $V^{(\delta_{bb},\delta^\infty_{aa})}(1) = (0,6)$ (OWR of 2.7) and $V^{(\delta_{bb},\delta^\infty_{ab})}(1) = (1,5)$ (OWR of 1.9). This means that now $(\delta_{bb},\delta^\infty_{aa}) \prec_1 (\delta_{bb},\delta^\infty_{ab})$, which is a preference reversal. The Bellman Optimality principle is thus violated.

As shown by Example 2, $\pi \succ_s \pi'$ does not imply $(\delta,\pi) \succsim_s (\delta,\pi')$ for every π,π',δ,s. So, in policy iteration, we cannot prune policy π' on the argument it is beaten by π since π' may lead to an optimal policy (δ,π'). Similar arguments explain that a direct adaptation of value iteration for OWR optimization may fail to find the optimal policy.

The above observations constitute the deadlock to overcome to be able to find efficiently OWR-optimal solutions. This motivates us to propose a solving method based on linear programming.

Solution Method. In order to use OWR in MMDPs, we first compute the ideal point I by setting I_i as the optimal value of $\mathcal{P}$ with reward function R_i.

Although OWR is not linear, its optimization in MMDPs does not impact the dynamic of the system, which thus remains linear. Therefore, OWR is optimized under the same constraints as Program $(v\mathcal{D})$, which gives the following program $(\mathcal{D}')$:

$$(\mathcal{D}')\begin{cases} \min \sum_{i\in O} w_i \eta^\lambda_{\langle i\rangle} \\ \text{s.t. } \eta^\lambda_i = \lambda_i\big(I_i - \sum_{s\in S}\sum_{a\in A} R_i(s,a)\,x_{sa}\big) \quad \forall i\in O \\ \sum_{a\in A} x_{sa} - \gamma \sum_{s'\in S}\sum_{a\in A} T(s',a,s)\,x_{s'a} = \mu(s) \quad \forall s\in S \\ x_{sa}\geq 0 \quad \forall s\in S, \forall a\in A \end{cases} \quad (\mathcal{C}')$$

where for all $i\in O$, I_i is computed by optimizing objective i with Program $(\mathcal{P})$ or Program $(\mathcal{D})$. Since OWR is not linear but only piecewise-linear (one piece per permutation of objectives), a linear reformulation of $(\mathcal{D}')$ can be written.

First, denoting $L_k(\eta^\lambda) = \sum_{i=1}^k \eta^\lambda_{\langle i\rangle}$ and $w'_i = w_i - w_{i+1}$ for $i = 1,\ldots,n-1$, $w'_n = w_n$, $(\mathcal{D}')$ can be rewritten as:

$$\min_{\eta^\lambda\in E} \sum_{k\in O} w'_k L_k(\eta^\lambda) \tag{6}$$

where E is defined by Constraints $(\mathcal{C}')$. Moreover, as shown by [14], the quantity $L_k(\eta^\lambda)$, for a given vector η^λ, can be computed by the following LP formulations:

$$L_k(\eta^\lambda) = \max_{(u_{ik})_{i\in O}} \{\sum_{i\in O} \eta^\lambda_i u_{ik} : \sum_{i\in O} u_{ik} = k,\ 0\leq u_{ik}\leq 1\} \tag{7}$$

$$= \min_{\substack{t_k \\ (d_{ik})_{i\in O}}} \{k t_k + \sum_{i\in O} d_{ik} : \eta^\lambda_i \leq t_k + d_{ik},\ d_{ik}\geq 0\} \tag{8}$$

where (7) follows from the definition of $L_k(\eta^\lambda)$ as the sum of the k largest values η^λ_i, while (8) is the dual LP with dual variable t_k corresponding to equation

$\sum_{i \in O} u_{ik} = k$ and variables d_{ik} corresponding to upper bounds on u_{ik}. Therefore, we have:

$$\begin{aligned}
&\min_{\eta^\lambda \in E} \sum_{k \in O} w'_k L_k(\eta^\lambda) \\
&= \min_{\eta^\lambda \in E} \sum_{k \in O} w'_k \min_{\substack{t_k \\ (d_{ik})_{i \in O}}} \{k t_k + \sum_{i \in O} d_{ik} : \eta^\lambda_i \leq t_k + d_{ik},\ d_{ik} \geq 0\} \quad (9) \\
&= \min_{\eta^\lambda \in E} \min_{\substack{(t_k)_{k \in O} \\ (d_{ik})_{i,k \in O}}} \{\sum_{k \in O} w'_k (k t_k + \sum_{i \in O} d_{ik}) : \eta^\lambda_i \leq t_k + d_{ik},\ d_{ik} \geq 0\} \quad (10)
\end{aligned}$$

where (9) derives from (8) and (10) derives from (9) as $w'_k > 0$. Together with the LP constraints $(\mathcal{C}')$ of set E. This leads to the following linearization of $(\mathcal{D}')$:

$$\begin{aligned}
\min \ & \sum_{k \in O} w'_k (k t_k + \sum_{i \in O} d_{ik}) \\
\text{s.t. } & \lambda_i \big(I_i - \sum_{s \in S} \sum_{a \in A} R_i(s,a)\, x_{sa}\big) \leq t_k + d_{ik} \quad \forall i, k \in O \\
& \sum_{a \in A} x_{sa} - \gamma \sum_{s' \in S} \sum_{a \in A} T(s', a, s)\, x_{s'a} = \mu(s) \quad \forall s \in S \\
& x_{sa} \geq 0 \quad \forall s \in S, \forall a \in A; \quad d_{ik} \geq 0 \quad \forall\, i, k \in O
\end{aligned}$$

Therefore, we get an exact LP formulation of the entire OWR problem $(\mathcal{D}')$. The randomized policy characterized by the x_{sa}'s at optimum is the OWR optimal policy. Our previous observation concerning the state-dependency of the OWR optimality tells us that the OWR-optimal solution might change with μ, which differs from the classical case. When the initial state is not known, distribution μ can be chosen as the uniform distribution over the possible initial states. When the initial state s_0 is known, $\mu(s)$ should be set to 1 when $s = s_0$ and to 0 otherwise. The solution found by the linear program does not specify which action to choose for the states that receive a null weight and that are not reachable from the initial state as they do not impact the value of the OWR-optimal policy.

5 Experimental Results

We tested our solving method on the navigation problem over a grid $N \times N$ ($N = 20, 50$ or 100 in our experiments). In this problem, a robot has four possible actions: Left, Up, Right, Down. The transition function models the fact that when moving, the robot may deviate from its trajectory with some fixed probability because it does not have a perfect control of its motor.

We ran four series of experiments with 100 instances each time. Unless otherwise stated, the parameters are chosen as follows. Rewards are two-dimensional vectors whose components are randomly drawn within interval $[0, 1]$. The discount factor is set to 0.9 and the initial state is set arbitrarily to the upper left corner of the grid. We set $w = (2/3, 1/3)$ (normalized vector obtained from $(1, 1/2)$) and $\lambda = (1, 1)$.

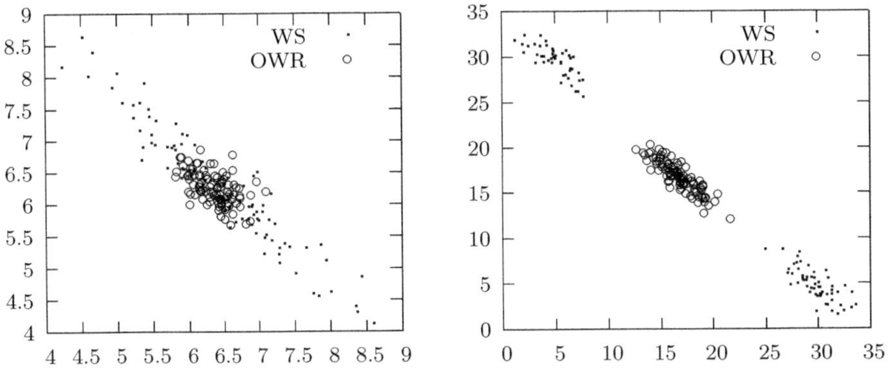

Fig. 5. 1st series (left), 2nd series (right) of experiments

As criteria are generally conflicting in real problems, for the first set of experiments, to generate realistic random instances, we simulate conflicting criteria with the following procedure: we pick one criterion randomly for each state and action and its value is drawn uniformly in $[0, 0.5]$ and the value of the other is drawn in $[0.5, 1]$. The results are represented on Figure 5 (left). One point (a dot for WS and a circle for OWR) represents the optimal value function in the initial state for one instance. Naturally, for some instances, WS provides a balanced solution but in most cases, WS gives a bad compromise solution. Figure 5 (left) shows that we do not have any control on tradeoffs obtained with WS. On the contrary, when using OWR, the solutions are always balanced.

To confirm the effectiveness of our approach, we ran a second set of experiments on pathological instances of the navigation problem. All the rewards are drawn randomly as for the first set of experiments. Then, in the initial state, for each action that does not move to a wall, we choose randomly one of the criteria and add a constant (here, arbitrarily set to 5). Then by construction, the value functions of all non-dominated deterministic policies in the initial state are unbalanced. The results are shown on Figure 5 (right). Reassuringly, we can see that OWR continues to produce fair solutions on the contrary to WS.

Our approach is still effective in higher dimensions. We ran a third set of experiments with three objectives as in higher dimensions, the experimental results would be difficult to visualize and as in dimension three, one can already show that OWR can be more effective than Minmax Regret or Augmented Tchebycheff. This last point could not have been shown in dimension two. In this third set of experiments, we set $w = (9/13, 3/13, 1/13)$ (normalized vector obtained from $(1, 1/3, 1/9)$) and $\lambda = (1, 1, 1)$. The random rewards are generated in order to obtain pathological instances in the spirit of the previous series of experiments. We set the initial state in the middle of the grid as we need to change the rewards of three actions. First, all rewards are initialized as in the first series of experiments (one objective drawn in $[0.5, 1]$, the other two in $[0, 0.5]$). In the initial state, for a first action, we add a constant C (here, $C = 5$) to the first component of its reward and a smaller constant c (here, $c = \frac{4}{5}C$) to its second

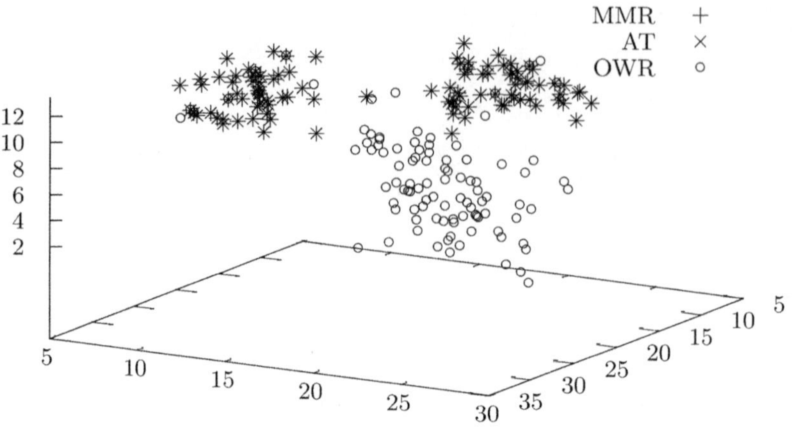

Fig. 6. Experiments with 3 objectives

one. For a second action, we do the opposite. We add c to its first component and C to its second one. For a third action, we add 5 to its third component and we subtract $2C$ from one of its first two ones chosen randomly. In such an instance, a policy choosing the third action in the initial state would yield a very low regret for the third objective, but the regrets for the first two objectives would not be balanced. In order to obtain a policy which yields a balanced profile on regrets, one needs to consider the first two actions.

The results of this set of experiments are shown on Figure 6. MMR stands for Minmax Regret and AT for Augmented Tchebycheff. Each point corresponds to the value of the optimal (w.r.t. MMR, AT or OWR) value function in the initial state of a random instance. One can notice that MMR and AT give the same solutions as both criteria are very similar. In our instances, it is very rare that one needs the augmented part of AT. Furthermore, one can see that the OWR-optimal solutions are between those optimal for MMR and AT. Although the OWR-optimal solutions are weaker on the third dimension, they fairly take into account potentialities on each objective and are better on at least one of the first two objectives.

For the last series of experiments, we tested our solution method with different scaling factors on the same instances as in the second series. With $\lambda = (1.75, 1)$ (resp. $\lambda = (1, 1.75)$), one can observe on the left (resp. right) hand side of Figure 7 that the obtained optimal tradeoffs with OWR now slightly favor the first (resp. second) objective as it could be expected.

We also perform experiments with more than three objectives. In Table 2, we give the average execution time in function of the problem size. The experiments were run using CPLEX 12.1 on a PC (Intel Core 2 CPU 2.66Ghz) with 4GB of RAM. The first row (n) gives the number of objectives. Row Size gives the number of states of the problem. Row TW gives the execution time for WS approach while row TO gives the execution time for OWR. All the times are given in

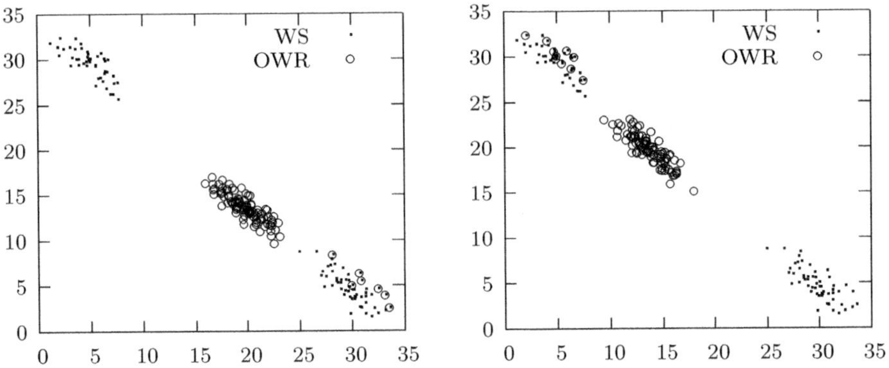

Fig. 7. 4th series of experiments (left: $\lambda = (1.75, 1)$, right: $\lambda = (1, 1.75)$)

Table 2. Average execution time in seconds

n	2			4			8			16		
Size	400	2500	10000	400	2500	10000	400	2500	10000	400	2500	10000
TW	0.2	5.2	147.6	0.10	5.1	143.7	0.1	4.7	146.0	0.12	4.9	143.6
TO	0.4	13.6	416.2	0.65	27.6	839.4	1.4	55.4	1701.7	3.10	111.5	3250.4

seconds as averages over 20 experiments. The OWR computation times increase proportionally to the number of criteria. Nevertheless, due to the huge number of variables x_{sa}'s, one may need to apply some column generation techniques [4] for larger problems.

6 Conclusion

We have proposed a method to generate fair solutions in MMDPs with OWR. Although this scalarizing function is not linear and cannot be optimized using value and policy iterations, we have provided an LP-solvable formulation of the problem. In all the experiments performed, OWR significantly outperforms the weighted sum concerning the ability to provide policies having a well-balanced valuation vector, especially on difficult instances designed to exhibit conflicting objectives. Moreover, introducing scaling factors λ_i in OWR yields deliberately biased tradeoffs within the set of Pareto-optimal solutions, thus providing full control to the decision maker in the exploration of policies.

Acknowledgements. The research by W. Ogryczak was partially supported by European Social Fund within the project Warsaw University of Technology Development Programme. The research by P. Perny and P. Weng was supported by the project ANR-09-BLAN-0361 GUaranteed Efficiency for PAReto optimal solutions Determination (GUEPARD).

References

1. Altman, E.: Constrained Markov Decision Processes. CRC Press, Boca Raton (1999)
2. Boutilier, C.: Sequential optimality and coordination in multiagent systems. In: Proc. IJCAI (1999)
3. Chatterjee, K., Majumdar, R., Henzinger, T.: Markov decision processes with multiple objectives. In: Durand, B., Thomas, W. (eds.) STACS 2006. LNCS, vol. 3884, pp. 325–336. Springer, Heidelberg (2006)
4. Desrosiers, J., Luebbecke, M.: A primer in column generation. In: Desaulniers, G., Desrosier, J., Solomon, M. (eds.) column generation, pp. 1–32. Springer, Heidelberg (2005)
5. Furukawa, N.: Vector-valued Markovian decision processes with countable state space. In: Recent Developments in MDPs, vol. 36, pp. 205–223 (1980)
6. Geoffrion, A.: Proper efficiency and the theory of vector maximization. J. Math. Anal. Appls. 22, 618–630 (1968)
7. Guestrin, C., Koller, D., Parr, R.: Multiagent planning with factored MDPs. In: NIPS (2001)
8. Hansen, P.: Bicriterion Path Problems. In: Multiple Criteria Decision Making Theory and Application, pp. 109–127. Springer, Heidelberg (1979)
9. Kostreva, M., Ogryczak, W., Wierzbicki, A.: Equitable aggregations and multiple criteria analysis. Eur. J. Operational Research 158, 362–367 (2004)
10. Littman, M.L., Dean, T.L., Kaelbling, L.P.: On the complexity of solving Markov decision problems. In: UAI, pp. 394–402 (1995)
11. Llamazares, B.: Simple and absolute special majorities generated by OWA operators. Eur. J. Operational Research 158, 707–720 (2004)
12. Marshall, A., Olkin, I.: Inequalities: Theory of Majorization and its Applications. Academic Press, London (1979)
13. Mouaddib, A.: Multi-objective decision-theoretic path planning. IEEE Int. Conf. Robotics and Automation 3, 2814–2819 (2004)
14. Ogryczak, W., Sliwinski, T.: On solving linear programs with the ordered weighted averaging objective. Eur. J. Operational Research 148, 80–91 (2003)
15. Puterman, M.: Markov decision processes: discrete stochastic dynamic programming. Wiley, Chichester (1994)
16. Steuer, R.: Multiple criteria optimization. John Wiley, Chichester (1986)
17. Viswanathan, B., Aggarwal, V., Nair, K.: Multiple criteria Markov decision processes. TIMS Studies in the Management Sciences 6, 263–272 (1977)
18. White, D.: Multi-objective infinite-horizon discounted Markov decision processes. J. Math. Anal. Appls. 89, 639–647 (1982)
19. Yager, R.: On ordered weighted averaging aggregation operators in multi-criteria decision making. IEEE Trans. on Syst., Man and Cyb. 18, 183–190 (1988)
20. Yager, R.: Decision making using minimization of regret. Int. J. of Approximate Reasoning 36, 109–128 (2004)

Scaling Invariance and a Characterization of Linear Objective Functions

Saša Pekeč

Fuqua School of Business, Duke University
100 Fuqua Drive, Durham, NC 27708-0120, USA

Abstract. A decision-maker who aims to select the "best" collection of alternatives from the finite set of available ones might be severely restricted in the design of the selection method. If the representation of valuations of available alternatives is subject to invariance under linear scaling, such as the choice of the unit of measurement, a sensible way to compare choices is to compare weighted sums of individual valuations corresponding to these choices. This scaling invariance, in conjunction with additional reasonable axioms, provides a characterization of linear 0-1 programming objective functions.

The problem of finding an optimal subset of available data to be aggregated, allowing for use of different aggregation methods for different subsets of data, is also addressed. If the input data in the optimal aggregation problem are measured on a ratio scale and if the aggregation must be unanimous and symmetric, the arithmetic mean is the only sensible aggregation method.

Keywords: Choice, Invariance, Linear scaling, Meaningfulness, Linear 0-1 programming.

1 Introduction

The problem of selecting an optimal subset of alternatives from the set of n alternatives has been studied in a wide variety of contexts ranging from psychology and economics (choice models) to management science and theoretical computer science (combinatorial optimization models). In this paper it is shown that, independent of the context and actual data, basic properties of the information associated to the set of alternatives dictate which method of selection of an optimal subset of alternatives should be used.

In a generic choice problem that is under consideration in this paper a decision-maker has to choose a subset of alternatives from the finite set of available ones. Available alternatives are numerated and the set of n available alternatives is denoted by $[n] := \{1, 2, \ldots, n\}$. Information about (or decision-maker's valuations of) available alternatives is represented by real numbers w_i that are associated to each available alternative $i \in [n]$. The decision-maker has to choose a subset of alternatives keeping in mind that some subsets of alternatives are not feasible and the list of non-feasible subsets is known. Furthermore, the decision-maker

R.I. Brafman, F. Roberts, and A. Tsoukiàs (Eds.): ADT 2011, LNAI 6992, pp. 205–218, 2011.

can valuate each feasible subset of alternatives, i.e., a possible candidate for the optimal choice. This valuation of $S \subset [n]$ is a real number that depends on the weights of alternatives from S, w_i, $i \in S$. It is possible that the decision-maker uses completely different valuation methods for different feasible subsets of alternatives. The decision-maker will choose the feasible subset with the highest (lowest) value.

For example, a production problem of choosing a collection of products to be produced from the set of n possible ones, where some combinations of products cannot be produced at the same time (for technological or some other reasons), can be modeled as a choice problem described above. The weight of each product and any combination of products could be its market value (or the production cost). It should be noted that the valuations of combinations of products could be combination specific, taking into account all possible synergetic values present in a particular combination of products (e.g., offering a complete product line; e.g., reduction in production costs,...) or negative effects (e.g., offering two similar products might effect market value of both products). A similar example could be a customer choice of optional equipment in a car or optional upgrades in a computer. While market values of each of computer upgrades (e.g., faster processor, better CPU board, larger hard disk, more RAM, better graphics card,...) are known, not all combinations of upgrades are mutually feasible nor are equally effective (e.g., the effect of a graphics card upgrade is nil if the processor is not fast enough; the effect of extra RAM is negligible if there is already plenty of RAM available). Another example is the problem of choosing a team or a committee from the pool of n candidates. The decision-maker could have a valuation function for the effectiveness of each team (e.g., expected time for completing given set of tasks) and could know which teams cannot be formed (i.e. which teams are not feasible for whatever reasons, e.g., scheduling constraints of some candidates).

The main object of the analysis in this paper will be the type of information described by weights w_i of the alternatives. These weights are in the same units of measurement for all alternatives and are often unique up to some assumption about at least the unit of measurement. For example, monetary values can be described in US dollars but could also be described in thousands of dollars or in any other currency or any other unit of measurement of monetary amounts. Similarly, if weights represent time (say, to complete the task), these weights can be represented in seconds, minutes,... The same conclusion goes for almost any type of information described by w_i (e.g., length, volume, mass,...). Given multiple acceptable ways to represent the weights in the form $\lambda w_1, \ldots, \lambda w_n$, for any $\lambda > 0$, a desirable choice model or optimization model property is that the structure of the optimal solution or choice is invariant to the representation choice. For example, if all weights w_i represent monetary value in Euros, the model solution should point to the same decision (structure of the optimal solution or the optimal choice) as if all weights w_i were represented in US dollars. (The value of the objective function could change since units of measurement changed, but there was no structural change in the problem inputs.)

As mentioned above, whenever it is allowable to pick the unit of measurement of data, weights $w_1, w_2, \ldots, w_n$ can be replaced by weights $\lambda w_1, \lambda w_2, \ldots, \lambda w_n$, $\lambda > 0$. (This corresponds to a change of the unit of measurement where the old unit of measurement is multiplied by $1/\lambda$.) In the language of measurement theory (a theoretical framework for studying allowable transformations of data), data that allows such transformations is said to be measured on a scale weaker or equal to ratio scale. Data representing monetary amounts (a.k.a. cardinal utility), time, length, ... are all examples of ratio scale data. In such situations, i.e. when input data is measured on a ratio scale or a weaker scale, any method that is used to select an optimal subset of available alternatives should have a property that the choice proposed by this method is invariant to positive linear scalings of weights associated with available alternatives (This statement will be made precise in the next section.)

The central result in this paper is that, when weights corresponding to alternatives are invariant under simple linear scaling (i.e., in the language of measurement theory, are measured on a scale weaker or equal to ratio scale), the decision-maker has little freedom in designing the methods of valuation of feasible subsets of alternatives. Under certain additional conditions, taking linear combination of weights associated with chosen alternatives are the only valuation methods of feasible subsets of alternatives that yield an optimal choice that is invariant under positive linear scaling of weights. In other words, even a simple invariance of input data puts very stringent limits on the objective function choice.

The choice model, invariance under linear scaling, as well as the main result are formulated and stated precisely in the next section. Section 3 contains the proof of the main theorem and a discussion related to possible modifications of the conditions of the theorem. The problem of optimal aggregation and its connections to the optimal choice is addressed in Section 4.

The final section of the paper is devoted to some closing remarks.

2 The Choice Model

As already stated in the Introduction, the set of the available alternatives is denoted by $[n]$. The collection of all feasible subsets of $[n]$ is denoted by $\mathcal{H}$.

Throughout we will assume that $[n] \notin \mathcal{H}$. (This assumption is not so restrictive since if the choice of all alternatives is a feasible option, the decision-maker could always compare the optimal choice among all feasible alternatives but $[n]$ with the choice $[n]$ to determine the final optimal choice.)

We will use boldfaced letters to denote vectors. Thus, $\mathbf{w}$ denotes the vector of weights $(w_1, \ldots, w_n)^T$ associated to alternatives $1, 2, \ldots, n$.

We will also utilize the obvious one-to-one correspondence between vectors $\mathbf{x} \in \{0,1\}^n$ and subsets $S \subseteq [n]$: $\mathbf{x}$ is the incidence vector of the set $S \subseteq [n]$ if and only if

$$x_i = 1 \Leftrightarrow i \in S.$$

Thus, the set of all feasible subsets of $[n]$ can be represented by the set of incidence vectors of elements of $\mathcal{H}$, i.e., without fear of ambiguity we can abuse the notation by writing $\mathcal{H} \subset \{0,1\}^n$ whenever incidence vectors of subsets will be more handy for notation purposes than the subsets themselves.

The problem of finding the optimal choice among n alternatives where $\mathcal{H}$ is the set of feasible choices is an optimization problem

$$\max\{P(\mathbf{x};\mathbf{w}) : \mathbf{x} \in \mathcal{H} \subset \{0,1\}^n\}. \tag{1}$$

where $P : \{0,1\}^n \times \mathbf{R}^n \to \mathbf{R}$.

Alternately, the problem of finding an optimal choice is

$$\max\{f_S(\mathbf{w}) : S \in \mathcal{H}\} \tag{2}$$

where f_S is a real valued function ($f_S : \mathbf{R}^n \to \mathbf{R}$) defined by $f_S(\mathbf{w}) = P(\mathbf{x}_S;\mathbf{w})$ where x_S stands for the incidence vector of the set S. The collection of functions f_S is denoted by $\mathcal{F}(P) := \{f_S : \mathbf{R}^n \to \mathbf{R} : S \subset [n]\}$ (f_S are defined as above).

Note that any collection of $2^n - 1$ functions $\{f_S : \mathbf{R}^n \to \mathbf{R} : S \subset [n]\}$ defines an objective function P and, hence, the problem (1). Thus formulations (1) and (2) are equivalent.

Remark. Note that the problem (2) with the family of $2^n - 1$ functions $\{f_S^{(L)} : \mathbf{R}^n \to \mathbf{R} : S \subset [n]\}$ defined by $f_S^{(L)}(\mathbf{w}) = \sum_{i \in S} w_i$ is equivalent to problem (1) with the objective function $P(\mathbf{x};\mathbf{w}) = \mathbf{w}^T\mathbf{x}$, i.e., one of the central problems of combinatorial optimization, the linear 0-1 programming problem:

$$\max\{\mathbf{w}^T\mathbf{x} : \mathbf{x} \in \mathcal{H} \subset \{0,1\}^n\}. \tag{3}$$

Linear 0-1 programming problem is the dominant optimization model in almost every quantitative aspect of management sciences and widely used in practice. Thus, it should not be surprising that, among all possible formulations of problem (1), the linear 0-1 programming problems are the most studied ones. Even this simple case (simple compared to general formulation (1)) is not completely understood. The reason for this is that the computational complexity of actually finding the maximum in (3) critically depends on the structure of the set of feasible solutions $\mathcal{H}$. (For example, choosing $\mathcal{H}$ to be the set of all Hamiltonian cycles of the complete graph on k vertices, $N = k(k-1)/2$, formulates the celebrated traveling salesman problem with edge weights given by $\mathbf{w} = (w_1, \ldots, w_n)^T$ - a canonical example of an NP-complete problem.)

What seems a bit more surprising is that linear 0-1 programming formulation (3) is used (almost exclusively) as a mathematical model for optimization problems over discrete structures. Choosing an objective function for a problem is a modeling issue and there is no a-priori reason that the objective function must be linear. This paper provides one argumentation for use of linear objective functions. We will show that invariance to linear scaling of weights w_i constrains the format of the objective function. ∎

The least one should expect from a satisfactory model is that conclusions that can be drawn from the model are invariant with respect to the choice of an acceptable way to represent problem parameters. For example, if $w_1, \ldots, w_n$ represent monetary amounts, then $w_1, \ldots, w_n$ can be expressed in any currency and denomination. In fact, whenever $w_1, \ldots, w_n$ are numerical representations of problem data, it is likely that, for any $\lambda > 0$, $\lambda w_1, \ldots, \lambda w_n$ are also acceptable numerical representations of data. This amounts to changing the unit of measurement (e.g., $\lambda = 1/1000$ describes the change from dollars to thousands of dollars, λ describes the change from currency x to currency y if the current exchange rate is λ units of y for one unit of x, etc). Hence, it is reasonable to assume that problem (1) satisfies the following property:

$$\begin{gathered} \forall \mathbf{w} \in \mathbf{R}^n, \forall \lambda > 0 : \\ P(\mathbf{x}^*; \mathbf{w}) = \max\{P(\mathbf{x}; \mathbf{w}) : \mathbf{x} \in \mathcal{H}\} \\ \Leftrightarrow \\ P(\mathbf{x}^*; \lambda\mathbf{w}) = \max\{P(\mathbf{x}; \lambda\mathbf{w}) : \mathbf{x} \in \mathcal{H}\} \end{gathered} \tag{4}$$

In other words, the conclusion of optimality ("$\mathbf{x}^*$ is an optimal solution") should be invariant under positive linear scaling of problem parameters $\mathbf{w}$ (that is, replacing $\mathbf{w}$ by $\lambda\mathbf{w}$, $\lambda > 0$).

Remark. As already stated in the Introduction, measurement theory provides a mathematical foundation for analysis of how data is measured and how the way data is measured might affect conclusions that can be drawn from a mathematical model. Scales of measurement where everything is determined up to the choice of the unit of measurement (e.g., measurement of mass, time, length, monetary amounts,...) are called *ratio scales.* In measurement theory terminology, requirement (4) is the requirement that the conclusion of optimality for problem (1) is *meaningful* if $w_1, \ldots, w_n$ are measured on a ratio scale. Informally, a statement involving scales of measurement is meaningful if its truth value does not depend on the choice of an acceptable way to measure data related to the statement. (More about measurement theory can be found in [4,13,6,9]. More about applying the concept of meaningfulness to combinatorial optimization problems can be found in [10] and [8].) ∎

A central question that motivates the work in the paper is whether there exists an objective function P with the following property:

Invariance under Linear Scaling (ILS). For any choice of a nonempty set of feasible solutions $\mathcal{H} \subset \{0,1\}^n$, requirement (4) is satisfied.

Clearly, the answer is: Yes. For example, the linear objective function $P(\mathbf{x}, \mathbf{w}) = \mathbf{w}^T\mathbf{x}$ has property **(ILS)**.

Are there any other objective functions having property **(ILS)**? There are plenty of degrees of freedom for the objective function choice; recall that the objective function form can vary over feasible subsets S. On the other hand, property **(ILS)** through the invariance requirement (4) is essentially one-dimensional and completely defined by $\lambda > 0$, but it does allow for unbounded one-dimensional scaling.

It will be shown that, provided that the objective function has some other *reasonable* properties, the linear objective function is essentially the only objective function having property **(ILS)**. Of course, the key word here is "reasonable". In order to describe these "reasonable" properties we again turn to the representation of an objective function P by the corresponding family $\mathcal{F}(P) = \{f_S : \mathbf{R}^n \to \mathbf{R} : S \subset [n]\}$:

Locality (L). It is reasonable to assume that the value $f_S(\mathbf{w})$ depends only on the weights corresponding to the elements from S. In other words, changing the weight w_j corresponding to any element $j \notin S$, will not change the value of f_S. More precisely, if

$$\forall S \subset [n], \forall j \notin S : \frac{\partial f_S}{\partial w_j} = 0$$

we will say that the family $\mathcal{F}(P)$ (or P) is **local** (has property **L**).

Normality (N). The weights $\mathbf{w}$ should (in a transparent way) indicate the value of f_S for all singletons S. We will say that the family $\mathcal{F}(P)$ (or P) is **normalized** (has property **(N)**) if, for any singleton $\{i\}$ and any $\mathbf{w} \in \mathbf{R}^n$ $f_{\{i\}}(\mathbf{w}) = w_i$ (i.e.,$f_{\{i\}}$ restricted to the i-th coordinate is the identity function).

The property **(N)** should not be considered restrictive: if $\mathcal{F}(P)$ were not normalized, it would make sense to reformulate the problem by introducing new weights $\bar{\mathbf{w}}$ defined by $\bar{w}_i := f_{\{i\}}(w_i)$. Of course, all other f_S would need to be redefined: $\bar{f}_S(\bar{\mathbf{w}}) := f_S(\mathbf{w})$.

Completeness (C). For any nonempty S, unbounded change in $\mathbf{w}$ should result in unbounded change in $f_S(\mathbf{w})$. In fact, we will require that $f_S(\mathbf{R}^n) = \mathbf{R}$. In other words, if for every nonempty $S \subset [n]$, $f_S \in \mathcal{F}(P)$ is surjective, we say that $F(P)$ (or P) is **complete** (has property **(C)**).

The property **(C)** is rather strong but it can be substantially relaxed as will be demonstrated in Theorem 2.

Separability (S). The rate of change of $f_S(\mathbf{w})$ with respect to changing w_i should depend only on w_i (and not on the values of w_j, $j \neq i$). Furthermore, this dependence should be "smooth". More precisely, f is **separable** (has property **(S)**) if for any $i \in [n]$, there exists a function $g_i : \mathbf{R} \to \mathbf{R}$, $g_i \in C^1(\mathbf{R})$, such that

$$\frac{\partial f}{\partial w_i}(\mathbf{w}) = g_i(w_i).$$

We say that $\mathcal{F}(P)$ (or P) is separable (has property **(S)**) if every function $f_S \in \mathcal{F}(P)$ is separable.

The separability is arguably the most restrictive of the properties from the point of view of modeling (in the sense that one might argue that there are many problems for which any optimization model with the objective function that has property **(S)** would not be satisfactory). Also, the property **(S)** plays a crucial role in obtaining the main characterization result in this paper. (One could argue that **(S)** is at least as critical as **(ILS)**.)

Possible variations of all these properties are briefly addressed in the next section after the proof of Theorem 1.

3 The Main Theorem

The main result of this paper is a characterization theorem:

Theorem 1. *Let P be the objective function for the problem (1). Suppose that $\mathcal{F}(P)$ satisfies* **(L)**, **(N)**, **(C)**, *and* **(S)**. *Then P has property* **(ILS)** *if and only if every $f_S \in \mathcal{F}(P)$ is linear, that is, if and only if for every $S \subset [n]$ there exist constants $C_{S,i}$, $i \in S$, such that*

$$f_S(\mathbf{w}) = \sum_{i \in S} C_{S,i} w_i. \tag{5}$$

We first give a "workable" reformulation of property **(ILS)**.

Proposition 1. *P satisfies* **(ILS)** *if and only if*

$$\begin{gathered} \forall S, T \subset [n], \forall \mathbf{w} \in \mathbf{R}^n, \forall \lambda \in \mathbf{R}_+ : \\ f_S(\mathbf{w}) \geq f_T(\mathbf{w}) \Leftrightarrow f_S(\lambda \mathbf{w}) \geq f_T(\lambda \mathbf{w}) \end{gathered} \tag{6}$$

Proof: Note that (4) can be rewritten as

$$\begin{gathered} \forall \mathbf{w} \in \mathbf{R}^n, \forall \lambda > 0 : \\ f_{S^*}(\mathbf{w}) = \max\{f_S(\mathbf{w}) : S \in \mathcal{H}\} \\ \Leftrightarrow \\ f_{S^*}(\lambda \mathbf{w}) = \max\{f_S(\lambda \mathbf{w}) : S \in \mathcal{H}\} \end{gathered} \tag{7}$$

Obviously, (6) $\Rightarrow$ **(ILS)**. Conversely, for any $S, T \subset [n]$, we define $\mathcal{H} = \{S, T\}$ which gives **(ILS)** $\Rightarrow$ (6). ∎

Homogeneous functions play a central role in the proof of Theorem 1. We say that $f : \mathbf{R}^n \to \mathbf{R}$ is a r-**homogeneous** function if for every $\lambda > 0$ and every $\mathbf{w}$, $f(\lambda \mathbf{w}) = \lambda^r f(\mathbf{w})$.

The plan of the proof is as follows: we will first show that properties **(L)**, **(N)**, **(C)**, and **(ILS)** imply that every f_S in $\mathcal{F}(P)$ is 1-homogeneous. Then we will use a well known result about homogeneous functions (Euler's homogeneity relation) to show that **(L)** and **(S)** imply that every f_S must be a linear function.

Lemma 1. *Let P satisfy* **(L)** *and* **(ILS)**. *Suppose that $f_{S_0} \in \mathcal{F}(P)$ is an r-homogeneous function. Then, for any $T \subset [n]$ such that $S_0 \cap T = \emptyset$ and such that $f_T(\mathbf{R}^n) \subseteq f_{S_0}(\mathbf{R}^n)$, f_T is also r-homogeneous.*

Proof: We need to show that for any $\mathbf{w} \in \mathbf{R}^n$ and any $\lambda \in \mathbf{R}_+$

$$f_T(\lambda \mathbf{w}_T) = \lambda^r f_T(\mathbf{w}_T).$$

Since $f_T(\mathbf{R}^n) \subseteq f_{S_0}(\mathbf{R}^n)$, there exists $\mathbf{w}'$ such that

$$f_{S_0}(\mathbf{w}') = f_T(\mathbf{w}).$$

Note that $S_0 \cap T = \emptyset$ implies that we can choose $\mathbf{w}'$ such that $w'_j = w_j$ for every $j \in T$ (because f_{S_0} has property **(L)**). Let $\mathbf{w}''$ be such that $w''_i = w'_i$ for every $i \in S$, $w''_j = w_j$ for every $j \notin S$. Then, we have

$$f_T(\mathbf{w}'') = f_T(\mathbf{w}) = f_{S_0}(\mathbf{w}') = f_{S_0}(\mathbf{w}'') \tag{8}$$

where the first and last equality hold because of locality for f_T and f_{S_0}, respectively. Hence, for any $\lambda > 0$,

$$f_T(\lambda\mathbf{w}) = f_T(\lambda\mathbf{w}'') = f_{S_0}(\lambda\mathbf{w}'') = \lambda^r f_{S_0}(\mathbf{w}'') = \lambda^r f_T(\mathbf{w}'') = \lambda^r f_T(\mathbf{w}).$$

The first and the last equality holds because of locality of f_T and the construction of $\mathbf{w}''$, the second one follows from (6), applied to S_0, T and $\mathbf{w}''$, the third one by r-homogeneity of f_{S_0}, and the fourth one is just (8). ∎

Lemma 2. *Let P satisfy* **(L)**, **(C)**, *and* **(ILS)**. *Then for any two non-empty $S, T \subset [n]$, $f_S \in \mathcal{F}(P)$ is r-homogeneous if and only if $f_T \in \mathcal{F}(P)$ is r-homogeneous.*

Proof: If $S \cap T = \emptyset$, then this is a direct consequence of Lemma 1 (since $f_S(\mathbf{R}^n) = f_T(\mathbf{R}^n)$ by property **(C)**.

If $S \cap T \neq \emptyset$, then we use the disjoint case above repeatedly as follows: f_S is r-homogeneous if and only if $f_{T \setminus S}$ is r-homogeneous if and only if $f_{S \setminus T}$ is r-homogeneous if and only if f_T is r-homogeneous. ∎

Finally, before proving Theorem 1, we need to prove several facts about r-homogeneous functions.

Lemma 3 (Euler's homogeneity relation, [3]). *Let $f : \mathbf{R}^n \to \mathbf{R}$ be r-homogeneous and differentiable on the open and connected set $D \subseteq \mathbf{R}^n$. Then for any $\mathbf{w} \in D$*

$$rf(\mathbf{w}) = \frac{\partial f(\mathbf{w})}{\partial w_1} w_1 + \frac{\partial f(\mathbf{w})}{\partial w_2} w_2 + \ldots + \frac{\partial f(\mathbf{w})}{\partial w_k} w_n. \tag{9}$$

Proof: Let $G : \mathbf{R}_+ \times \mathbf{R}^n \to \mathbf{R}$ and $H : \mathbf{R}^n \to \mathbf{R}$ be defined by:

$$G(\lambda, \mathbf{w}) := f(\lambda\mathbf{w}) - \lambda^r f(\mathbf{w}) = 0,$$

$$H(\mathbf{w}) := \frac{\partial f(\mathbf{w})}{\partial w_1} w_1 + \frac{\partial f(\mathbf{w})}{\partial w_2} w_2 + \ldots + \frac{\partial f(\mathbf{w})}{\partial w_n} w_n - rf(\mathbf{w}).$$

Since

$$\frac{\partial G(\lambda, \mathbf{w})}{\partial \lambda} = \frac{\partial f(\lambda\mathbf{w})}{\partial w_1} w_1 + \frac{\partial f(\lambda\mathbf{w})}{\partial w_2} w_2 + \ldots + \frac{\partial f(\lambda\mathbf{w})}{\partial w_n} w_n - r\lambda^{r-1} f(\mathbf{w}) = \frac{1}{\lambda} H(\lambda\mathbf{w})$$

we conclude (by setting $\lambda = 1$) that $H(\mathbf{w}) = 0$ for all $\mathbf{w} \in D$, which is exactly (9). ∎

Lemma 4. *Let $f : \mathbf{R}^n \to \mathbf{R}$ be an r-homogeneous function satisfying property ($\mathbf{S}$). Then there exist constants C_i such that*

$$f(w_1, \ldots, w_n) = \sum_{i=1}^{n} C_i w_i^r.$$

Proof: By property ($\mathbf{S}$), there exist functions $g_i \in C^1(\mathbf{R})$, so that Euler's homogeneity relation (9) can be written as

$$rf(\mathbf{w}) = g_1(w_1)w_1 + g_2(w_2)w_2 + \ldots + g_n(w_n)w_n. \tag{10}$$

Taking the partial derivative with respect to the i-th variable we get:

$$rg_i(w_i) = r\frac{\partial f}{\partial w_i}(\mathbf{w}) = g_i'(w_i)w_i + g_i(w_i)$$

which must hold for every w_i. Hence,

$$w_i g_i'(w_i) - (r-1)g_i(w_i) = 0, \ \forall w_i \in \mathbf{R}.$$

The general solution of this linear homogeneous ordinary differential equation is $g_i(t) = C_i t^{r-1}$ Hence, from (10) we get

$$f(\mathbf{w}) = C_1 w_1^r + C_2 w_2^r + \ldots + C_n w_n^r.$$ ∎

Proof of Theorem 1:
Obviously, any family $\mathcal{F}(P)$ where all f_S are of the form (5) satisfies relation (6). Hence, by Proposition 1, P has property **(ILS)**.
Conversely, suppose that P has property **(ILS)**. Note that **(N)** implies that f_S is 1-homogeneous for any singleton S. Hence, by Lemma 2, we conclude that every $f_T \in \mathcal{F}(P)$ is 1-homogeneous ($f_\emptyset = 0$ by **(L)** and Lemma 1). Finally, (5) follows from Lemma 4. ∎

Theorem 1 demonstrates that, if we require that the model satisfy some reasonable criteria (i.e., invariance of the conclusion of optimality under linear scalings of the problem parameters, locality, normality, completeness, and separability), the choice of the objective function is limited to the choice among linear objective functions.

It should be noted that full strength of normality **(N)** and completeness **(C)** were not necessary for the proof of the theorem. In fact, one can replace these two properties by the requirement for the existence of an r-homogenous function $f_S \in \mathcal{F}(P)$ and by requiring that

$$f_S(\mathbf{R}^n) = f_{\{1\}}(\mathbf{R}^n) = f_{\{2\}}(\mathbf{R}^n) = \cdots = f_{\{n\}}(\mathbf{R}^n) = \bigcup_{T \subset [n]} f_T(\mathbf{R}^n) \tag{11}$$

holds. Thus we have the following straightforward generalization of Theorem 1:

Theorem 2. *Let P be the objective function for the problem (1). Suppose that $\mathcal{F}(P)$ satisfies* **(L)**, *and* **(S)**. *Furthermore suppose that there exists an r-homogeneous function $f_S \in \mathcal{F}(P)$ and that relation (11) holds. Then P has property* **(ILS)** *if and only if for every $S \subset [n]$ there exist constants $C_{S,i}$, $i \in S$, such that*

$$f_S(\mathbf{w}) = \sum_{i \in S} C_{S,i} w_i^r. \tag{12}$$

Locality **(L)** and Separability **(S)** imply that the objective function is smooth (has continuous second partial derivatives). The smoothness was essential in the presented proofs of both Lemma 3 and Lemma 4. It is quite possible that the properties **(L)** and **(S)** can be reformulated so that smoothness is not required and that Theorem 2 still holds. As already mentioned, the essence of locality **(L)** is the requirement that the value of the function f_S is independent of the values of w_i corresponding to $j \notin S$, and the essence of separability **(S)** is that the rate of change of f_S with respect of changing w_i depends only on the value of that w_i. For example, for any odd p, the function

$$P(\mathbf{x}, \mathbf{w}) = (x_1 w_1^p + \ldots + x_n w_n^p)^{1/p}$$

does satisfy locality **(L)**, normality **(N)**, completeness **(C)**, and invariance under linear scaling **(ILS)** but is not separable. So, separability is a necessary property for characterization of linear objective functions.

Remark. The objective function defined by (5) is linear, but it is not the objective function of the linear 0-1 programming problem (3) unless $C_{S,i} = C_{T,i}$ for all $i \in S, T$ and $S, T \in \mathcal{H}$. Additional (symmetry) properties are needed to ensure that. ∎

4 Optimal Aggregation

There is a vast literature on aggregating data or aggregating expert opinions. For example, the issues of aggregation are central in multiple criteria decision making and in multiattribute utility theory ([2] provides a survey of the field). Similarly, combining expert judgments or forecasts is another area where data aggregation plays a central role; see [1] for a survey. Finally, social welfare functions can be viewed as data aggregation methods; e.g., see [11,12].

Here we consider a generic aggregation problem where the input consists of the set of real numbers representing data to be aggregated. The decision-maker decides which data should be aggregated and which data should be ignored. Furthermore, the decision-maker might use different aggregation methods for different subsets of data that were selected for aggregation. For example, a company might attempt to obtain estimates, from various sources using diverse methods, on the added value (expressed in monetary amounts) of the prospective acquisition. Once the estimates are collected, a pro-acquisition manager could choose which estimates to present to the board. It is plausible that the choice of the estimates might dictate the choice of the aggregation method (for example if a particular collection of estimates was aggregated repeatedly using the same method in the past, the argument for using a different aggregation method this time might not be a convincing one and could reveal pro-acquisition opinion).

Formally, the optimal aggregation problem has the same formulation as the optimal choice problem (2) with $[n]$ denoting the index set of data to be aggregated (e.g., the experts, data sources), $w_1, w_2, \ldots, w_n$ denoting values of the data to be aggregated, $\mathcal{H}$ denoting collections of data that are feasible for aggregation (it might not be allowed to aggregate some combinations of data), and f_S denoting the aggregation method used when data from set S are chosen to be aggregated.

Thus, all statements from Section 2 and Section 3 apply to the optimal aggregation. In other words, if data to be aggregated are measured on a ratio scale or weaker, the objective function P from the optimal aggregation problem (1) has to satisfy property (ILS). If, in addition, (L), (N), (C) and (S) also hold, Theorem 1 implies that all aggregation methods f_S could only be linear combinations of the values corresponding to the elements of S.

The following property is almost universally considered as a desired property of any aggregation method:

Unanimity(U). If all data to be aggregated have equal value, the result of aggregation should be that value. In other words, f_S is unanimous if whenever there exists a u such that $w_i = u$ for all $i \in S$, then $f_S(\mathbf{w}) = u$. We say that the objective function P from (1) satisfies (U) if and only if all functions f_S from $\mathcal{F}(P)$ are unanimous.

Note that (U) is a stronger property than (N): if P satisfies (U) it trivially satisfies (N).

Theorem 3. *Let P be the objective function for the problem (1). Suppose that $\mathcal{F}(P)$ satisfies* **(L)**, **(C)**, **(S)**, *and* **(U)**. *Then P has property* **(ILS)** *if and only if every $f_S \in \mathcal{F}(P)$ is linear, that is, if and only if for every $S \subset [n]$ there exist constants $C_{S,i}$, $i \in S$, such that*

$$f_S(\mathbf{w}) = \sum_{i \in S} C_{S,i} w_i.$$

In addition, for every $S \subset [n]$,

$$\sum_{i \in S} C_{S,i} = 1. \tag{13}$$

Proof: As already noted (U) implies (N). Hence, Theorem 1 implies the linearity of all $f_S \in \mathcal{F}(P)$. The coefficients $C_{S,i}$ must sum to one by unanimity. Take $u \neq 0$ and set $w_i = u$ for all $i \in S$. Then,

$$u = f_S(\mathbf{w}) = \sum_{i \in S} C_{S,i} u = u \sum_{i \in S} C_{S,i}$$

where first equality follows by (U) and the second by linearity of f_S. Since $u \neq 0$, (13) follows. ∎

Many aggregation methods are symmetric, that is, invariant to permutations of data being aggregated. This property ensures that all expert opinions are equally valued.

In order to define symmetry precisely, let Π_S denote the set of permutations of $[n]$ for which all elements from $[n] \setminus S$ are fixed. In other words, $\pi \in \Pi_S$ if and only if $\pi(i) = i$ for all $i \notin S$. For a vector $\mathbf{w} \in R^n$ and a permutation π, let $\pi(\mathbf{w})$ denote the vector defined by $[\pi(\mathbf{w})]_i = w_{\pi(i)}$.

Symmetry (Sym). f_S is symmetric if for any $\mathbf{w}$ and any $\pi \in \Pi(S)$, $f_S(\mathbf{w}) = f_S(\pi(\mathbf{w}))$. The objective function P from (1) satisfies (U) if and only if all functions f_S from $\mathcal{F}(P)$ are symmetric.

Theorem 4. *Let P be the objective function for the problem (1). Suppose that $\mathcal{F}(P)$ satisfies* **(L)**, **(C)**, **(S)**, **(U)**, *and* **(Sym)**. *Then P has property* **(ILS)** *if and only if every $f_S \in \mathcal{F}(P)$ is the arithmetic mean of $\{w_i : i \in S\}$.*

Proof: By Theorem 3. It only remains to show that (Sym) also implies that $C_{S,i} = \frac{1}{|S|}$ for every $S \subset [n]$ and every $i \in S$. Since every $f_S(\mathbf{w}) = \sum_{i \in S} C_{S,i} w_i$ is symmetric, there exists C_S such that $C_S = C_{S,i}$ for every $i \in S$. Thus, by (13), $C_S = \frac{1}{|S|}$. Hence,

$$f_S(\mathbf{w}) = \frac{1}{|S|} \sum_{i \in S} w_i.$$

In other words, every f_S is the arithmetic mean of the weights corresponding to elements of S. ∎

In conclusion, the optimal aggregation problem can be formulated as an optimal choice problem. Thus, if representation of the data to be aggregated is invariant under linear scaling, aggregation methods that can be used for aggregating subsets of available data are limited. As shown by Theorem 3, if unanimity of aggregation is required, these aggregation methods must be convex combinations of data to be aggregated. If in addition, symmetry of aggregation is required, the arithmetic mean is the only possible aggregation method yielding to meaningful conclusions about optimal choice of data to be aggregated (as shown by Theorem 4).

5 Closing Remarks

The choice/optimization model studied here encompasses a large class of choice and decision models (e.g., 0-1 programming is a very special case). However, the model does have obvious limitations. For example, in many situations it is not possible to give a valuation of an alternative in the form of a single number (e.g., valuations of risky prospects, e.g. stock investments, often include standard deviation in addition to expected value). Another limitation of the presented model is its deterministic nature. An analysis of a simple model such as the one presented here is a necessary step toward an analysis of more complex models that are able to capture multidimensional input data and inherent stochastic nature of input data valuations. In fact, in many situations when complex models of choice are considered, a first run through the model would attempt to assign a single number to each of the available alternatives. For example, one could use (a best guess for) the expected value of a particular piece of input data instead of its (unknown) probability distribution. Similarly, when data corresponding to an available alternative is multidimensional, the decision-maker could try to collapse all that information into a single number. Whenever such simplifications are being made, a decision-maker essentially simplifies his/her sophisticated choice model into a choice model studied here. The limitations of the model presented here should not necessarily be viewed as negative: enriching the model could possibly add further constraints on the objective function choice. In other words, the simple structure of our model is already sufficient to let the property of invariance to scaling of input data, "force" linearity onto the objective function.

The prescriptive flavor of our analysis opens it to criticism of "reasonable" assumptions that were utilized in the presented proofs. Keeping in mind that the invariance under linear scalings **(ILS)** is central to our analysis, it should be noted that we tried to avoid requiring any "nice" behavior with respect to additivity of $\mathbf{R}^n$ since such property together with **(ILS)** would strongly indicate that the objective function must have the form of the linear functional on $\mathbf{R}^n$. In our characterization, the additivity is a consequence of 1-homogeneity and separability. It is important to note that it is the separability condition, and not scaling invariance, that eliminates large classes of objective/aggregation functions such as ordered weighted averaging operators (for which simple objectives like max and min are special cases) [14] and fuzzy measure-based aggregation operators (e.g., like those based on Choquet capacity)[7,5]. However, our goal was not to present yet another characterization theorem based on the set of more or less reasonable conditions, but to point out the importance of information about the type of input data and its implications in the model design and construction of the choice method. Thus, the approach presented here differs from a standard prescriptive approach since the main driving force towards narrowing possible methods of choice is not the set of some desirable conditions that have to be satisfied but the very fact that the input data of the choice model are of a certain type. Hence the main message of this work is not contained in specific forms of choice and aggregation methods as prescribed by characterization theorems, but in a claim that a decision-maker should pay close attention to the type of input data when designing the methods of choice and aggregation.

References

1. Clemen, R.T.: Combining Forecasts: A Review and Annotated Bibliography. Intl. J. Forecasting 5, 559–583 (1989)
2. Dyer, J.S., Fishburn, P.C., Steuer, R.E., Wallenius, J., Zionts, S.: Multiple Criteria Decision Making, Multiattribute Utility Theory: The Next Ten Years. Management Science 38(5), 645–654 (1992)

3. Eichhorn, W.: Functional Equations in Economics. Addison-Wesley, Reading (1978)
4. Krantz, D.H., Luce, R.D., Suppes, P., Tversky, A.: Foundations of Measurement, vol. I. Academic Press, New York (1971)
5. Labreuche, C., Grabisch, M.: The Choquet Integral for the Aggregation of Interval Scales in Multicriteria Decision Making. Fuzzy Sets and Systems 137(1), 11–26 (2003)
6. Luce, R.D., Krantz, D.H., Suppes, P., Tversky, A.: Foundations of Measurement, vol. III. Academic Press, New York (1990)
7. Marichal, J.-L.: On Choquet and Sugeno Integrals as Aggregation Functions. In: Grabisch, M., Murofushi, T., Sugeno, M. (eds.) Fuzzy Measures and Integrals, pp. 247–272. Physica Verlag, Heidelberg (2000)
8. Pekeč, A.: Limitations on Conclusions from Combinatorial Optimization Models. Ph.D. Dissertation, Rutgers University (1996)
9. Roberts, F.S.: Measurement Theory. Addison-Wesley, Reading (1979)
10. Roberts, F.S.: Limitations of Conclusions Using Scales of Measurement. In: Pollock, S.M., Rothkopf, M.H., Barnett, A. (eds.) Handbooks in OR & MS, vol. 6, pp. 621–671. North-Holland, Amsterdam (1994)
11. Sen, A.K.: Collective Choice and Social Welfare. North-Holland, Amsterdam (1984)
12. Sen, A.K.: Choice, Welfare and Measurement. Harvard University Press, Cambridge (1987)
13. Suppes, P., Krantz, D.H., Luce, R.D., Tversky, A.: Foundations of Measurement, vol. II. Academic Press, New York (1989)
14. Yager, R.R.: On Ordered Weighted Averaging Aggregation Operators in Multicriteria Decision Making. IEEE Trans. Systems Man Cybernet. 18, 183–190 (1988)

Learning the Parameters of a Multiple Criteria Sorting Method

Agnès Leroy[1], Vincent Mousseau[2], and Marc Pirlot[1]

[1] MATHRO, Faculté Polytechnique, Université de Mons
9, Rue de Houdain, Mons, Belgium
marc.pirlot@umons.ac.be

[2] Laboratoire Génie Industriel, Ecole Centrale Paris
Grande Voie des Vignes 92295 Châtenay Malabry, France
vincent.mousseau@ecp.fr

Abstract. Multicriteria sorting methods aim at assigning alternatives to one of the predefined ordered categories. We consider a sorting method in which categories are defined by profiles separating consecutive categories. An alternative a is assigned to the lowest category for which a is at least as good as the lower profile of this category, for a majority of weighted criteria. This method, that we call MR-Sort, corresponds to a simplified version of ELECTRE Tri. To elicit the values for the profiles and weights, we consider a learning procedure. This procedure relies on a set of known assignment examples to find parameters compatible with these assignments. This is done using mathematical programming techniques.

The focus of this study is experimental. In order to test the mathematical formulation and the parameters learning method, we generate random samples of simulated alternatives. We perform experiments in view of answering the following questions: (a) assuming the learning set is generated using a MR-Sort model, is the learning method able to restore the original sorting model? (b) is the learning method able to do so even when the learning set contains errors? (c) is MR-Sort model able to represent a learning set generated with another sorting method, i.e. can the models be discriminated on an empirical basis?

Keywords: Multicriteria Decision Aiding, Sorting, Preference Elicitation, Learning Methods.

1 Introduction

In this paper we deal with multiple criteria sorting methods that assign each alternative to a category selected in a set of ordered categories. We consider assignment rules of the following type. Each category is associated with a "lower profile" and an alternative is assigned one of the categories above this profile as soon as the alternative is at least as good as the profile for a (weighted) majority of criteria.

R.I. Brafman, F. Roberts, and A. Tsoukiàs (Eds.): ADT 2011, LNAI 6992, pp. 219–233, 2011.

Such a procedure is a simplified version of ELECTRE Tri, an outranking sorting procedure in which the assignment of an alternative is determined using a more complex concordance non-discordance rule [16]. Several papers have recently been devoted to the elicitation by learning of the parameters of the ELECTRE Tri method. These learning procedures usually rely on a set of known assignment examples and use mathematical programming techniques to find parameters compatible with these assignments (see e.g. [13], [11], [14], [6]). Unfortunately, the number of parameters involved is rather high and the mathematical formulation of the constraints resulting from the assignment examples are nonlinear so that the proposed methods do not try in general to determine all parameters at the same time. They generally assume that some of these parameters are known and determine the remaining ones accordingly.

To better tackle these difficulties, we have decided to work with a simplified version of ELECTRE Tri, essentially that characterized by [1,2]. In this version, an alternative is assigned above a limit profile if this alternative is at least as good as the profile for a sufficient coalition of criteria. We assume in addition that additive weights can be assigned to all criteria in such a way that a coalition is sufficient if the sum of the associated weights passes some majority threshold. In such a method, the parameters to be determined are the limit profiles of the categories, the criteria weights and the majority threshold.

The set of constraints on the parameters expressing the assignment of the examples, as well as other constraints, form a nonlinear mixed integer program that can be solved using CPLEX for realistic problems. Learning sets composed of up to 100 assignment examples and involving up to 5 criteria and 3 categories have been solved to optimality in a few seconds.

The interest of this study is experimental. In order to test the mathematical formulation and the parameters learning method, we have generated random samples of simulated alternatives represented by normalized performance vectors (values uniformly drawn from the [0,1] interval). We have then performed series of experiments in view of answering the following questions:

- Q1 **Model retrieval:** assuming that the examples have been assigned by means of a simulated sorting procedure based on a majority rule, does the learning method allow to elicit values of the parameters that are close to those of the original procedure used for their assignment? What size of a learning set is needed in order to obtain a "good approximation‘" of these parameters?
- Q2 **Tolerance for error:** assuming that the examples have only been "approximately" assigned using a simulated sorting model, i.e. that a certain proportion of assignment errors (5 to 15%) have been introduced, to what extent do these errors perturb the elicitation of the assignment model?
- Q3 **Idiosyncrasy:** we generate an assignment model that is not based on a majority rule but on an additive value function. We assign the alternatives in the learning set according with the latter rule. The question we try to answer is whether the change in the model can be easily detected by the elicitation procedure. In other words, can the models be discriminated on an empirical basis, i.e., on the sole evidence of assignment examples?

We present the results of our experiments as well as the conclusions that we draw from them (for more detail the interested reader can refer to [10]). Further research perspectives are outlined.

2 MR-Sort: A Sorting Method Based on a Majority Rule

As announced in the introduction, we depart from the usual Electre Tri sorting model that appears too complex (too many parameters) for our purpose of experimenting with a learning method. In addition, the precise procedure used for assigning alternatives to categories has not been characterized in an axiomatic manner. These are the reasons why we have turned to a simpler version of Electre Tri that has been characterized by [1,2].

At this stage, let us assume that an alternative is just a n-tuple of elements which represent its evaluations on a set of n criteria. We denote the set of criteria by $N = \{1, \ldots, n\}$ and assume that the values of criterion i range in the set X_i. Hence the set of alternatives can be identified with the Cartesian product $X = \Pi_{i=1}^n X_i$.

According to Bouyssou and Marchant, a *non-compensatory sorting method* (NCSM) is a procedure for assigning any alternative $x \in X$ to a particular category, in a given ordered set of categories. For simplicity, assume that there are only two categories. They thus form an ordered bipartition (X^1, X^2) of X, X^1 (resp. X^2) being interpreted as the set of "bad" (resp. "good") alternatives. A sorting method (in two categories) is *non-compensatory*, in Bouyssou-Marchant sense, if the following conditions hold:

- for each criterion i, there is a partition (X_i^1, X_i^2) of X_i; X_i^1 (resp. X_i^2) is interpreted as the set of "bad" (resp. "good") levels in the range of criterion i;
- there is a family $\mathcal{F}$ of "sufficient" coalitions of criteria (i.e. subsets of N), with the property that a coalition that contains a sufficient coalition is itself sufficient;
- the set of "good" levels $\mathcal{X}_i^2$ on each criterion and the set of sufficient coalitions $\mathcal{F}$ are such that alternative $x \in X$ belongs to the set of "good" alternatives X^2 iff the set of criteria on which the evaluation of x belongs to the set of "good" levels is a sufficient coalition, i.e.:

$$x = (x_1, \ldots, x_i, \ldots, x_n) \in X^2 \quad \text{iff} \quad \{i \in N | x_i \in X_i^2\} \in \mathcal{F}. \qquad (1)$$

Non compensatory sorting models have been fully characterized by a set of axioms in the case of two categories [1]. [2] extends the above definition and characterization to the case of more than two categories. These two papers also contain definitions and characterizations of NCSM with vetoes.

In the present paper we consider a special case of the NCSM model (with two or more categories and no veto). The Bouyssou-Marchant models are specialized in the following way:

1. We assume that X_i is a subset of $\mathbb{R}$ (e.g. an interval) for all $i \in N$ and the partitions (X_i^1, X_i^2) of X_i are compatible with the order on the real numbers $<$, i.e., for all $x_i \in X_i^1$, $x_i' \in X_i^2$, we have $x_i < x_i'$. We assume furthermore that X_i^2 has a smallest element b_i, which implies that $x_i < b_i \leq x_i'$.
2. There is a weight w_i associated with each criterion and a threshold λ such that a coalition is sufficient iff the sum of the weights of the criteria belonging to the coalition passes threshold λ: for all subset F of N, $F \in \mathcal{F}$ iff $\sum_{i \in N} w_i \geq \lambda$; we may assume w.l.o.g. that the weights are normalized $(\sum_{i \in N} w_i = 1)$.

Rule (1) can thus be rephrased as:

$$x = (x_1, \dots, x_i, \dots, x_n) \in X^2 \quad \text{iff} \quad \sum_{i \in N: x_i \geq b_i} w_i \geq \lambda. \tag{2}$$

To bridge the gap with the classical ELECTRE TRI model, let us consider that A is the set of alternatives and $g_i : A \rightarrow \mathbb{R}$ are functions associating each alternative $a \in A$ its evaluation on criterion i. Alternative a is hence represented by the n-tuple $(g_1(a), \dots, g_i(a), \dots, g_n(a)) \in X = \Pi_{i=1}^n X_i$. A is partitioned into two categories (A^1, A^2), with A^1 (resp. A^2) the set of "bad" (resp. "good") alternatives. We extend rule (2) to sets of alternatives having vectors in X as their evaluation on the n criteria and we assume that (A^1, A^2) satisfies the extension of rule (2), namely:

$$a \in A^2 \quad \text{iff} \quad \sum_{i \in N: g_i(a) \geq b_i} w_i \geq \lambda. \tag{3}$$

Clearly, (3) is also a particular case of the classical ELECTRE TRI (pessimistic) assignment rule.

In rules (2) or (3), the b_i's compose a vector $b \in \mathbb{R}^n$, which is the (lower) limit profile of category A^2. An alternative a belongs to A^2 iff its evaluations $g(a_i)$ are at least as good as b_i on a subset of criteria that has sufficient weight.

In the sequel, we call a model that assigns alternatives to (two) categories according to rule (3) a *Majority Rule Sorting Model* (MR-Sort). The parameters of such a model are the n components of the limit profile b, the weights of the criteria $w_1, \dots w_n$ and the majority threshold λ, in all $2n + 1$ parameters.

This setting can easily be generalized to sorting in k categories $(A^1, \dots, A^h, \dots, A^k)$ forming an ordered partition of A. The MR-Sort assignment rule is the following. Alternative $a \in A$ is assigned to category A^h, for $h = 2, \dots k - 1$ if

$$\sum_{i \in N: g_i(a) \geq b_i^{h-1}} w_i \geq \lambda \quad \text{and} \quad \sum_{i \in N: g_i(a) \geq b_i^h} w_i < \lambda \tag{4}$$

where $b^h = (b_1^h, \dots b_n^h)$ is the lower limit profile of category A^h. Alternative a is assigned to category A^1 if $\sum_{i \in N: g_i(a) \geq b_i^1} w_i < \lambda$. It is assigned to category A^k if $\sum_{i \in N: g_i(a) \geq b_i^{k-1}} w_i \geq \lambda$. A MR-Sort model with k categories involves $kn + 1$ parameters ($k - 1$ limit profiles, the weight vector and the majority threshold).

3 Learning a MR-Sort Model

Assuming that the decision maker is able to provide us with a number of a priori assignments of alternatives to categories, we may take advantage of this information to restrict the set of MR-Sort models compatible with such an information and possibly select one or several typical ones among them. Let $A^* \subseteq A$ be the subset of alternatives assigned to categories by the decision maker. A^* will be referred to as the *learning set*. Let $A^* = \{a_1, \ldots, a_j, \ldots, a_{na}\}$, where na is the number of alternatives in the learning set. In the case of two categories the DM's assignments result in a bipartition (A^{*1}, A^{*2}) of the learning set into a set of "bad" and "good" alternatives, respectively. These assignments generate constraints on the parameters of the MR-Sort models. Below, these constraints receive a linear formulation and are integrated into a mixed integer linear program (MIP) that is designed to select a particular feasible set of parameters.

3.1 The Case of Two Categories

We consider the case involving two categories separated by a frontier denoted b (for more than two categories see section 3.2). For $a_j \in A^*$, let us define the binary variables δ_{ij} $(i = 1, \ldots, n)$ such that $\delta_{ij} = 1 \Leftrightarrow g_i(a_j) \geq b_i$ and $\delta_{ij} = 0 \Leftrightarrow g_i(a_j) < b_i$. For δ_{ij} to be defined consistently, we impose the following constraints (M being an arbitrary large positive value):

$$\begin{cases} M(\delta_{ij} - 1) \leq g_i(a_j) - b_i < M \cdot \delta_{ij} \\ \delta_{ij} \in \{0, 1\} \end{cases} \tag{5}$$

Using the δ_{ij} binary variables, we define continuous variables c_{ij} such that $c_{ij} = 0 \Leftrightarrow \delta_{ij} = 0$ and $c_{ij} = w_i \Leftrightarrow \delta_{ij} = 1$ (where w_i denotes the weight of criterion g_i). To do so, we impose that $\delta_{ij} - 1 + w_i \leq c_{ij} \leq \delta_{ij}$ and $0 \leq c_{ij} \leq w_i$.

As we consider two categories, the set of assignment examples is defined by two subsets $A^{*1} \subset A$ and $A^{*2} \subset A$; A^{*1} (A^{*2}, respectively) is composed of alternatives which the DM intuitively assigns to the good (resp. bad) category. In order for these assignment examples to be reproduced by the MR-sort model, the constraints (6) should be posed.

$$\begin{cases} \sum_{i \in N} c_{ij} < \lambda, \forall a_j \in A^{*1} \\ \sum_{i \in N} c_{ij} \geq \lambda, \forall a_j \in A^{*2} \end{cases} \tag{6}$$

In order to discriminate among the MR-sort models compatible with the preference information provided by the DM (assignment examples A^{*1} and A^{*2}), we consider the objective function which maximize the robustness of assignment examples, as defined in (6). To do so, we introduce additional continuous variables, x_j and y_j, for each $a_j \in A^*$, and α defined in (7). Hence, maximizing $z = \alpha$ amounts to maximizing the value of the minimal slack in the constraints (6). Strict inequalities are transformed into non-strict ones by introducing an arbitrary small positive quantity ε.

$$\begin{cases} \sum_{i \in N} c_{ij} + x_j + \varepsilon = \lambda & \forall a_j \in A^{*1} \\ \sum_{i \in N} c_{ij} = y_j + \lambda & \forall a_j \in A^{*2} \\ \alpha \leq x_j & \forall a_j \in A^* \\ \alpha \leq y_j & \forall a_j \in A^* \end{cases} \tag{7}$$

This leads us to the following mathematical program :

$$\begin{cases} max\ \alpha \\ \quad \sum_{i \in N} c_{ij} + x_j + \varepsilon = \lambda & \forall a_j \in A^{*1} \\ \quad \sum_{i \in N} c_{ij} = \lambda + y_j & \forall a_j \in A^{*2} \\ \quad \alpha \leq x_j,\ \alpha \leq y_j & \forall a_j \in A^* \\ \quad c_{ij} \leq w_i & \forall a_j \in A^*, \forall i \in N \\ \quad c_{ij} \leq \delta_{ij} & \forall a_j \in A^*, \forall i \in N \\ \quad c_{ij} \geq \delta_{ij} - 1 + w_i & \forall a_j \in A^*, \forall i \in N \\ \quad M\delta_{ij} + \varepsilon \geq g_i(a_j) - b_i & \forall a_j \in A^*,\ \forall i \in N \\ \quad M(\delta_{ij} - 1) \leq g_i(a_j) - b_i & \forall a_j \in A^*,\ \forall i \in N \\ \quad \sum_{i \in N} w_i = 1,\ \lambda \in [0.5, 1] \\ \quad w_i \in [0, 1] & \forall i \in N \\ \quad c_{ij} \in [0, 1],\ \delta_{ij} \in \{0, 1\} & \forall a_j \in A^*,\ \forall i \in N \\ \quad x_j, y_j \in \mathbb{R} & \forall a_j \in A^* \\ \quad \alpha \in \mathbb{R} \end{cases} \tag{8}$$

3.2 More Than 2 Categories

It is not difficult to modify program (8) in order to deal with more than two categories. We consider the general case in which k categories are defined by $k-1$ limit profiles b^1, b^2, ..., b^h, ..., b^{k-1} (where $b^h = (b^h_1, ...b^h_n)$). For each alternative a_j in category A^{*h} of the learning set A^* (for $h = 2, \ldots, k-1$), we introduce $2n$ binary variables δ^h_{ij} and δ^{h-1}_{ij}, for $i = 1, \ldots, n$. We force δ^l_{ij} to be equal to 1 iff $g_i(a_j) \geq b^l_i$ for $l = h-1, h$ and $\delta^h_{ij} = 0 \Leftrightarrow g_i(a_j) < b^h_i$. We introduce $2n$ continuous variables c^l_{ij} $(l = h-1, h)$ constrained to be equal to w_i if $\delta^l_{ij} = 1$ and to 0 otherwise (as is done in (8)). Finally, we express that a_j is at least as good as profile b_{h-1} on a subset of criteria that has sufficient weight while this is not true w.r.t. profile b_h; we write constraints similar to (6) to express this. The case in which a_j belongs to one of the extreme categories (A^{*1} and A^{*k}) is simpler. It requires the introduction of only n binary variables and n continuous variables. Indeed if a_j belongs to A_1 we just have to express that the subset of criteria on which a_j is at least as good as b_1 has not sufficient weight. In a dual way, when a lies in A^k, the best category, we have to express that it is at least as good as the upper profile b^k on a subset of criteria that has sufficient weight.

3.3 Infeasible Learning Sets

The MIP programs presented in the two previous subsections may prove infeasible in case the assignments of the alternatives in the learning set are incompatible with all MR-sort models. In order to be able to tackle such problems we formulate a MIP that finds a MR-sort model maximizing the number of alternatives in the learning set that the model correctly assigns.

In the two categories case, for each $a_j \in A^*$, we introduce a binary variable γ_j which is equal to one if alternative a_j is correctly assigned by the MR-Sort model, and equal to zero otherwise. To ensure that the γ_j variables are correctly defined, we modify the constraints (6) in the following way:

$$\begin{cases} \sum_{i \in N} c_{ij} < \lambda + M(1-\gamma_j), \forall a_j \in A^{*1} \\ \sum_{i \in N} c_{ij} \geq \lambda - M(1-\gamma_j), \forall a_j \in A^{*2} \end{cases} \tag{9}$$

Starting from (8) and substituting constraints (6) by (9), and the objective function by the new objective $z = \sum_{a_j \in A^*} \gamma_j$, we obtain a MIP that yields a subset $A^{*\prime} \subseteq A^*$ of maximal cardinality that can be represented by an MR-Sort model. A generalization to more than two categories is obtained by bringing similar changes to the model described in section 3.2. These models will be used in the second and third experiments below (sections 4.1 and 4.2).

4 Empirical Design and Results

Our goal is to test the "learnability" of the MR-Sort model based on the previous MIP formulation. The three issues raised in the introduction, namely, model retrieval, tolerance for error, and idiosyncrasy are investigated through simulations. Such simulations involve generating alternatives, simulating a DM assigning these alternative, and learning an MR-Sort model from this information.

4.1 Experiment 1: Model Retrieval

Our strategy is the following. We generate a set of alternative, and a hypothetical MR-sort model, denoted M. Then we simulate the behavior of a DM assigning the generated alternatives, while having this MR-Sort model in mind. Hence, we constitute a learning set A^* by assigning the generated alternatives using the MR-Sort model. We infer an MR-Sort model M' compatible with the learning set using the MIP formulation presented in section 3. Although these models may be quite different, they coincide on the way they assign elements of A^*, by construction. In order to compare models M and M', we randomly generate a large set of other alternatives and we compute the percentage of "assignment errors", i.e. the proportion of these alternatives that models M and M' assign to different categories.

For small learning sets, it is expected that the inferred model is rather arbitrary. Therefore, our first experiment aims at investigating the following two issues:

- What is the typical size of a learning set which would lead to an inferred model "close" to the original one, i.e. yielding a small percentage of assignment errors?
- Does the MIP inference program remain tractable when the size of the problem increases ?

Generating alternatives. A set of na alternatives is generated. Each alternative is identified with a vector drawn from the unit hypercube $[0,1]^n$ (uniform distribution). Such a vector represents the evaluations of an alternative along n criteria. Note that these evaluations are independent random variables. Drawing the evaluations from the [0,1] interval is not restrictive, since the assignment rule is invariant up to a strictly increasing transformation of the scale of each criterion.

Simulating an MR-Sort model. n numbers are randomly drawn from the unit interval, then normalized, yielding weights w_i associated to each criterion i. A majority threshold λ is is randomly drawn from the $[0.5, 1]$ interval. To generate the $k-1$ vector profiles $b_h = (b_1^h, , \ldots, b_i^h, \ldots, b_n^h)$, we proceed as follows. For $i = 1, ..., n$, b_i^1 is randomly drawn from the $[0, \frac{2}{k}]$ interval, b_i^2 is randomly drawn from the $[b_i^1, \frac{3}{k}]$ interval; generally, b_i^h is randomly drawn from the $[b_i^{h-1}, \frac{h+1}{k}]$ interval, for $h = 2, ..., k-1$. In this way, we guarantee that each vector profile b^h dominates the previous one b^{h-1}. Moreover, for $h = 1, ..., k-1$, b_i^h divide the $[0,1]$ interval scale of criterion i in sub-interval that are of similar length (roughly speaking $\frac{1}{k}$ on average).

Empirical design. We run 10 instances of each of the problems obtained by varying the following parameters:

- Two, three categories,
- Three, four, five criteria,
- Learning sets containing 10 to 100 alternatives.

We use CPLEX to solve the MIP model (for several categories) and infer an MR-Sort model. We record the CPU time used. We also compute the proportion of vectors from a set B of 10 000 randomly generated alternatives that are assigned to the same category by the initial and inferred models. The lower this proportion, the greater the discrepancy between the original and inferred models.

Results. The main results of these experiments are summarized in figure 1. Figure 1a shows the average percentage of assignment errors as a function of the size of the learning set (from 10 to 100). The percentage of assignment errors is the percentage of the 10 000 alternatives in B that are not assigned to the same category by the original and the inferred model. The percentage represented is the *average* of the percentages observed on 10 learning sets instances. Figure 1b represents the computing time used to learn the inferred model as a function of the size of the learning set (all computations have been performed on a single standard PC).

Comments. On figure 1a we see that the assignment errors tend to decrease when the size of the learning set increases, in each simulation setting. This obviously means that the higher the cardinality of the learning set, the more determined the model. For a given size of the learning set, the error rate grows with the number

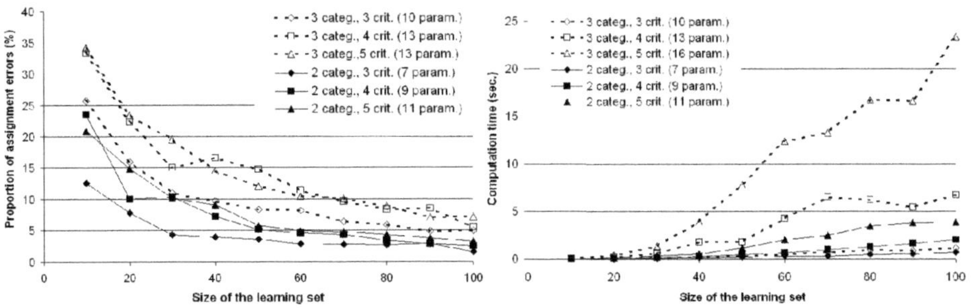

(a) Assignment errors (%) (b) CPU time

Fig. 1. Results of experiment 1

of model parameters (number of criteria, number of categories). To guarantee an error rate errors inferior to 10%, we typically need learning sets consisting of 40 (resp. 70) alternatives in the case of 2 (resp. 3) categories, 5 criteria). In other words, small learning sets (e.g. 10 or 20), do not allow to retrieve the original model parameters with good precision. Experiments of this type allow to estimate the size of the learning sets that yield a given level of the error rate.

Computing times (figure 1b) remain under 10 seconds on average for 2 categories and for 3 categories up to 4 criteria, even for the largest learning sets we consider (100 alternatives). However, the case of 3 categories and 5 criteria suggests that computing time could soon become unacceptable when the number of categories and/or criteria takes larger values (about 25 seconds for 100 alternatives in the learning set in the case of 3 categories and 5 criteria).

4.2 Experiment 2: Tolerance for Error

In this second experiment we study to which extent the model is "learnable" when the assignment examples are "approximately" simulated by an MR-Sort model, i.e. that a proportion of assignment errors has been introduced in the learning set. To do so, we randomly generate sets of na alternatives and an MR-Sort model as described in section 4.1. We then modify the obtained learning set by introducing a proportion of 5%, 10% and 15% of errors. These errors correspond to alternatives in the learning set which are randomly assigned to a category which differs from the one computed by the MR-Sort model.

As these learning sets include a proportion of erroneous assignments, it is possible that no MR-Sort model can restore the whole learning set. Therefore we compute the maximum number of alternatives of the learning set whose assignments is compatible with an MR-Sort model. This is done using the MIP formulation given in section 3.3. As in the first experiment, we run 10 instances of each of the problems obtained by varying the size of the learning sets (from 10 to 100 alternatives), with a fixed number of categories (two) and criteria (three).

Figure 2a represents the ratio of the maximal value of the objective function of the MIP by the size of the learning set, i.e. the maximal proportion of alternatives in the learning set whose assignments are compatible with a MR-Sort model (as a function of the percentage of assignment errors made in assigning the alternatives in the learning set to a category). Figure 2b represents the proportion of randomly generated evaluation vectors that are assigned to the same category by the initial and the inferred MR-Sort model. Figure 3 shows how the CPU time evolves with the size of the learning set and the proportion of errors.

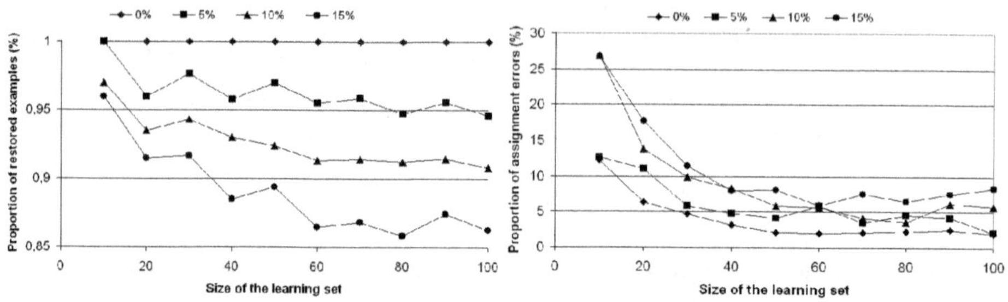

(a) Optimal objective value (%) (b) Assignment errors (%)

Fig. 2. "Learnability" of an MR-Sort model (2 categories, 3 criteria) using learning sets involving errors

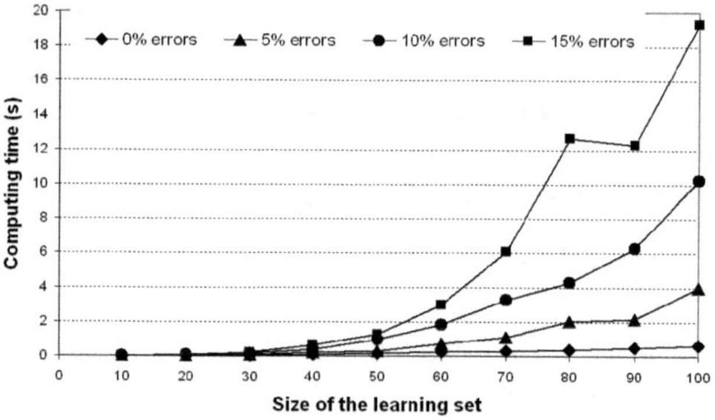

Fig. 3. Computing time in Experiment 2

Comments. On figure 2a we observe that the the maximal proportion of alternatives in the learning set whose assignments are compatible with a MR-Sort model decrease from a high value to reach asymptotically a minimum, when the size of the learning set increases. Moreover, it should be noted that, when the learning set is large, the proportion of restored examples in learning sets containing 5% (10%, 15%, respectively) errors approximately corresponds to 95% (90%, 85%, respectively). This means that, when the learning set is small, the MR-Sort model is flexible enough to reproduce almost all the learning set despite the errors; however, when the size of the learning set is large, as the MR-Sort model is more specific, the proportion of alternatives in the learning set whose assignment is not reproduced by the inferred model corresponds to the proportion of errors introduced in the learning set. Note however that alternatives in the learning set that are excluded when inferring the model do not necessarily correspond to the errors introduced in the learning set. However, the proportion of alternatives excluded when inferring the model is at most equal to the proportion of introduced errors.

On figure 2b we see that the proportion of randomly generated evaluation vectors that are assigned to different categories by the initial and the inferred MR-Sort model decreases with the size of the learning set, independently of the proportion of errors in the learning set. For sufficiently large learning sets (40 alternatives or more), the presence of errors in the learning set deteriorates the ability of the model to restore the assignment of random alternatives, but in a limited way only. For instance, a model inferred using a learning set of 100 alternatives with 15% errors induces 8% incorrect assignments, while a model inferred using a learning set of the same size with no error induce 2% incorrect assignments. It appears that the presence of a limited number of errors in the learning set does not strongly impact the "learnability" of the model.

Figure 3 shows that the CPU time increases with the size of the learning set, for all proportion of errors in the learning set. Moreover, for large learning sets (more than 50 alternatives) the proportion of errors in the learning set significantly impacts the CPU time. Although this experiment considers datasets with two categories and four criteria only, the average computing time with a learning set of 100 alternatives and 15% errors is approximately 20 seconds. Moreover, it should be recalled that the CPU time also increases in the case of error-free learning sets with the number of categories and criteria. This suggests that the inference program using a learning set with error might become intractable when the number of criteria and categories increases.

4.3 Experiment 3: Idiosyncratic Behavior

In the third experiment, we have tried to see to what extent an MR-Sort model is able to account for assignments made by another–definitely different–sorting model. In view of this, we have generated a sorting model based on an additive value function (AVF-Sort model). Such a model is used e.g. in the UTADIS method [5,17]. We generate such a model by slightly modifying the procedure designed for generating a MR-Sort model (see section 4.1).

Simulating an AVF-Sort model. We generate weights and profiles as for the MR-Sort model. For each vector profile $b_h = (b_1^h, , \dots, b_i^h, \dots, b_n^h)$, we compute an associated threshold $\beta^h = \sum_{i=1}^n w_i b_i^h$. Then we assign alternatives to categories by means of the following rule. Alternative $a = (g_1(a), \dots, g_n(a))$ is assigned to category A^h, for $h = 2, \dots, k$ if

$$\beta^{h-1} \leq \sum_{i \in N} w_i g_i(a) < \beta^h; \tag{10}$$

alternative a is assigned to category A^1 if $\sum_{i \in N} w_i g_i(a) < \beta^1$; it is assigned to category A^k if $\beta^{k-1} \leq \sum_{i \in N} w_i g_i(a)$.

As in the previous experiments, we assign 10 to 100 alternatives considered as forming a learning set to categories, using an AVF-Sort model. Then we run the

MIP described in section 3.3 to learn an MR-Sort model that assigns as many as possible of these alternatives to the same category as the AVF-Sort model.

Results. Figure 4 shows the proportion of alternatives in the learning set that a learned MR-Sort model has been able to assign to the same category as the original AVF-Sort model. Figure 4a shows the results for two categories and 3 to 5 criteria. Figure 4b shows similar results for three categories and 3 and 4 criteria. In the latter case, the maximal size of learning set is 80 (for larger size, computing times become excessive).

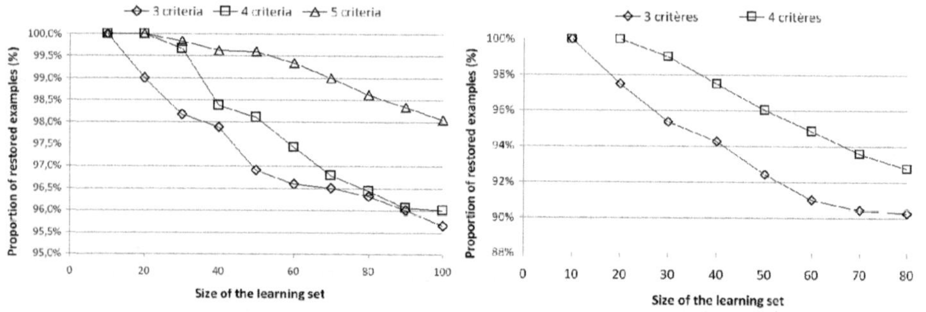

(a) Two categories (b) Three categories

Fig. 4. Results of experiment 3: maximal proportion of alternatives in learning set compatible with a MR-Sort model

Comments. It may come as a surprise that MR-Sort models are flexible enough to accommodate more than 95% (resp. 90%) of the alternatives assigned by the AVF-Sort model when there are two (resp. three) categories. Hence, it seems uneasy to detect, on the sole basis of the assignment of the alternatives in a learning set, which sorting model has been used to generate the learning set.

Another observation is the following. The larger the number of criteria, the higher the proportion of alternatives in the learning set that can be assigned consistently by the two models. This is surely due to the higher number of degrees of freedom (parameters) in the models when there are more criteria.

No extensive experimentation has been performed sofar on the way the learned MR-Sort model behaves when its assignments are compared to those of the original AVF-model on a large sample of generated alternatives. This has only be checked in the case of two categories and three criteria on a set of 100,000 generated alternatives. The proportion of these alternatives assigned to different classes by the two models amounts to 15.4%, a proportion significantly more important than that observed on the learning set.

5 Conclusion

This paper has experimentally investigated the feasibility of eliciting the parameters of an MR-Sort model based on a set of assignment examples. It has explored how the mathematical programs involved in the elicitation respond when exposed to three different learning settings.

The main insight resulting from these experiments is that eliciting an MR-Sort model is highly demanding in terms of information. A large number of assignment examples are required to reliably elicit such sorting models: from around 20 examples for the 7 parameters model with 2 categories and 3 criteria to around 75 examples for the 16 parameters model with 3 categories and 5 criteria; learning sets of that size guarantee less than 10% assignment errors when using the learned model. The need for much information is also attested by the fact that the MR-Sort model is able to accommodate both learning sets with errors and sets of assignments made using a different model.

This has implications for practice, which go beyond the learning of an MR-Sort model. We develop some of these implications below as well as some further research issues.

Parsimony in the choice of a model. The scarcity of available information may (should ?) drive the analyst from considering models involving many parameters (such as the classical ELECTRE TRI model) to simpler ones (such as MR-Sort). Our experiments tend to show that the expressive power of MR-Sort is more than sufficient when the learning set contains a few dozens of assignment examples. In such circumstances, the learning of models involving more parameters is likely to yield highly arbitrary results, since the learned model will be one among many possible models which equally well reproduce the assignment examples. Even in the context of the simple MR-Sort model, it could be advisable to model the decision problem using few categories when the preference information is scarce. Indeed reducing the number of categories–provided this is an option–will strongly influences the number of parameters to be elicited. The degree of refinement of the categorizations should be related to the number of assignment examples that the DM is expected to provide.

Questioning about parameters. In this work, we have only used assignment examples for learning the parameters of the model. To reduce the complexity of the learning process, one may think of directly asking the DM for the value of some of the parameters, such as the profiles or the weights, and learn the others. This approach has been proposed for ELECTRE TRI in [14,12]. It is however exposed to criticism since one cannot assume that the DM is aware of the meaning of the technical parameters of a method[1]; the cognitive value of spontaneous answers, for instance about the weights of criteria, is questionable (see [3], section 4.4). In principle, questions about parameters, asked to the DM, should preferably be formulated in terms of comparisons of alternatives, assignment to categories, etc, i.e. in terms of objects and issues related to the real decision problem.

Working with all models compatible with the assignment examples. In view of the vastly undetermined character of the parameters in preference models learned

[1] For instance, weights of criteria represent tradeoffs in an additive value function model, while they are used to measure the strength of coalitions of criteria in the ELECTRE methods.

on the basis of examples (not only in ordered assignment problems but also in ranking problems), it has been advocated ([8,9,7]) to work with all models compatible with the available information (assignment examples in a sorting problem, pairs belonging to the preference relation in a ranking problem, restriction on the range of parameters, etc). Valid recommendations hence are basically those shared by all models compatible with the information. Our experiments challenge the operational character of such an approach, also referred to as "Robust Ordinal Regression". It is likely that, unless the available information is very rich or the domain of variation of the parameters severely restricted, the conclusions compatible with all possible models will be very poor. In any case, this approach calls for empirical validation of its operational character. Note that experimental results similar to those presented in the present paper have been obtained for ranking problems under the additive value function model (see [15]).

Introducing vetoes in the MR-Sort model. As we have seen with our second experiment, considering sets of assignment examples that imperfectly follow an MR-Sort model leads to learned models that incorrectly assign some examples. In some cases, these examples have not been correctly assigned due to the fact that they are too much below the level of their category bottom profile on some criteria. Introducing vetoes in the MR-Sort model is a simple way of fixing such situations. Although it is possible indeed to learn both the parameters of a MR-Sort model and the veto threshold at a time (a mathematical program doing that is proposed in [10]), one could think to proceeding in steps. First elicit the MR-Sort model that best fits the assignment examples. Then examine the incorrectly assigned examples and see whether such incorrect assignments may be caused by veto effects. Finally, estimate the veto thresholds. An alternative approach whose objective is to minimize the number of criteria on which veto occurs is proposed in [4]. Further work should be devoted to developing the appropriate tools for estimating veto thresholds.

Selecting informative assignment examples. In our experiments, the assignment examples were generated randomly. Since the amount of information contained in the examples in view of determining the model is a crucial issue, one may want in practice to select as informative as possible assignment examples. An example is all the more informative that it strongly reduces the set of parameters compatible with its assignment. Developing a methodology for efficiently (i.e. by means of questions the answer of which is as informative as possible) eliciting sorting or ranking models by learning is an interesting research challenge.

In view of the issues raised above, we hope to have convinced the reader that the experimental analysis of learning methods in MCDA is a subject that has not received enough attention sofar. We believe that it deserves further efforts and we have tried to suggest a few new research directions.

Acknowledgment. We thank two anonymous referees for helpful comments. The usual caveat applies.

References

1. Bouyssou, D., Marchant, T.: An axiomatic approach to noncompensatory sorting methods in MCDM, I: The case of two categories. European Journal of Operational Research 178(1), 217–245 (2007)
2. Bouyssou, D., Marchant, T.: An axiomatic approach to noncompensatory sorting methods in MCDM, II: More than two categories. European Journal of Operational Research 178(1), 246–276 (2007)
3. Bouyssou, D., Marchant, T., Pirlot, M., Tsoukiàs, A., Vincke, P.: Evaluation and decision models with multiple criteria: Stepping stones for the analyst, Boston. International Series in Operations Research and Management Science, vol. 86 (2006)
4. Cailloux, O., Meyer, P., Mousseau, V.: Eliciting Electre Tri category limits for a group of decision makers. Tech. rep., Laboratoire Génie Industriel, Ecole Centrale Paris (June 2011), Cahiers de recherche (2011-09)
5. Devaud, J., Groussaud, G., Jacquet-Lagrèze, E.: UTADIS: Une méthode de construction de fonctions d'utilité additives rendant compte de jugements globaux. In: European Working Group on MCDA, Bochum, Germany (1980)
6. Dias, L., Mousseau, V.: Inferring ELECTRE's veto-related parameters from outranking examples. European Journal of Operational Research 170(1), 172–191 (2006)
7. Greco, S., Kadziński, M., Mousseau, V., Słowiński, R.: ELECTRE-GKMS: Robust ordinal regression for outranking methods. European Journal of Operational Research 214(10), 118–135 (2011)
8. Greco, S., Mousseau, V., Słowiński, R.: Ordinal regression revisited: multiple criteria ranking using a set of additive value functions. European Journal of Operational Research 191(2), 415–435 (2008)
9. Greco, S., Mousseau, V., Słowiński, R.: Multiple criteria sorting with a set of additive value functions. European Journal of Operational Research 207(3), 1455–1470 (2010)
10. Leroy, A.: Apprentissage des paramètres d'une méthode multicritère de tri ordonné, Master Thesis, Université de Mons, Faculté Polytechnique (2010)
11. Mousseau, V., Figueira, J., Naux, J.: Using assignment examples to infer weights for ELECTRE TRI method: Some experimental results. European Journal of Operational Research 130(2), 263–275 (2001)
12. Mousseau, V., Figueira, J., Naux, J.: Using assignment examples to infer weights for ELECTRE TRI method: Some experimental results. European Journal of Operational Research 130(2), 263–275 (2001)
13. Mousseau, V., Slowinski, R.: Inferring an ELECTRE TRI model from assignment examples. Journal of Global Optimization 12(2), 157–174 (1998)
14. Ngo The, A., Mousseau, V.: Using Assignment Examples to Infer Category Limits for the ELECTRE TRI Method. Journal of Multiple Criteria Decision Analysis 11(1), 29–43 (2002)
15. Pirlot, M., Schmitz, H., Meyer, P.: An empirical comparison of the expressiveness of the additive value function and the Choquet integral models for representing rankings. In: 25th Mini-EURO Conference Uncertainty and Robustness in Planning and Decision Making, URPDM 2010 (2010)
16. Roy, B., Bouyssou, D.: Aide multicritère à la décision: méthodes et cas. Economica Paris, Paris (1993)
17. Zopounidis, C., Doumpos, M.: PREFDIS: a multicriteria decision support system for sorting decision problems. Computers & Operations Research 27(7-8), 779–797 (2000)

Handling Preferences in the "Pre-conflicting" Phase of Decision Making Processes under Multiple Criteria

Dmitry Podkopaev* and Kaisa Miettinen

Department of Mathematical Information Technology,
University of Jyväskylä, Jyväskylä, Finland
{dmitry.podkopaev,kaisa.miettinen}@jyu.fi

Abstract. Multiple criteria decision making (MCDM) literature concentrates on the concept of conflicting objectives, which is related to focusing on the need of trading off. Most approaches to eliciting preferences of the decision maker (DM) are built accordingly on contradistinguishing different attainable levels of objectives. We propose to pay attention to the non-conflicting aspects of decision making allowing the DM to express preferences as a desirable direction of consistent improvement of objectives. We show how such preference information combined with a dominance relation principle results in a Chebyshev-type scalarizing model, which can be used in early stages of decision making processes for deriving preferred solutions without trading off.

Keywords: multiobjective optimization, preference expressing, scalarizing function, direction of improvement, trade-off coefficients.

1 Introduction and Problem Statement

The main task of multiple criteria decision making (MCDM) is usually understood as helping the decision maker (DM) in finding the most preferred solution in the presence of conflicting objectives. Most interactive methods of multiobjective optimization (see e. g. Steuer 1986; Miettinen 1999; Branke et al. 2008; Ruiz et al. 2011) concentrate on dealing with Pareto optimal solutions only, which means that improvement in some objective function value is possible only by allowing some other objective(s) to impair. The DM is typically asked (Miettinen et al. 2008) to express preferences either by comparing different Pareto optimal outcomes as e. g. in the algorithm by Steuer (1986), or by establishing aspiration levels of objective function values as in the reference point method by Wierzbicki (1981, 1986). Thus, the DM is accustomed to contradistinguishing different attainable levels of objectives and trading off.

In real-life problems formulated as multiobjective optimization problems, it is not always possible to obtain information about attainable objective function

* On leave from the System Research Institute, Polish Academy of Sciences, Warsaw, Poland.

R.I. Brafman, F. Roberts, and A. Tsoukiàs (Eds.): ADT 2011, LNAI 6992, pp. 234–246, 2011.

values or the structure of the Pareto optimal set. One can easily imagine at least two kinds of such situations: making decisions on something new (e. g. designing a new product/construction), and dealing with a problem where the Pareto optimal set exploration is associated with high computational cost. Expressing preferences in terms of attainable objective function values or Pareto optimal solution comparison becomes especially difficult on early stages of decision making processes, before the structure of the Pareto optimal set is revealed. Our research is focused on interactive multiobjective optimization problems and is aimed in overcoming such difficulties.

As identified in Miettinen et al. (2008), interactive solution approaches can often be characterized by first having a learning phase when the DM gets to know the attainable objective function values and his/her own preferences. Once the DM has identified an interesting region of solutions, the learning phase is followed by a decision phase where the final decision is made. In this paper we challenge the established practice of studying only Pareto optimal solutions in both phases and approach to enabling freer search in the learning phase by turning to the non-conflicting nature of objectives.

We claim that the DM perceives multiple objectives in decision making problems not as conflicting, but as mutually supportive. Indeed, it follows directly from the MCDM problem statement that all the objectives are to be optimized simultaneously, rather than some of them have to be improved at the expense of deteriorating other ones. Therefore we propose to represent DM's preferences as a direction of simultaneous improvement of objectives. Expressing such kind of aspiration does not require any knowledge about the solution set and, thereby, can be used in the learning phase before any Pareto optimal solutions are available. Once DM's preferences are expressed in this way, one should combine them with the Pareto dominance relation in order to enable deriving Pareto optimal solutions satisfying DM's preferences. This makes possible passing from the learning phase to the decision phase.

We develop an approach to handling with such preferences in combination with a dominance relation principle. To make this approach applicable we present a scalarizing function involving DM's preferences, as scalarization is in practice a very popular way of deriving solutions to multiobjective problems (see e. g. Miettinen 1999; Miettinen and Mäkelä 2002). To be more specific, we use a modified Chebyshev-type scalarizing function to characterize solutions satisfying DM's preferences.

Let us use the following general formulation of the multiobjective optimization problem:

$$\max_{x \in X} f(x), \tag{1}$$

where

X is the set of feasible solutions;
$k \geq 2$ is the number of objectives;
$f = (f_1, f_2, \ldots, f_k)$;
$f_i : X \to \mathbf{R},\ i \in N_k := \{1, 2, \ldots, k\}$, are objective functions.

Solving this problem means finding the most preferred solution, i. e. an element of X which is the most preferred from the DM's point of view. Assuming that the DM prefers more to less in each objective, we state by the operator "max" that the DM aims at maximizing all the objective function values simultaneously.

For each feasible solution x, we have the corresponding vector of objective function values $y = f(x)$ called *objective vector* or *outcome*. We assume that when choosing the most preferred solution, the DM takes into account only these values. Therefore, we consider problem (1) to be equivalent to the following problem of finding the most preferred outcome:

$$\max_{y \in Y} y, \tag{2}$$

where $Y = \{f(x) : x \in X\}$, $Y \subseteq \mathbf{R}^k$, is the outcome set, and $\mathbf{R}^k$ is called the objective space.

For any two vectors $v,\ w \in \mathbf{R}^q$, $q \in \mathbf{N}$, we define

$$v \geq w \text{ if and only if } v_i \geq w_i \text{ for any } i \in N_q \text{ and}$$

$$v > w \text{ if and only if } v_i > w_i \text{ for any } i \in N_q.$$

Given a set $Z \subseteq \mathbf{R}^k$, the subset of Pareto optimal objective vectors is defined by

$$P(Z) = \{z \in Z : \nexists z' \in Z \ (z' \geq z \text{ and } z' \neq z)\}.$$

A feasible solution is called Pareto optimal, if its outcome belongs to $P(Y)$. We share the widely accepted assumption that the most preferred solution of problem (1) should be Pareto optimal.

The paper is organized as follows. In Section 2 we define the direction of simultaneously improved objectives representing DM's preferences and in Section 3 we present an extension of Pareto dominance relation involving some DM's preference information in terms of relative importance of objectives. Section 4 combines both types of preference information into one preference model based on a Chebyshev-type scalarizing function. In Section 5 we discuss applicability of our approach and compare it with existing techniques of DM's preference elicitation. Finally we conclude in Section 6.

2 Direction of Proportional Improvement of Objectives

It is intuitively obvious that if a decision problem involves multiple goals, from the DM's point of view there is no sense in achieving one goal without achieving or with insufficient achievement of the other goals. For example designing a passenger car with extremely low fuel consumption but the maximum speed of 1 km/h; or investing to a portfolio with zero risk but vanishingly small profit or no profit at all do not have much sense. Moreover, in many practical decision making problems there are certain proportions in which the objectives should be improved to achieve the most intensive synergy effect. The idea of the most promising direction of simultaneous improvement of objectives agrees with the well-known assumption of concavity of the utility function (Guerraggio and Molho 2004), implying that this function grows faster in certain directions of simultaneous increase of objective function values.

The preference specification describing the direction of consistent improvement of objectives consists of a *starting point* in the objective space, and a vector representing a *direction of improvement*. In terms of problem (2), the starting point is defined by $s \in \mathbf{R}^k$ and the direction by $\delta \in \mathbf{R}^k$. Although it is not required for the starting point to be an outcome, it is assumed that s is meaningful for the DM. In other words, s represents some *hypothetical outcome*, which can be evaluated by the DM on the basis of his/her preferences. We emphasize the fact that the DM wants to improve all the objectives by setting $\delta > 0$.

The information represented by s and δ is interpreted as follows: *the DM wants to improve the hypothetical outcome s as much as possible, increasing the objective function values in proportions δ.*

The DM selects the starting point keeping in mind that it has to be improved then with respect to all objectives, i. e. the final solution outcome should have greater values of all components. Observe that the smaller are starting point components, the more likelihood that any outcome which is interesting for the DM can be obtained by increasing the starting point components. Taking into account this observation, we propose the following approaches to selecting s.

- Many real-life MCDM problems arise from the desire to improve the existing solution. The outcome of that solution can serve as the starting point.
- The DM may provide worst imaginable values of objective functions to use them as the starting point components.
- The nadir point defined by $y^{\text{nad}} = (y_1^{\text{nad}}, y_2^{\text{nad}}, \ldots, y_k^{\text{nad}})$, where $y_i^{\text{nad}} = \min\{y_i : \ y \in P(Y)\}$ (see for example Miettinen 1999) is a good candidate for the starting point. In the case of a computationally costly problem, evolutionary algorithms can be used to estimate components of y^{nad} (Deb, Miettinen and Chaudhuri 2010).

From the given starting point the DM defines the improvement direction by one of the following ways (or their combination).

- The DM sets the values $\delta_1, \delta_2, \ldots, \delta_k$ directly. This is possible when the DM understands the idea of the improvement direction and can operate with objective function values in his/her mind.
- The DM says that the improvement of objective i by one unit (the unitary increase of the i-th objective function value) should be accompanied by improvement of each other objective j, $j \neq i$, by a value θ_j. Thereby, the improvement direction is defined by $\delta_i = 1$ and $\delta_j = \theta_j$, $j \neq i$.
- The DM defines the above proportions freely for any pairs of objective functions. This can be implemented as an interactive procedure allowing the DM

to pick up any pair of objective functions i and j, $i \neq j$, and set the desirable ratio of improvement between them as θ_{ij}. A mechanism ensuring that $k(k-1)$ values θ_{ij} fully and consistently define k values $\delta_1, \delta_2, \ldots, \delta_k$ should then be used.
- The DM defines a reference point $r \in \mathbf{R}^k$, $r > s$ (not necessary $r \in Y$) representing a (hypothetical) outcome (s)he would like to achieve. Then the direction of improvement is defined by $r - s$.

Once DM's preferences are expressed as the improvement direction, a solution satisfying them can be determined. It is easy to explain to the DM the geometrical interpretation of such a solution outcome as the outcome which is farthest from s along the half-line $\{s + h\delta,\ h \geq 0\} \subset \mathbf{R}^k$, or in other words, the outcome solving the following single objective optimization problem:

$$\max\{s + h\delta : \quad h \in \mathbf{R},\ h > 0,\ s + h\delta \in Y\}. \tag{3}$$

We assume that the DM is aware of the possibility of situation depicted in Figure 1, where such an solution is not Pareto optimal. On the other hand, the DM is interested in Pareto optimal solutions only. This justifies inclusion of the Pareto optimality condition into the preference model.

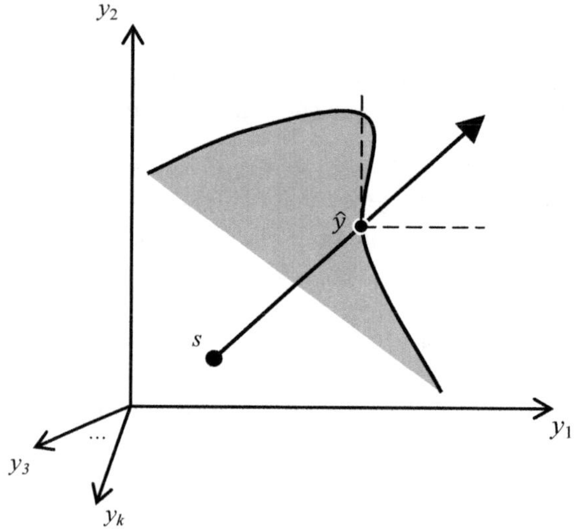

Fig. 1. Outcome $\hat{y}$ satisfying DM's preferences is not Pareto optimal, because it is dominated by other outcomes (outlined by dashed lines)

In the next section we present an extension of the Pareto optimality condition, which enables the DM to express some additional preference information.

3 Bounding Trade-Off Coefficients

Expressing preferences as a direction of simultaneous improvement of objectives allows the DM not to think in terms of Pareto optimal solutions and trading off, which can be useful in the learning phase of the decision making process, before any information about the Pareto optimal solution set is available. But even in this early phase, the DM may have some a priory judgments about relative importance of objectives. Let us describe a model based on bounding trade-off coefficients, which enables the DM to express such kind of preferences.

The idea of using bounds on trade-off coefficients for representing DM's preference information can be outlined as follows. Each Pareto optimal outcome y is characterized by $k(k-1)$ trade-off coefficients $t_{ij}(y)$, $i, j \in N_k$, $i \neq j$, where $t_{ij}(y)$ is defined as the ratio of increasing the i-th objective function value to decreasing the j-th objective function value when passing from y to other outcomes. The preferences of the DM are represented by values α_{ij} for some $i, j \in N_k$, $i \neq j$ where α_{ij} serves as the upper bound of $t_{ij}(y)$ for any $y \in Y$. The value α_{ij} is interpreted as follows: the DM agrees with a loss in value of the j-th objective function, if the value of i-th objective function will increase more than α_{ij} times the value of the loss. An outcome $y \in P(Y)$ cannot be considered as preferred by the DM, if there exist i and j, $i \neq j$, such that $t_{ij}(y) > \alpha_{ij}$. Indeed, the latter inequality means the existence of an outcome y' such that when moving from y to y', the DM receives gain in value of the i-th objective function which is greater than α_{ij} times the loss in value of the j-th objective function. Then y' is regarded as more preferred than y, thereby y cannot be considered as a candidate to be the most preferred outcome.

Summing up, the outcomes satisfying DM's preferences are only those Pareto optimal outcomes $y \in Y$, for which no one trade-off coefficient $t_{ij}(y)$ exceeds its upper bound α_{ij} whenever the latter is defined. Such outcomes are called trade-off outcomes of problem (2). Let us emphasize that the DM can define bounds on trade-off coefficients for all $k(k-1)$ pairs of different objective functions, as well as for only some of them.

In the next subsection we describe the approach to defining trade-off coefficients and deriving trade-off outcomes developed by Wierzbicki (1990), Kaliszewski (1994), and Kaliszewski and Michalowski (1997). In Subsection 3.2 we introduce its modification described in Podkopaev (2010), which allows the DM to express preferences more freely.

3.1 Global Trade-Off Approach

For any $y^* \in Y$ and $j \in N_k$, we define

$$Z_j(y^*, Y) = \{y \in Y : y_j < y_j^* \text{ and } y_s \geq y_s^* \text{ for all } s \in N_k \setminus \{j\}\}.$$

Definition 1. *Let $i, j \in N_k$, $i \neq j$. If $Z_j(y^*, Y) \neq \emptyset$, then the number*

$$T_{ij}(y^*, Y) = \sup_{y \in Z_j(y^*, Y)} \frac{y_i - y_i^*}{y_j^* - y_j} \tag{4}$$

is called a global trade-off coefficient *between the i-th and the j-th objective functions for outcome y^*. If $Z_j(y^*, Y) = \emptyset$, then $T_{ij}(y^*, Y) = -\infty$ by definition.*

The value $T_{ij}(y^*, Y)$ indicates, how much at most the outcome y^* can be improved in i-th objective relatively to its deteriorating in j-th objective when passing from y^* to any other outcome, under the condition that the other objectives are not impaired.

The DM defines bounds on trade-off coefficients α_{ij} for some $i, j \in N_k$, $i \neq j$. The bounds which are not defined by the DM are set to be infinite. A Pareto optimal outcome is called a *global trade-off outcome* of problem (1), if the following inequalities hold:

$$T_{ij}(y^*, Y) \leq \alpha_{ij} \text{ for any } i, j \in N_k, i \neq j. \tag{5}$$

The next result by Kaliszewski and Michalowski (1997) can be used for deriving global trade-off outcomes.

Theorem 1. *Let $y^0 \in \mathbf{R}^k$, $y_i^0 > y_i$ for all $y \in Y, i \in N_k$ and let $\rho_i > 0, i \in N_k$. If for some $\lambda_i > 0, i \in N_k$, outcome y^* is a solution to*

$$\min_{y \in Y} \max_{i \in N_k} \lambda_i \left((y_i^0 - y_i) + \sum_{j \in N_k} \rho_j \left(y_j^0 - y_j \right) \right), \tag{6}$$

then $y^* \in P(Y)$ and

$$T_{ij}(y^*, Y) \leq \frac{1 + \rho_j}{\rho_i} \quad \textit{for all } i, j \in N_k,\ i \neq j. \tag{7}$$

Parameters ρ_i, $i \in N_k$, introduced in Theorem 1, are used to implicitly define upper bounds on trade-off coefficients via (7). Thus problem (6) allows imposing upper bounds $\alpha_{ij}, i, j \in N_k$, $i \neq j$, on trade-off coefficients only if there exist ρ_i, $i \in N_k$, such that

$$\alpha_{ij} = \frac{1 + \rho_j}{\rho_i} \quad \text{for all } i, j \in N_k,\ i \neq j. \tag{8}$$

In the case of more than two objectives, this implies limiting the DM in expressing his/her preferences in the sense that among all possible combinations of bounds on trade-off coefficients defined by $(\alpha_{ij} > 0 :\ i, j \in N_k,\ i \neq j) \in \mathbf{R}^{k(k-1)}$ only those ones are available, which belong to the k-dimensional subset of $\mathbf{R}^{k(k-1)}$, defined by (8) for some ρ_i, $i \in N_k$.

3.2 B-Efficiency Approach

We apply a modification which allows the DM to define bounds on trade-off coefficients explicitly, with $k(k-1)$ degrees of freedom. The only restriction imposed on these bounds is the inequality system

$$\alpha_{is}\alpha_{sj} \geq \alpha_{ij} \text{ for any } i, j, s \in N_k, i \neq j, j \neq s. \tag{9}$$

These inequalities follow from assumptions of asymmetricity and transitivity of the DM's strict preference relation (Podkopaev 2008) and being explained to and accepted by the DM, do not actually restrict him/her in expressing preferences.

Let us transform the objective space with the following transformation matrix: $\mathrm{B} = [\beta_{ij}]_{k\times k} \in \mathbf{R}^{k\times k}$, where

$$\beta_{ij} = \frac{1}{\alpha_{ji}} \text{ for any } i, j \in N_k. \tag{10}$$

The transformed outcome set is defined by

$$\mathrm{B}Y = \{\mathrm{B}y : \ y \in Y\}.$$

For any set $Z \subseteq \mathbf{R}^k$, we define the subset of weakly Pareto optimal objective vectors:

$$W(Z) = \{z \in Z : \text{ for any } z' \in Z \text{ there exists } p \in N_k \text{ such that } z_p \geq z'_p\}.$$

We call elements of $W(\mathrm{B}Y)$ B-*efficient outcomes*. An outcome y^* is B-efficient, if no other outcome y dominates it in the following sence:

$$\mathrm{B}y > \mathrm{B}y^*. \tag{11}$$

It has been proved in Podkopaev (2007) that whenever bounds on trade-off coefficients are finite and inequalities (9) hold, any element of $W(\mathrm{B}Y)$ is a Pareto optimal outcome of problem (1) satisfying bounds on trade-off coefficients (5), i. e. it is a global trade-off outcome of problem (1) (defined in Subsection 3.1). The converse is not generally true, i. e. not every global trade-off outcome belongs to $W(\mathrm{B}Y)$.

To explain the difference between global trade-off outcomes and B-efficient outcomes, we need to represent DM's preferences in terms of values β_{ij} instead of α_{ij} in order to enable giving interpretation of the B-efficiency concept. For any $i, j \in N_k$, $i \neq j$, the value β_{ij} has the clear meaning as the highest price in terms of the i-th objective function loss, which the DM agrees to pay for the unitary gain in value of the j-th objective function.

Let $y^* \in P(Y)$. It follows from the definition that y^* is not a global trade-off outcome, if some other outcome $y \in Z_i(y^*, Y)$ dominates it in the following sence:

$$y_i^* - y_i < \beta_{ij}\left(y_j - y_j^*\right) \text{ for some } j \in N_k \setminus \{i\}.$$

It is proved in Podkopaev (2008) that y^* is not a B-efficient outcome, if for some $y \in Z_i(y^*, Y)$ we have

$$y_i^* - y_i < \sum_{j \in N_k \setminus \{i\}} \beta_{ij}\left(y_j - y_j^*\right). \tag{12}$$

Thus, in terms of bounds on global trade-off coefficients, the DM considers y better than y^* if the amount of decreasing i-th objective function (when passing from y^* to y) is small enough to be accepted by the DM in exchange for increasing any of the other objective functions. In the approach based on B-efficient solutions the amount of decreasing the i-th objective function is compared to the weighted sum of amounts of increasing all the other objective functions. In other words, all the gains from increasing the other objective functions are taken into account simultaneously.

Provided that the idea of trade-off coefficients and the meaning of values α_{ij} or β_{ij} are explained to the DM, (s)he can express preferences by defining either of these two sets of values. Let us remind that it is not necessary to get information about all $k(k-1)$ bounds on trade-off coefficients. The DM can set or modify bounds on trade-off coefficients for selected pairs of objectives one-by-one. The issue of tracking down that conditions (9) are satisfied during such a process is addressed in Podkopaev (2010).

4 Preference Model

We are now in a position to construct the model of DM's preferences from the two types of preference information described in two previous sections. In order to make the model applicable we address the following two issues. At first, the DM has to be aware of how his/her preference information is used. We explain how a solution satisfying both types of preference information is selected from DM's perspective. Secondly, a mathematical technique for deriving such a solution is to be provided. We construct a scalarization model for this purpose.

The preference information obtained from the DM consists of the following parts:

- the starting point defined as a (hypothetical) outcome s;
- the direction of consistent improvement of objectives defined as a positive vector δ in the outcome space;
- (optional) the bounds of trade-off coefficients defined as positive numbers β_{ij} for all or some of pairs of objective functions $i, j \in N_k, i \neq j$.

We assume that the DM agrees with the idea of applying this preference information for selecting a solution as follows: searching for the outcome which is farthest from s in the direction δ and if this outcome is dominated[1] by some other outcome, trying to improve it even more applying the domination principle. Let us explain this selection process in detail from the DM's perspective.

As stated in Section 2, the DM aspires at improving objective function values, moving from the starting point $s \in Y$ along the consistent improvement direction $\delta \in \mathbf{R}^k$ as far as possible inside the outcome set. Let y^0 denote the farthest outcome in this direction (defined as the solution to (3)). If y^0 is B-efficient, then it cannot be further improved based on the available information and thereby is considered as satisfying DM's preferences. If y^0 is not B-efficient, then there exists an outcome dominating it. In this case an outcome dominating y^0 is selected as detailed below.

Given a point z on the line defined by the consistent improvement direction, let us call *superior to z* any outcome dominating z. If y^0 is not B-efficient, then it

[1] Hereinafter we use the notion of domination only in the sense of the domination relation related to bounding trade-off coefficients and defined by (11).

has a superior. Let us continue moving from y^0 along the improvement direction until we find the farthest point in this direction having a superior. Denote this farthest point by $\bar{y}$. The outcome satisfying DM's preferences can be selected among any superiors of $\bar{y}$.

Denote by $\hat{y}$ the outcome selected in the above described way. To show that $\hat{y}$ can be considered to satisfy DM's preferences (in the case where y^0 is not B-efficient), it is enough to observe that $\hat{y}$ dominates $\bar{y}$, and $\bar{y}$ is more preferred than y^0 (since it is located farther from s in the direction of improvement). Thus $\hat{y}$ is more preferred than y^0. Besides that, as follows from Theorem 2 below, there does not exist an outcome dominating $\hat{y}$ in the sence of B-efficiency.

Figure 2 illustrates how the solution selection rule based on DM's preferences can be explained to the DM in the case where y^0 is not B-efficient. The dashed lines represent borders of the sets of vectors in the objective space which dominate y^0 and $\bar{y}$.

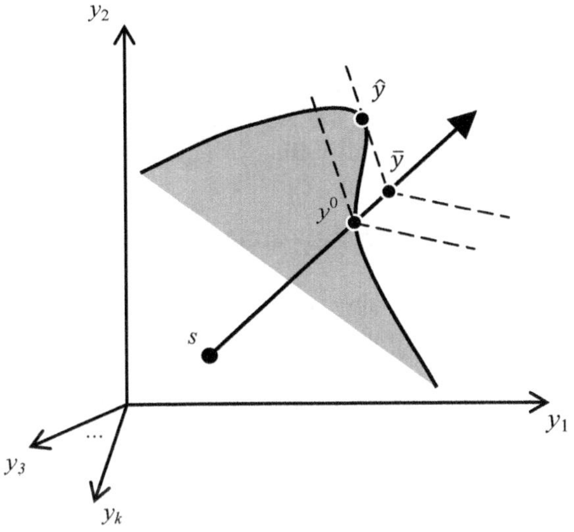

Fig. 2. Selecting solution $\hat{y}$ satisfying DM's preferences

The next theorem provides mathematical technique for deriving solutions based on DM's preferences according to the described above rules.

Theorem 2. *Suppose that $\alpha_{ij} < \infty$, $i, j \in N_k$, $i \neq j$, and inequalities (9) hold. Let y^* be a solution of*

$$\min_{y \in Y} \max_{i \in N_k} \frac{1}{\delta_i} \left((s_i - y_i) + \sum_{\substack{j \in N_k \\ j \neq i}} \frac{s_j - y_j}{\alpha_{ji}} \right). \tag{13}$$

Then the following three statements are true.

1) Solution y^* is a B*-efficient outcome of problem (1).*

2) If the solution of (3) is B*-efficient, then it coincides with y^*.*

3) If the solution of (3) is not B*-efficient, then y^* is a superior to the farthest point having superiors along the half-line $s + \delta h,\ h \geq 0$.*

The theorem is proved easily based on the fact that the level curves of the scalarizing function are borders of domination cones with apexes lying on the half-line $s + \delta h,\ h \geq 0$.

Theorem 2 states that an outcome satisfying DM's preferences expressed as starting point s, direction δ and bounds on trade-off coefficients $\alpha_{ij},\ i, j \in N_k,\ i \neq j$, can be obtained as a solution of Chebyshev-type scalarized problem (13).

Remark 1. Earlier we mentioned that those bounds on trade-off coefficients α_{ij} which are not defined by the DM should be set to infinity. But in Theorem 2 we require that all of them are finite. This condition is necessary for ensuring that a solution obtained from (13) is Pareto optimal (see for example Wierzbicki 1986), otherwise only weak Pareto optimality is guaranteed. Therefore we propose to assign large enough numbers to all undefined bounds on trade-off coefficients, so that they will have negligibly small influence on the preference model.

5 Application of the Preference Model

Based on Theorem 2 we can suggests the following procedure for deriving a solution satisfying DM's preferences:

- The DM expresses preferences in the form of a direction of consistent improvement of objectives and possible bounds on trade-off coefficients.
- The preference information is presented as values of parameters $(s_1, s_2, \ldots, s_k)$ (the starting point), $(\delta_1, \delta_2, \ldots, \delta_k) \in \mathbf{R}^k$ (the direction of improvement of objectives), and possible $\alpha_{ij},\ i, j \in N_k,\ i \neq j$ (bounds on trade-off coefficients).
- Problem (13) is solved providing a solution which satisfies DM's preferences.

This procedure can be incorporated in any decision making method, whenever there is need for eliciting preference information in terms of desirable proportions of simultaneous improvements, and possibility to solve scalarized problem (13). As an example let us mention the interactive method NAUTILUS developed by Miettinen et al. (2010), where the exploration of the outcome set is entirely based on gradual improvement of non-Pareto optimal outcomes with respect to all objectives simultaneously. Although NAUTILUS utilizes different ways of DM's preference elicitation, our technique can be incorporated there without changes.

Observe that problem (13) is very similar to (6) and other scalarized problems, used for deriving solutions in reference-point-based methods (see for example Wierzbicki 1981, 1986, 1990; Kaliszewski 1994; Kaliszewski and Michalowski

1997). The main difference of our approach is the way how DM's preferences are elicited and the solution selection process is interpreted. In reference-point-based methods a solution closest (in some sense) to the reference point is searched for and therefore the absolute position of the reference point has a crucial meaning. In our approach, setting the reference point is one of many ways to define the desired proportions of objective function improvement. At that only the direction in which the reference point is located with respect to the starting point is important.

The concept of proportional improvement of objectives is very similar to (and to a large degree inspired by) the consensus direction technique of deriving preferred solutions, which was developed by Kaliszewski (2006). That technique is based on specifying a direction in the objective space, but in contrast to our approach, it is interpreted as a direction of proportional deterioration of objectives starting from a reference point.

6 Conclusions

We have presented an approach to expressing preference information as proportions, in which the DM wishes to improve objectives. It can be applied when attainable levels of objective function values are unknown and other methods of expressing preference relying on such knowledge cannot be used. To derive solutions satisfying DM's preferences, one can use the scalarized problem based on a modification of the widely used Chebyshev-type scalarization. This technique can be incorporated into any MCDM method, where the DM's preference can be expressed in an appropriate way.

The presented technique of eliciting DM's preferences and deriving preferred solutions is very simple. The main purpose of describing it is drawing attention to non-conflicting aspects of MCDM and showing that one can easily operate with preference information based on the idea of mutually supportive objectives.

References

1. Branke, J., Deb, K., Miettinen, K., Slowinski, R. (eds.): Multiobjective Optimization: Interactive and Evolutionary Approaches. Springer, Heidelberg (2008)
2. Deb, K., Miettinen, K., Chaudhuri, S.: Towards an Estimation of Nadir Objective Vector Using a Hybrid of Evolutionary and Local Search Approaches. IEEE Transactions on Evolutionary Computation 14(6), 821–841 (2010)
3. Guerraggio, A., Molho, E.: The origins of quasi-concavity: a development between mathematics and economics. Historia Mathematica 31, 62–75 (2004)
4. Kaliszewski, I.: Qualitative Pareto analysis by cone separation technique. Kluwer Academic Publishers, Boston (1994)
5. Kaliszewski, I.: Multiple criteria decision making: selecting variants along compromise lines. Techniki Komputerowe 1, 2006, 49–66 (2006)
6. Kaliszewski, I., Michalowski, W.: Efficient solutions and bounds on trade-offs. Journal of Optimization Theory and Applications 94, 381–394 (1997)

7. Miettinen, K.: Nonlinear Multiobjective Optimization. Kluwer Academic Publishers, Boston (1999)
8. Miettinen, K., Eskelinen, P., Ruiz, F., Luque, M.: NAUTILUS method: An interactive technique in multiobjective optimization based on the nadir point. European Journal of Operational Research 206, 426–434 (2010)
9. Miettinen, K., Mäkelä, M.M.: On scalarizing functions in multiobjective optimization. OR Spectrum 24, 193–213 (2002)
10. Miettinen, K., Ruiz, F., Wierzbicki, A.P.: Introduction to Multiobjective Optimization: Interactive Approaches. In: Branke, J., Deb, K., Miettinen, K., Słowiński, R. (eds.) Multiobjective Optimization. LNCS, vol. 5252, pp. 27–57. Springer, Heidelberg (2008)
11. Podkopaev, D.: An approach to finding trade-off solutions by a linear transformation of objective functions. Control and Cybernetics 36(2), 347–356 (2007)
12. Podkopaev, D.: Representing partial information on preferences with the help of linear transformation of objective space. In: Trzaskalik, T. (ed.) Multiple Criteria Decision Making 2007, pp. 175–194. The Karol Adamiecki University of Economics in Katowice Scientific Publications (2008)
13. Podkopaev, D.: Incorporating Explicit Tradeoff Information to Interactive Methods Based on the Chebyshev-type Scalarizing Function. Reports of the Department of Mathematical Information Technology. Series B: Scientific Computing. No. B9/2010. University of Jyväskylä, Jyväskylä (2010)
14. Ruiz, F., Luque, M., Miettinen, K.: Improving the computational efficiency in a global formulation (GLIDE) for interactive multiobjective optimization. Annals of Operations Research (2011), `http://dx.doi.org/10.1007/s10479-010-0831-x`
15. Steuer, R.E.: Multiple Criteria Optimization: Theory, Computation and Application. Wiley Series in Probability and Mathematical Statistics. John Wiley, New York (1986)
16. Wierzbicki, A.P.: A mathematical basis for satisficing decision making. In: Morse, J.N. (ed.) Organizations: Multiple Agents with Multiple Criteria. LNEMS, vol. 190, pp. 465–485. Springer, Berlin (1981)
17. Wierzbicki, A.P.: On the completeness and constructiveness of parametric characterization to vector optimization problems. OR Spectrum 8, 73–87 (1986)
18. Wierzbicki, A.P.: Multiple criteria solutions in noncooperative game theory, part III: theoretical foundations. Discussion Paper No. 288. Kyoto Institute of Economic Research (1990)

Bribery in Path-Disruption Games*

Anja Rey and Jörg Rothe

Institut für Informatik, Universität Düsseldorf, 40225 Düsseldorf, Germany

Abstract. Bachrach and Porat [1] introduced path-disruption games. In these coalitional games, agents are placed on the vertices of a graph, and one or more adversaries want to travel from a source vertex to a target vertex. In order to prevent them from doing so, the agents can form coalitions, and a coalition wins if it succeeds in blocking all paths for the adversaries. In this paper, we introduce the notion of bribery for path-disruption games. We analyze the question of how hard it is to decide whether the adversaries can bribe some of the agents such that no coalition can be formed that blocks all paths for the adversaries. We show that this problem is NP-complete, even for a single adversary. For the case of multiple adversaries, we provide an upper bound by showing that the corresponding problem is in Σ_2^p, the second level of the polynomial hierarchy, and we suspect it is complete for this class.

1 Introduction

Consider the following scenario that might occur in a network application. An intruder wants to send data from a source computer to a target computer and a security system has the task to prevent this from happening. Situations like this can be modeled in game-theoretic terms. For example, Bachrach and Porat [1] introduced path-disruption games, cooperative games where agents are located on the vertices of a graph and one or more adversaries want to travel from a source vertex to a target vertex. To stop them, the agents might form coalitions that block all paths for the adversaries. If a coalition of agents succeeds in doing so, it wins the game.

We will focus on path-disruption games here, but mention that such situations can be modeled in terms of a noncooperative game as well. For example, Jain et al. [2] considered zero-sum security games on graphs, motivated by a real-life scenario where the Mumbai police located a limited number of inspection checkpoints on the road network of the city to prevent what had happened in the Mumbai attacks of 2008: The attackers entered the city on certain entrance points (corresponding to the source vertices) and then tried reach certain target locations (corresponding to the target vertices) to launch their attacks.

As the above example shows, path-disruption games do not only have applications in network security but also in other settings whenever an adversarial player

* This work was supported in part by DFG grant RO 1202/12-1 and the European Science Foundation's EUROCORES program LogICCC.

R.I. Brafman, F. Roberts, and A. Tsoukiàs (Eds.): ADT 2011, LNAI 6992, pp. 247–261, 2011.

may wish to travel through a graph and agents want to prevent that. In computer science, such situations may also occur in the field of multiagent systems. The computational analysis of social-choice-theoretic scenarios (a field known as *computational social choice*, see, e.g., [3,4,5]) and game-theoretic scenarios (known as *algorithmic game theory*) have become fields of increasing interest in recent years. In particular, coalitional games (such as weighted voting games [6,7], network flow games [8,9,10], etc.) have been analyzed from a computational complexity point of view.

In cooperative game theory, a key question is to analyze the stability of games, that is, to determine which coalition will form and how to divide the payoff within a coalition (see, e.g., Bachrach et al. [11] for the cost of stability in coalitional games). Path-disruption games combine the ideas of cooperative game theory, where agents have common interests and collaborate, with an aspect from noncooperative game theory by also considering an adversary who can actively interfere with the situation in order to achieve his or her individual goals in opposition to the agents. Inspired by bribery in the context of voting (see Faliszewski et al. [12]), we introduce the notion of bribery in path disruption games. Here, the adversary breaks into the setting and tries to change the outcome to his or her advantage by paying a certain amount of money, without exceeding a given budget.

In particular, we analyze the complexity of the problem of whether the adversaries in a path-disruption game can bribe some of the agents such that no coalition will be formed preventing the adversaries from reaching their targets. We show that this problem is NP-complete, even for a single adversary. For the case of multiple adversaries, we provide an upper bound by showing that the corresponding problem is in Σ_2^p, the second level of the polynomial hierarchy [13,14], and we suspect it is complete for this class. Besides this we leave new approaches and related problems open for further discussion.

Section 2 gives the needed notions from complexity theory, coalitional game theory, and graph theory. In Section 3, path-disruption games are formally defined. Bribery is introduced in Section 4. We present our complexity results in Section 5. Finally, a conclusion and future work can be found in Section 6.

2 Preliminaries

Let $\mathbb{R}$, $\mathbb{R}_{\geq 0}$, and $\mathbb{Q}_{\geq 0}$ denote the set of real numbers, nonnegative real numbers, and nonnegative rational numbers, respectively. Let $\mathbb{N}^+ = \{1, 2, \dots\}$ denote the set of positive integers.

A *coalitional game* consists of a set of *players* N and a *coalitional function* $v : \mathfrak{P}(N) \rightarrow \mathbb{R}$. When considering a multiagent application, *players* in a coalitional game are often referred to as *agents*. Here, the terms *agent* and *player* are used synonymously. A *simple game* is a coalitional game, where $v(S) \leq v(T)$ for $S \subseteq T \subseteq N$ (monotonicity) and a coalition $C \subseteq N$ either wins or loses the game, i.e., the coalitional function is the characteristic function $v : \mathfrak{P}(N) \rightarrow \{0, 1\}$. Further basics on game theory can be found, e.g., in the textbook by Osborne and Rubinstein [15].

A graph $G = (V, E)$ can be either directed or undirected. We analyze path-disruption games on undirected graphs, as this is the more demanding case regarding the computational hardness results. Given an undirected graph, we can simply reduce the problem to the more general case of a directed graph by substituting each undirected edge $\{u, v\}$ by the two directed edges (u, v) and (v, u). Given a graph $G = (V, E)$, we denote an *induced subgraph restricted to a subset of edges* $E' \subseteq E$ by

$$G|_{E'} = (V, E')$$

and an *induced subgraph restricted to a subset of vertices* $V' \subseteq V$ by

$$G|_{V'} = (V', \{\{v, u\} \in E \mid v \in V' \wedge u \in V'\}).$$

We assume the reader is familiar with the basic notions of complexity theory, such as the complexity classes P, NP, and $\Sigma_2^p = \mathrm{NP}^{\mathrm{NP}}$ (which is the second level of the polynomial hierarchy [13,14]) and the notion of (polynomial-time many-one) reducibility, denoted by $\leq_{\mathrm{m}}^{\mathrm{p}}$, and hardness and completeness with respect to $\leq_{\mathrm{m}}^{\mathrm{p}}$. For further reading we refer to the textbooks by Papadimitriou [16] and Rothe [17].

Two well-known NP-complete problems (see, e.g., [18]) that will be used in this paper are defined as follows. In the first problem, Partition, we ask whether a sequence of positive integer weights can be partitioned into two subsequences of equal weight.

Partition	
Given:	A nonempty sequence of positive integers $A = (a_1, \ldots, a_n)$ such that $\sum_{i=1}^{n} a_i$ is even.
Question:	Is there a subset $A' \subseteq A$ such that $\sum_{a_i \in A'} a_i = \sum_{a_i \in A \setminus A'} a_i$?

The second problem is also a partitioning problem, but now the question is whether the vertex set of a given graph with edge weights can be partitioned into two vertex sets such that the total weight of the edges crossing this cut is at least as large as a given value.

MaxCut	
Given:	A graph $G = (V, E)$, a weight function $w : E \to \mathbb{N}^+$, and a bound $K \in \mathbb{N}^+$.
Question:	Is there a partition of the vertex set V into two disjoint subsets $V_1, V_2 \subseteq V$ such that $\sum_{\{u,v\} \in E, u \in V_1, v \in V_2} w(\{u, v\}) \geq K$?

Our results are also based on a further decision problem mentioned by Bachrach and Porat [1]:

MultipairCut with Vertex Costs (MCVC)	
Given:	A graph $G = (V, E)$, m vertex pairs (s_j, t_j), $1 \leq j \leq m$, a weight function $w : V \to \mathbb{R}_{\geq 0}$, and a bound $k \in \mathbb{R}_{\geq 0}$.
Question:	Is there a subset $V' \subseteq V$ such that $\sum_{v \in V'} w(v) \leq k$ and $G\|_{V \smallsetminus V'}$ contains no path linking a pair (s_j, t_j), $1 \leq j \leq m$?

Proposition 1. MCVC *belongs to* P *for problem instances with* $m < 3$, *yet is* NP*-complete for problem instances with* $m \geq 3$.

The related optimization problem for $m < 3$ can be solved in polynomial time using the same algorithm as the decision problem with a corresponding output.

3 Path-Disruption Games

Following Bachrach and Porat [1], we define several path-disruption games (for short, PDGs) on graphs. Given a graph $G = (V, E)$ with $n = \|V\|$ vertices, each agent $i \in N = \{1, \ldots, n\}$ represents vertex v_i. Moreover, there are several adversaries who want to travel from a source vertex s to a target vertex t in V. We say a coalition $C \subseteq N$ *blocks a path from* s *to* t if there is no path from s to t in the induced subgraph $G|_{V \smallsetminus \{v_i \mid i \in C\}}$ or if s or t are not even in $V \smallsetminus \{v_i \mid i \in C\}$.

Bachrach and Porat [1] distinguish four types of path-disruption games: PDGs with a single adversary and with multiple adversaries, and for both with and without costs. We denote path-disruption games with costs by PDGC, and path-disruption games without costs by PDG. The most general game is the model with several adversary players and costs for each vertex to be blocked.

PDGC-Multiple	
Domain:	A graph $G = (V, E)$, $n = \|V\|$, a cost function $c : V \to \mathbb{R}_{\geq 0}$, a reward $r \in \mathbb{R}_{\geq 0}$, and adversaries $(s_1, t_1), \ldots, (s_m, t_m)$.
Agents:	$N = \{1, \ldots, n\}$, i represents v_i, $1 \leq i \leq n$.
Coal. Fcn.:	$v(C) = \begin{cases} r - m(C) & \text{if } m(C) < \infty, \\ 0 & \text{otherwise} \end{cases}$ with $m(C) = \begin{cases} \min\{c(B) \mid B \subseteq C \wedge \tilde{v}(B) = 1\} & \text{if } \tilde{v}(C) = 1, \\ \infty & \text{otherwise,} \end{cases}$ where $c(B) = \sum_{i \in B} c(v_i)$ and $\tilde{v}(C) = \begin{cases} 1 & \text{if } C \text{ blocks each path from } s_j \text{ to } t_j \text{ for each } j, 1 \leq j \leq m, \\ 0 & \text{otherwise.} \end{cases}$

Letting $m = 1$, we have a restriction to a single adversary, namely PDGC-Single. Letting $c(v_i) = 0$ for all i, $1 \leq i \leq n$, $r = 1$, and $v(C) = \tilde{v}(C)$, the simple games without costs, PDG-Multiple and PDG-Single, are defined. We say a coalition $C \subseteq N$ *wins the game* if $\tilde{v} = 1$, and *loses* otherwise.

In the definition of path-disruption games, weights and bounds are real numbers. However, to make the problems for these games suitable for computer processing (and to define their complexity in a reasonable way), we will henceforth assume that all weights and bounds are rational numbers. The same holds for MCVC as defined in Section 2 and the bribery problems for path-disruption games to be introduced in the following section.

4 Bribery

Given a PDG or PDGC, can an adversary (s,t) bribe a coalition $B \subseteq N$ of agents such that no coalition $C \subseteq N$ will be formed that blocks each path from s to t? There are several possibilities to define such a decision problem. Considering the simplest form of PDG, single adversary without costs and with constant prices for each agent and an infinite budget for the adversary, the answer is yes if and only if $(G, s, t) \in$ GAP, where GAP is the graph accessibilty problem (see, e.g., [17]): Given a graph G and two distinct vertices, a source vertex s and a target vertex t, can t be reached via a path from s? This problem can be solved in nonlogarithmic space (and thus in polynomial time). Since bribery of all agents on a path from s to t will guarantee the adversary a safe travel, the equivalence holds. In the following we consider bribery on a PDG with costs.

PDGC-Single-Bribery	
Given:	A PDGC with $m = 1$, a price function $b : V \to \mathbb{Q}_{\geq 0}$, and a budget $k \in \mathbb{Q}_{\geq 0}$.
Question:	Is there a coalition $B \subseteq N$ such that $\sum_{i \in B} b(v_i) \leq k$, and no coalition $C \subseteq N \setminus B$ has a value $v(C) > 0$?

Analogously, the multiple-adversary case PDGC-Multiple-Bribery can be defined.

5 Complexity Results

In this section, we give complexity results for the bribery problems in path-disruption games. Theorem 1 classifies PDGC-Single-Bribery in terms of its complexity.

Theorem 1. PDGC-Single-Bribery *is* NP*-complete.*

Proof. First we show that the problem is in NP. Given a PDG consisting of

- a graph $G = (V, E)$,
- a cost function $c : V \to \mathbb{Q}_{\geq 0}$,
- a reward $r \in \mathbb{Q}_{\geq 0}$,
- a source and a target vertex, $s, t \in V$,
- a price function $b : V \to \mathbb{Q}_{\geq 0}$, and
- a bound $k \in \mathbb{Q}_{\geq 0}$,

we can nondeterministically guess a coalition $B \subseteq N$, $N = \{1, \ldots, n\}, n = \|V\|$. Obviously, it can be tested in polynomial time whether $\sum_{i \in B} b(v_i) \leq k$. If this inequality fails to hold, bribery of B is not possible. Otherwise, we need to test whether it holds for all coalitions $C \subseteq N \setminus B$ that $v(C) \leq 0$. That is the case if and only if

- either $\tilde{v}(C) = 0$ or
- $r \leq m(C) < \infty$.

We can test this property by the following algorithm. Let $c' : V \to \mathbb{Q}_{\geq 0}$ be a new cost function with

$$c'(v_i) \quad = \quad \begin{cases} c(v_i) & \text{if } i \notin B \\ r & \text{if } i \in B. \end{cases}$$

Note that c' can be constructed in polynomial time. Determine the minimal cost K needed to seperate s from t regarding c'. This can be done by means of the algorithm solving the MCVC-problem for $m = 1$, which runs in polynomial time.

If $K \geq r$, we have that for all $C \subseteq N \setminus B$,

$$v(C) = \begin{cases} r - K \leq 0 & \text{if } \tilde{v} = 1 \\ 0 & \text{if } \tilde{v} = 0. \end{cases}$$

Thus, for all $C \subseteq N \setminus B$, the coalitional function is at most 0 and bribery is possible.

If, on the other hand, $K < r$, there exists a minimal winning coalition $C \subseteq N$ with $m(C) = K$, $v(C) = r - K > 0$. Since we defined $c(v_i) = r$ for all $i \in B$, C is a subset of $N \setminus B$. Therefore, bribery of B is not possible.

Next we show that PDGC-Single-Bribery is NP-hard. We prove this by means of a reduction from Partition that is based on the reduction Partition $\leq_{\mathrm{m}}^{\mathrm{p}}$ MaxCut by Karp [19]. Given an instance $A = (a_1, a_2, \ldots, a_m)$ of Partition, create the following MaxCut instance:

– $G' = (V', E')$, where $V' = \{v_1, v_2, \dots, v_m\}$ and

$$E' = \{\{v_i, v_j\} \mid v_i, v_j \in V, i \neq j\},$$

– $w : E' \to \mathbb{N}^+$ with $w(\{v_i, v_j\}) = a_i \cdot a_j$, and

– $K = S^2/4$ with $S = \sum_{i=1}^{m} a_i$.

Obviously, the MaxCut property is satisfied if and only if A belongs to Partition.

Next, given A and G', we create the following instance X of PDGC-Single-Bribery. The path-disruption game consists of $G = (V, E)$, where

$$\begin{aligned} V &= V' \cup \{v_{m+1}, v_{m+2}\} \cup \{v_{m+2+i}, v_{2m+2+i} \mid 1 \leq i \leq m\} \\ &\quad \cup \{v_{3m+2+j} \mid e_j \in E', 1 \leq j \leq m(m-1)/2\}, \\ E &= \{\{u, v_{3m+2+j}\}, \{v_{3m+2+j}, v\} \mid \{u, v\} = e_j \in E'\} \\ &\quad \cup \{\{v_{m+1}, v_{m+2+i}\}, \{v_{m+2+i}, v_i\} \mid 1 \leq i \leq m\} \\ &\quad \cup \{\{v_i, v_{2m+2+i}\}, \{v_{2m+2+i}, v_{m+2}\} \mid 1 \leq i \leq m\} \end{aligned}$$

and furthermore of source vertex $s = v_{m+1}$, target vertex $t = v_{m+2}$, reward

$$r = \frac{S^2}{2} + S,$$

and cost function $c : V \to \mathbb{Q}_{\geq 0}$,

$$c(v_i) = \begin{cases} r & \text{if } 1 \leq i \leq m+2 \\ a_j & \text{if } m+3 \leq i \leq 2m+2,\ i = m+2+j \\ a_j \cdot \left(\frac{S}{2} + 1\right) & \text{if } 2m+3 \leq i \leq 3m+2,\ i = 2m+2+j \\ w(e_j) & \text{if } 3m+3 \leq i \leq n,\ i = 3m+2+j, \end{cases}$$

with $n = 3m + 2 + m(m-1)/2$.

Moreover, let $k = S/2$ and let the price function be $b : V \to \mathbb{Q}_{\geq 0}$,

$$b(v_i) = \begin{cases} k+1 & \text{if } 1 \leq i \leq m+2 \\ a_j & \text{if } m+3 \leq i \leq 2m+2,\ i = m+2+j \\ k+1 & \text{if } 2m+3 \leq i \leq n. \end{cases}$$

Figure 1 illustrates this construction.

We claim that

$$A \in \text{Partition} \iff X \in \text{PDGC-Single-Bribery}. \tag{1}$$

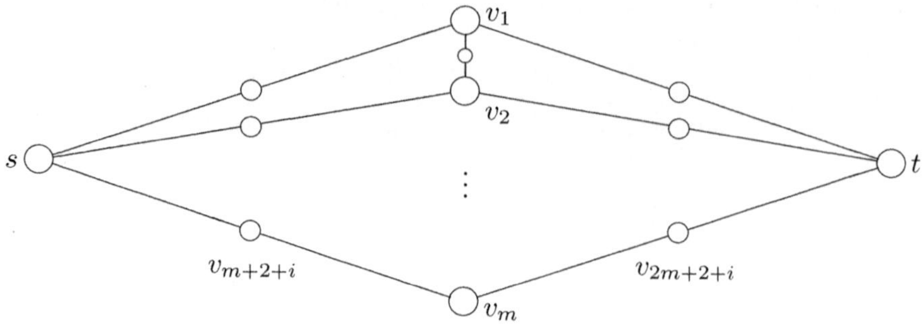

Fig. 1. Construction of the PDGC-SINGLE-BRIBERY instance X

From left to right, suppose $A \in$ PARTITION. Then there is a subset $A' \subseteq A$ with

$$\sum_{a_i \in A'} a_i = \sum_{a_i \in A \smallsetminus A'} a_i = \frac{S}{2}.$$

We show that bribery is possible for coalition

$$B = \{m + 2 + i \mid a_i \in A'\} \subseteq N.$$

First, note that

$$\sum_{m+2+i \in B} b(v_{m+2+i}) = \sum_{a_i \in A'} b(v_{m+2+i}) = \sum_{a_i \in A'} a_i = \frac{S}{2} = k.$$

Second, we need to prove that for each coalition $C \subseteq N \smallsetminus B$, $v(C) \leq 0$. Let C be an arbitrary coalition of $N \smallsetminus B$. If $\tilde{v}(C) = 0$, then $v(C) = 0$ by definition. Otherwise, C contains a minimal winning subcoalition $C' \subseteq C$ with $\tilde{v}(C') = 1$ and $m(C) = \sum_{i \in C'} c(v_i)$.

If C' contains an agent situated on a vertex in $\{v_1, \ldots, v_{m+2}\}$, then $m(C) \geq r$, so $v(C) \leq 0$. Thus, we may assume that $C' \cap \{1, \ldots, m + 2\} = \emptyset$.

C' must contain $\{2m + 2 + i \mid a_i \in A'\}$; otherwise, a path from $s = v_{m+1}$ over v_{m+2+i}, v_i, and v_{2m+2+i} to $t = v_{m+2}$ for an i, $a_i \in A'$, is not blocked.

For all i, $i \in A \smallsetminus A'$, we have that $m + 2 + i$ or $2m + 2 + i$ has to be in C'. Define

$$\tilde{A}_1 = \{a_i \mid a_i \in A \smallsetminus A',\ 2m + 2 + i \in C'\},$$
$$x = \sum_{a_i \in \tilde{A}_1} a_i \leq S/2,$$

and let $\tilde{A}_2$ be the set containing the remaining $a_i \notin A' \cup \tilde{A}_1$. Consequently, $\{m + 2 + i \mid a_i \in \tilde{A}_2\} \in C'$.

If $\tilde{A}_2 = \emptyset$, then $C' = \{2m+2+i \mid 1 \leq i \leq m\}$ with $\sum_{i \in C'} c(v_i) = S \cdot (S/2+1) = r$. Thus, assume that $\tilde{A}_2 \neq \emptyset$.

If $\tilde{A}_1 = \emptyset$, $\{m+2+i \mid a_i \in A \setminus A'\} \subseteq C'$. C' is a minimal winning coalition if and only if additionally $\{3m+2+j \mid e_j = \{v_{j_1}, v_{j_2}\} \in E', a_{j_1} \in A', a_{j_2} \notin A'\}$ are in C'. So,

$$\begin{aligned} m(C) &= \sum_{a_i \in A'} c(v_{2m+2+i}) + \sum_{a_i \notin A'} c(v_{m+2+i}) + \sum_{a_{j_1} \in A'} \sum_{\substack{a_{j_2} \notin A' \\ e_j = \{v_{j_1}, v_{j_2}\} \in E'}} c(v_{3m+2+j}) \\ &= \sum_{a_i \in A'} a_i \cdot \left(\frac{S}{2} + 1\right) + \sum_{a_i \notin A'} a_i + \sum_{a_{j_1} \in A'} \sum_{\substack{a_{j_2} \notin A' \\ e_j = \{v_{j_1}, v_{j_2}\} \in E'}} w(e_j) \\ &= \frac{S}{2} \cdot \left(\frac{S}{2} + 1\right) + \frac{S}{2} + \sum_{a_{j_1} \in A'} \sum_{a_{j_2} \notin A'} a_{j_1} \cdot a_{j_2} \\ &= \frac{S^2}{4} + S + \frac{S}{2} \cdot \frac{S}{2} \;=\; \frac{S^2}{2} + S \;=\; r. \end{aligned}$$

Assume that $\tilde{A}_1 \neq \emptyset$. In order to block all paths, it must be the case that

$$\{3m+2+j \mid e_j = \{v_{j_1}, v_{j_2}\} \in E', a_{j_1} \in A', a_{j_2} \in \tilde{A}_2\} \subseteq C'$$

and

$$\{3m+2+j \mid e_j = \{v_{j_1}, v_{j_2}\} \in E', a_{j_1} \in \tilde{A}_1, a_{j_2} \in \tilde{A}_2\} \subseteq C'.$$

C' is not minimal if it contains both $m+2+i$ and $2m+2+i$ for an i, $1 \leq i \leq m$. If this was the case for an i, $a_i \in \tilde{A}_1$, then

- either the same subset of $\{3m+2+j, e_j \in E'\}$ would be in C' which would make $m+2+i$ redundant;
- or we have

$$\begin{aligned} \{3m+2+j \mid e_j = \{v_{j_1}, v_{j_2}\} \in E', a_{j_1} \in A', a_{j_2} \in \tilde{A}_2\} &\subseteq C', \\ \{3m+2+j \mid e_j = \{v_{j_1}, v_i\} \in E', a_{j_1} \in A'\} &\subseteq C', \\ \{3m+2+j \mid e_j = \{v_{j_1}, v_{j_2}\} \in E', a_{j_1} \in \tilde{A}_2, a_{j_2} \in \tilde{A}_1, j_2 \neq i\} &\subseteq C', \\ \{3m+2+j \mid e_j = \{v_i, v_{j_2}\} \in E', a_{j_2} \in \tilde{A}_1\} &\subseteq C', \end{aligned}$$

  which makes blocking of $2m+3+i$ unnecessary and is the same case as $\tilde{A}_1 = \tilde{A}_1 \setminus \{a_i\}$.

Thus, we have

$$\begin{aligned}
m(C) - r &= \sum_{a_i \in A'} c(v_{2m+2+i}) + \sum_{a_i \in \tilde{A}_1} c(v_{2m+2+i}) + \sum_{a_i \in \tilde{A}_2} c(v_{m+2+i}) \\
&\quad + \sum_{a_{j_1} \in A'} \sum_{\substack{a_{j_2} \in \tilde{A}_2 \\ e_j = \{v_{j_1}, v_{j_2}\} \in E'}} c(v_{3m+2+j}) \\
&\quad + \sum_{a_{j_1} \in \tilde{A}_1} \sum_{\substack{a_{j_2} \in \tilde{A}_2 \\ e_j = \{v_{j_1}, v_{j_2}\} \in E'}} c(v_{3m+2+j}) \\
&\quad - \frac{S^2}{2} - S \\
&= \sum_{a_i \in A'} a_i \cdot \left(\frac{S}{2} + 1\right) + \sum_{a_i \in \tilde{A}_1} a_i \cdot \left(\frac{S}{2} + 1\right) + \sum_{a_i \in \tilde{A}_2} a_i \\
&\quad + \sum_{a_{j_1} \in A'} \sum_{\substack{a_{j_2} \in \tilde{A}_2 \\ e_j = \{v_{j_1}, v_{j_2}\} \in E'}} w(e_j) + \sum_{a_{j_1} \in \tilde{A}_1} \sum_{\substack{a_{j_2} \in \tilde{A}_2 \\ e_j = \{v_{j_1}, v_{j_2}\} \in E'}} w(e_j) \\
&\quad - \frac{S^2}{2} - S \\
&= \frac{S}{2} \cdot \left(\frac{S}{2} + 1\right) + x \cdot \left(\frac{S}{2} + 1\right) + \left(\frac{S}{2} - x\right) + \sum_{a_{j_1} \in A'} \sum_{a_{j_2} \in \tilde{A}_2} a_{j_1} a_{j_2} \\
&\quad + \sum_{a_{j_1} \in \tilde{A}_1} \sum_{a_{j_2} \in \tilde{A}_2} a_{j_1} a_{j_2} - \frac{S^2}{2} - S \\
&= \frac{S^2}{4} + S + x \cdot \frac{S}{2} + \frac{S}{2} \cdot \left(\frac{S}{2} - x\right) + x \cdot \left(\frac{S}{2} - x\right) - \frac{S^2}{2} - S \\
&= -x^2 + \frac{S}{2}x = -x\left(x - \frac{S}{2}\right),
\end{aligned}$$

so $m(C) - r$ is a function in x. For each x with $0 \leq x \leq S/2$, it holds that $m(C) - r \geq 0$. Therefore, bribery is possible.

To prove the direction from right to left in (1), suppose that X belongs to PDGC-Single-Bribery. Then there exists a coalition $B \subseteq N$ with

$$\sum_{i \in B} b(v_i) \leq k \tag{2}$$

and for all coalitions $C \subseteq N \setminus B$, we have that

$$\text{either } \tilde{v}(C) = 0 \text{ or } m(C) \geq r. \tag{3}$$

Since all other vertices have a price greater than k, B is a subset of

$$\{m + 3, \ldots, 2m + 2\}.$$

Assume that $B = \emptyset$. Then $C = \{m+3, \dots, 2m+2\} \subseteq N \setminus B$ is a minimal winning coalition with $\tilde{v}(C) = 1$ and

$$m(C) = \sum_{i=1}^{m} c(v_{m+2+i}) = \sum_{i=1}^{m} a_i = S < \frac{S^2}{2} + S = r.$$

That is a contradiction to Condition (3). On the other hand, Condition (2) implies that

$$\sum_{i \in B} b(v_i) = \sum_{m+2+i \in B} a_i \leq k = S/2,$$

and, in particular, $B \neq \{m+3, \dots, 2m+2\}$. This leads to the following two cases.

Case 1: $\sum_{i \in B} b(v_i) < k = S/2$. Denote $\sum_{i \in B} b(v_i) = \sum_{m+2+i \in B} a_i = \sum_{i \in B} c(v_i)$ by x, $0 < x < S/2$. Then,

$$\begin{aligned} C &= \{2m+2+i \mid m+2+i \in B\} \\ &\cup \{m+2+i \mid 1 \leq i \leq m, m+2+i \notin B\} \\ &\cup \{3m+2+j \mid e_j = \{v_{j_1}, v_{j_2}\} \in E', m+2+j_1 \in B, m+2+j_2 \notin B\} \end{aligned}$$

is a minimal winning coalition in $N \setminus B$ with

$$\begin{aligned} m(C) - r &= \sum_{i \in C} c(v_i) - \frac{S^2}{2} - S \\ &= \sum_{m+2+i \in B} c(v_{2m+2+i}) + \sum_{\substack{i=1 \\ m+2+i \notin B}}^{m} c(v_{m+2+i}) \\ &\quad + \sum_{m+2+j_1 \in B} \sum_{\substack{m+2+j_2 \notin B \\ e_j = \{v_{j_1}, v_{j_2}\} \in E'}} c(v_{3m+2+j}) - \frac{S^2}{2} - S \\ &= x \cdot \left(\frac{S}{2} + 1\right) + (S - x) - \frac{S^2}{2} - S \\ &\quad + \sum_{m+2+j_1 \in B} \sum_{\substack{m+2+j_2 \notin B \\ e_j = \{v_{j_1}, v_{j_2}\} \in E'}} w(e_j) \\ &= x \cdot \frac{S}{2} - \frac{S^2}{2} + x \cdot (S - x) \\ &= -x^2 + \frac{3S}{2} x - \frac{S^2}{2} = -(x - S)\left(x - \frac{S}{2}\right). \end{aligned}$$

For x with $0 < x < \frac{S}{2}$, it holds that $m(C) - r < 0$, which again is a contradiction to Condition (3).

Case 2: $\sum_{i \in B} b(v_i) = k$. That is,

$$\sum_{m+2+i \in B} a_i = \frac{S}{2}.$$

Thus, a partition into $A' = \{a_i \mid m + 2 + i \in B\}$ and $A \setminus A'$ with

$$\sum_{a_i \in A'} a_i = \sum_{a_i \in A \setminus A'} a_i = \frac{S}{2}$$

exists.

This concludes the proof of (1). The observation that the construction described can be done in polynomial time completes the proof proof of the theorem. ❑

Theorem 2 provides an upper bound for PDGC-Multiple-Bribery. The exact complexity of this problem remains open.

Theorem 2. PDGC-Multiple-Bribery *belongs to* $\Sigma_2^p = \mathrm{NP}^{\mathrm{NP}}$.

Proof. Given an instance $G, c, (s_j, t_j), r, b, K$, PDGC-Multiple-Bribery can be characterized as follows:

$$(\exists B \subseteq N)(\forall C \subseteq N \setminus B)(\forall D \subseteq C)\left[\sum_{i \in B} b(v_i) \leq k \text{ and } v(D) \leq 0\right],$$

which is equivalent to

$$(\exists B \subseteq N)(\forall D \subseteq N \setminus B)\left[\sum_{i \in B} b(v_i) \leq k \text{ and } \left(\tilde{v}(D) = 0 \text{ or } \sum_{i \in D} c(v_i) \geq r\right)\right].$$

The property in brackets can obviously be tested in polynomial time. Thus, the problem satisfies the quantifier characterization of Σ_2^p (see [13,14]). ❑

The NP-hardness result for the single-adversary case straightforwardly reduces to the more general multiple-adversary case. However, we suspect that PDGC-Multiple-Bribery is even hard for the complexity class Σ_2^p.

6 Conclusion and Future Work

We have introduced the model of bribery in path-disruption games, a class of cooperative games defined by Bachrach and Porat [1]. We have shown that bribery in path-disruption games is computationally hard even in the single-adversary-case. From an algorithmic point of view it might be interesting to analyze whether there are special instances that are tractable. In real life, networks are often graphs with certain properties, e.g., graphs with a small diameter and a high clustering coefficient. Is the complexity of bribery problems in path-disruption games on those graphs different from the general case?

Bachrach and Porat [1] analyze PDGs on trees with the result that very often problems that are hard in general become solvable in polynomial time. We suspect that PDGC-Multiple-Bribery is NP-complete when restricted to planar graphs, in contrast to the general problem for which we can show only membership in Σ_2^p (see Theorem 2). Still, this is computationally intractable.

NP-completeness is only a worst-case measure of complexity. Future work might also tackle the issue of typical-case complexity of these problems. In the context of voting problems, much work has been done recently in this regard, both theoretically (see, e.g., [20,21,22]) and experimentally (see, e.g., [23,24]).

Moreover, it would be interesting to vary the model of bribery and to study the resulting problems in terms of their complexity. In the context of voting, such variations of bribery in elections are, e.g., microbribery [25] and swap bribery [26]. In the context of path-disruption games, one variation might be to define the costs of blocking a vertex in a graph and the prices for bribing the corresponding agents in relation to each other. This might be analyzed in connection with the stability of the game and might lead to a new perspective on the topic.

Another idea for expanding the model of Bachrach and Porat [1] is the following: Consider a network, where the $m \geq 1$ adversaries are placed on a source vertex s each, but their target vertex is unknown. Letting $p_{i,j}$ be the probability that adversary i wants to reach target vertex v_j, $1 \leq i \leq m$, $1 \leq j \leq n$, define the following game.

Probabilistic PDG-Multiple	
Domain:	A graph $G = (V, E)$, $n = \|V\|$, and adversaries $s_1, \ldots, s_m$.
Agents:	$N = \{1, \ldots, n\}$, where i represents v_i.
Coal. Fcn.:	$v(C) = \prod_{i=1}^{m} \sum_{j=1}^{n} p_{i,j} \cdot w(C, i, j)$ with $w(C,i,j) = \begin{cases} 1 & \text{if } C \text{ blocks each path from } s_i \text{ to } v_j, \\ 0 & \text{otherwise.} \end{cases}$

The other cases (such as Probabilistic PDGC-Multiple) can be defined analogously. The special case where $p_{i,j} = 1$ if $t_i = v_j$, and equals 0 otherwise, is exactly the corresponding path-disruption game as defined in Section 3. In future work, we intend to analyze the complexity of related problems such as bribery of this probabilistic model in comparison to the original path-disruption game.

Acknowledgments. We thank the anonymous reviewers for their helpful reviews and literature pointers.

References

1. Bachrach, Y., Porat, E.: Path-disruption games. In: Proceedings of the 9th International Joint Conference on Autonomous Agents and Multiagent Systems, IFAAMAS, pp. 1123–1130 (May 2010)
2. Jain, M., Korzhyk, D., Vaněk, O., Conitzer, V., Pěchouček, M., Tambe, M.: A double oracle algorithm for zero-sum security games on graphs. In: Proceedings of the 10th International Joint Conference on Autonomous Agents and Multiagent Systems, IFAAMAS, pp. 327–334 (May 2011)
3. Endriss, U., Lang, J. (eds.): Proceedings of the 1st International Workshop on Computational Social Choice. Universiteit van Amsterdam, `staff.science.uva.nl/~ulle/COMSOC-2006/proceedings.html` (2006)
4. Endriss, U., Goldberg, P. (eds.): Proceedings of the 2nd International Workshop on Computational Social Choice. University of Liverpool (2008), `www.csc.liv.ac.uk/~pwg/COMSOC-2008/proceedings.html`
5. Conitzer, V., Rothe, J. (eds.): Proceedings of the 3rd International Workshop on Computational Social Choice. Universität Düsseldorf (2010), `http://ccc.cs.uni-duesseldorf.de/COMSOC-2010/proceedings.shtml`
6. Elkind, E., Goldberg, L., Goldberg, P., Wooldridge, M.: Computational complexity of weighted threshold games. In: Proceedings of the 22nd AAAI Conference on Artificial Intelligence, pp. 718–723. AAAI Press, Menlo Park (July 2007)
7. Aziz, H., Paterson, M.: False name manipulations in weighted voting games: Splitting, merging and annexation. In: Proceedings of the 8th International Joint Conference on Autonomous Agents and Multiagent Systems, IFAAMAS, pp. 409–416 (May 2009)
8. Bachrach, Y., Rosenschein, J.: Computing the Banzhaf power index in network flow games. In: Proceedings of the 6th International Joint Conference on Autonomous Agents and Multiagent Systems, IFAAMAS, pp. 323–329 (2007)
9. Bachrach, Y., Rosenschein, J.: Power in threshold network flow games. Journal of Autonomous Agents and Multi-Agent Systems 18(1), 106–132 (2009)
10. Rey, A., Rothe, J.: Merging and splitting for power indices in weighted voting games and network flow games on hypergraphs. In: Proceedings of the 5th European Starting AI Researcher Symposium, pp. 277–289. IOS Press, Amsterdam (2010)
11. Bachrach, Y., Elkind, E., Meir, R., Pasechnik, D., Zuckerman, M., Rothe, J., Rosenschein, J.: The cost of stability in coalitional games. In: Mavronicolas, M., Papadopoulou, V.G. (eds.) SAGT 2009. LNCS, vol. 5814, pp. 122–134. Springer, Heidelberg (2009)
12. Faliszewski, P., Hemaspaandra, E., Hemaspaandra, L.: How hard is bribery in elections? Journal of Artificial Intelligence Research 35, 485–532 (2009)
13. Meyer, A., Stockmeyer, L.: The equivalence problem for regular expressions with squaring requires exponential space. In: Proceedings of the 13th IEEE Symposium on Switching and Automata Theory, pp. 125–129 (1972)
14. Stockmeyer, L.: The polynomial-time hierarchy. Theoretical Computer Science 3(1), 1–22 (1977)
15. Osborne, M., Rubinstein, A.: A Course in Game Theory. MIT Press, Cambridge (1999)
16. Papadimitriou, C.: Computational Complexity. Addison-Wesley, Reading (1994)
17. Rothe, J.: Complexity Theory and Cryptology. An Introduction to Cryptocomplexity. EATCS Texts in Theoretical Computer Science. Springer, Heidelberg (2005)

18. Garey, M., Johnson, D.: Computers and Intractability: A Guide to the Theory of NP-Completeness. W. H. Freeman and Company, New York (1979)
19. Karp, R.: Reducibilities among combinatorial problems. In: Miller, R., Thatcher, J. (eds.) Complexity of Computer Computations, pp. 85–103. Plenum Press, New York (1972)
20. Procaccia, A., Rosenschein, J.: Junta distributions and the average-case complexity of manipulating elections. Journal of Artificial Intelligence Research 28, 157–181 (2007)
21. Erdélyi, G., Hemaspaandra, L., Rothe, J., Spakowski, H.: Generalized juntas and NP-hard sets. Theoretical Computer Science 410(38-40), 3995–4000 (2009)
22. Homan, C., Hemaspaandra, L.: Guarantees for the success frequency of an algorithm for finding Dodgson-election winners. Journal of Heuristics 15(4), 403–423 (2009)
23. Walsh, T.: Where are the really hard manipulation problems? The phase transition in manipulating the veto rule. In: Proceedings of the 21st International Joint Conference on Artificial Intelligence, IJCAI, pp. 324–329 (July 2009)
24. Walsh, T.: An empirical study of the manipulability of single transferable voting. In: Proceedings of the 19th European Conference on Artificial Intelligence, pp. 257–262. IOS Press, Amsterdam (2010)
25. Faliszewski, P., Hemaspaandra, E., Hemaspaandra, L., Rothe, J.: Llull and Copeland voting computationally resist bribery and constructive control. Journal of Artificial Intelligence Research 35, 275–341 (2009)
26. Elkind, E., Faliszewski, P., Slinko, A.: Swap bribery. In: Mavronicolas, M., Papadopoulou, V.G. (eds.) SAGT 2009. LNCS, vol. 5814, pp. 299–310. Springer, Heidelberg (2009)

The Machine Learning and Traveling Repairman Problem

Theja Tulabandhula, Cynthia Rudin, and Patrick Jaillet

Massachusetts Institute of Technology, Cambridge MA 02139, USA
{theja,rudin,jaillet}@mit.edu

Abstract. The goal of the *Machine Learning and Traveling Repairman Problem* (ML&TRP) is to determine a route for a "repair crew," which repairs nodes on a graph. The repair crew aims to minimize the cost of failures at the nodes, but the failure probabilities are not known and must be estimated. If there is uncertainty in the failure probability estimates, we take this uncertainty into account in an unusual way; from the set of acceptable models, we choose the model that has the lowest cost of applying it to the subsequent routing task. In a sense, this procedure agrees with a managerial goal, which is to show that the data can support choosing a low-cost solution.

Keywords: machine learning, traveling repairman, integer programming, uncertainty, generalization bound, constrained linear function classes.

1 Introduction

We consider the problem of determining a route for a "repair crew" on a graph, where each node on the graph has some probability of failure. These probabilities are not known and must be estimated from past failure data. Intuitively the nodes that are more prone to failure should be repaired first. But if those nodes are far away from each other, the extra time spent by the repair crew traveling between nodes might actually increase the chance of failures occurring at nodes that have not yet been repaired. In that sense, it is better to construct the route to minimize the possible cost of failures, taking into account the travel time between nodes and also the (estimated) failure probabilities at each of the nodes. We call this problem the *machine learning and traveling repairman problem* (ML&TRP). There are many possible applications of the ML&TRP, including the scheduling of safety inspections or repair work for the electrical grid, oil rigs, underground mining, machines in a factory, or airlines.

One key idea we present here concerns the way that uncertainty is handled in probabilistic modeling, and the way the uncertainty relates to how the models are used in applications. Namely, when there is uncertainty in modeling, our idea is to choose a model that has advantages for our specific application, when we act on the predictions made by the model. For estimation problems, it is possible for many predictive models to be equally plausible, given finite data. In standard statistical and machine learning practice, we choose one of these

R.I. Brafman, F. Roberts, and A. Tsoukiàs (Eds.): ADT 2011, LNAI 6992, pp. 262–276, 2011.

models, but the choice of model is oblivious to the way that the model will be used in the application. Our idea is that we choose a model that predicts well, but that also has the advantage that it has a low "operating cost," which is the cost to act on the predictions made by the model. In this work, among all equally good predictive models for failure probabilities, we choose the one that leads to the lowest failure cost.

We present two formulations for the ML&TRP. The first formulation is *sequential*: the failure probabilities are estimated in a way that is oblivious to the failure cost; then, the route is determined by minimizing failure cost (which depends on the chosen probabilistic model). The second formulation handles uncertainty as discussed above, by computing the failure probabilities and the route *simultaneously*. This means that the estimated failure probabilities and the route are chosen together in a way that the failure cost will be low if possible; when there is uncertainty, the simultaneous formulation chooses the model with the lowest failure cost. The simultaneous formulation is optimistic; it provides the best possible, but still reasonable, scenario described by the data. The company might wish to know whether it is at all possible that a low-failure-cost route can be designed that is realistically supported by the data; the simultaneous formulation finds such a solution.

We design the failure cost in two ways, where either can be used for the sequential and the simultaneous formulations. The first failure cost is proportional to the sum (over nodes) of the expected number of failures at each node. The second failure cost considers, for each node, the probability that the first failure is before the repair crew's visit to the node. The first cost applies when the failure probability of a node does not change until it is visited by the crew, regardless of whether a failure already occurred at that node, and the second cost applies when the node is completely repaired after the first failure, or when it is visited by the repair crew, whichever comes first. In either case, the failure cost reduces to a weighted *traveling repairman problem* (TRP) objective [1].

The ML&TRP relates to literature on both machine learning and optimization (time-dependent traveling salesman problems). In machine learning, the use of unlabeled data has been explored extensively in the semi-supervised learning literature [2]. The ML&TRP does not fall under the umbrella of semi-supervised learning, since the incorporation of unlabeled data is used solely for determining the failure cost, and is not used to provide additional *distributional* information. Our work is slightly closer to work on graph-based regularization [3,4,5], but their goal is to obtain probability estimates that are smoothed on a graph with suitably designed edge weights. On the other hand, our goal is to obtain, in addition to probability estimates, a low-cost route for traversing a very different graph with edge weights that are physical distances. Our work contributes to the literature on the TRP and related problems by adding the new dimension of probabilistic estimation at the nodes. We adapt techniques from [6,7,8] within our work for solving the TRP part of the ML&TRP.

One particularly motivating application for the ML&TRP is smart grid maintenance. Since 2004, many power utility companies are implementing new inspection and repair programs for preemptive maintenance, where in the past, repair work was mainly made reactively [9]. Con Edison, which is New York City's power company, services tens of thousands of manholes (access points to underground grid) through new inspection and repair programs. The scheduling of manhole inspection and repair in Manhattan, Brooklyn and the Bronx is assisted by a statistical model [10]. This model does not take into account the route of the repair crew. This leaves open the possibility that, for this and for many other domains, estimating failure probabilities with knowledge of the repair crew's route could lead to an improvement in operations.

In Section 2, we provide the two formulations and the two ways of modeling failure cost. In Section 3, we describe mixed-integer nonlinear programs (MINLP) and algorithms for solving the ML&TRP. Section 4 gives an example and some experiments on data from the NYC power grid. Section 5 states a generalization result, and Section 6 concludes the paper.

2 ML&TRP Formulations

Consider two sets of instances, $\{x_i\}_{i=1}^m$, $\{\tilde{x}_i\}_{i=1}^M$, with $x_i \in \mathcal{X}$, $\tilde{x}_i \in \mathcal{X}$ that are feature vectors with $\mathcal{X} \subset \mathbb{R}^d$. Let the x_i^j indicate the j-th coordinate of the feature vector x_i. For the first set of instances, we are also given labels $\{y_i\}_{i=1}^m$, $y_i \in \{-1, 1\}$. These instances and their labels are the set of training examples. For the maintenance application, each of the $\{x_i\}_{i=1}^m$ encode manhole information (e.g., number and types of cables, number and types of previous events, etc.) and the labels $\{y_i\}_{i=1}^m$ encode whether the manhole failed ($y_i = 1$) or not ($y_i = -1$). More details about the features and labels can be found in Section 4. The other instances $\{\tilde{x}_i\}_{i=1}^M$ (with M unrelated to m), are unlabeled data that are each associated with a node on a graph G. The nodes of the graph G indexed by $i = 1, ..., M$ represent manholes on which we want to design a route. We are also given physical distances $d_{i,j} \in \mathbb{R}_+$ between all pairs of nodes i and j. A route on G is represented by a permutation π of the node indices $1, \ldots, M$. Let Π be the set of all permutations of $\{1, ..., M\}$. A set of failure probabilities will be estimated at the nodes and these estimates will be based on a function of the form $f_\lambda(x) = \lambda \cdot x$. The class of functions $\mathcal{F}$ is chosen to be:

$$\mathcal{F} := \{f_\lambda : \lambda \in \mathbb{R}^d, ||\lambda||_2 \leq M_1\}, \tag{1}$$

where M_1 is a fixed positive real number.

The sequential formulation has a machine learning step and a traveling repairman step, whereas the simultaneous formulation has both ML and TRP together in the first step, and the second step uses the route from the first step.

Sequential Formulation
Step 1. (ML) Compute the values $f_\lambda^*(\tilde{x}_i)$:

$$f_\lambda^* \in \text{argmin}_{f_\lambda \in \mathcal{F}} \text{TrainingError}(f_\lambda, \{x_i, y_i\}_{i=1}^m).$$

Step 2. (TRP) Compute a route using estimated scores on $\{\tilde{x}_i\}_{i=1}^M$:

$$\pi^* \in \text{argmin}_{\pi \in \Pi} \text{FailureCost}(\pi, f_\lambda^*, \{\tilde{x}_i\}_{i=1}^M, \{d_{i,j}\}_{i,j=1}^M).$$

We will define the TrainingError and FailureCost shortly.

Simultaneous Formulation
Step 1. Compute the values $f_\lambda^*(\tilde{x}_i)$:

$$f_\lambda^* \in \text{argmin}_{f_\lambda \in \mathcal{F}} \Big[\text{TrainingError}(f_\lambda, \{x_i, y_i\}_{i=1}^m) + C_1 \min_{\pi \in \Pi} \text{FailureCost}\left(\pi, f_\lambda, \{\tilde{x}_i\}_{i=1}^M, \{d_{i,j}\}_{i,j=1}^M\right) \Big].$$

Step 2. Compute a route corresponding to the scores:

$$\pi^* \in \text{argmin}_{\pi \in \Pi} \text{FailureCost}(\pi, f_\lambda^*, \{\tilde{x}_i\}_{i=1}^M, \{d_{i,j}\}_{i,j=1}^M).$$

A transformation of $f_\lambda^*(x)$ yields an estimate of the probability of failure $P(y = 1|x)$ (we discuss this later, see (2)). In Step 1, f_λ^* is chosen to yield probability estimates that agree with the training data, but at the same time, yield lower failure costs. The user-defined constant C_1 is a tradeoff parameter, moving from "oblivious" estimation models to cost-aware estimation models. When C_1 is small, the algorithm essentially becomes sequential, ignoring the FailureCost. When it is large, the algorithm is highly biased towards low FailureCost solutions. One might want to choose C_1 large when there is a lot of uncertainty in the estimates and a strong belief that a very low cost solution exists. Or, one could choose a large C_1 to determine what policy would be chosen when the cost is underestimated. A small C_1 is appropriate when the number of training examples is large enough so that there is little flexibility (uncertainty) in the choice of model f_λ^*. Or one would choose low C_1 when we wish to choose, among equally good solutions, the one with the lowest cost. We now define the TrainingError and two options for the FailureCost.

Training Error. In learning, the unregularized error is a sum (or average) of losses over the training examples: $\sum_{i=1}^m l(f_\lambda(x_i), y_i)$, where the loss function $l(\cdot, \cdot)$ can be any monotonic smooth function bounded below by zero. We choose the logistic loss: $l(f_\lambda(x), y) := \ln\left(1 + e^{-y f_\lambda(x)}\right)$ so that the probability of failure $P(y = 1|x)$, is estimated as in logistic regression by

$$P(y = 1|x) \text{ or } p(x) := \frac{1}{1 + e^{-f_\lambda(x)}}. \tag{2}$$

The negative log likelihood is:

$$\sum_{i=1}^m -\ln\left[p(x_i)^{(1+y_i)/2}(1 - p(x_i))^{(1-y_i)/2}\right] = \sum_{i=1}^m \ln\left(1 + e^{-y_i f_\lambda(x_i)}\right).$$

We then add an ℓ_2 penalty over the parameters λ (with coefficient C_2) to get

$$\text{TrainingError}(f_\lambda, \{x_i, y_i\}_{i=1}^m) := \sum_{i=1}^{m} \ln\left(1 + e^{-y_i f_\lambda(x_i)}\right) + C_2||\lambda||_2^2. \quad (3)$$

The coefficient C_2 is inversely related to the constant M_1 in (1) and both represent the same constraint on the function class. C_2 is useful for algorithm implementations whereas M_1 is useful for analysis.

Two Options for Failure Cost. In the first option (denoted as Cost 1), for each node there is a cost for (possibly repeated) failures prior to a visit by the repair crew. In the second option (denoted as Cost 2), for each node, there is a cost for the first failure prior to visiting it. There is a natural interpretation of the failures as being generated by a continuous random process at each of the nodes. When discretized in time, this is approximated by a Bernoulli process with parameter $p(\tilde{x}_i)$. Both Cost 1 and Cost 2 are appropriate for power grid applications. Cost 2 is also appropriate for delivery truck routing applications, where perishable items can fail (once an item has spoiled, it cannot spoil again). For many applications, neither of these two costs apply, in which case, it is possible to design a more appropriate or specialized cost and use that in place of the two we present here, using the same general idea of combining this cost with the training error to produce an algorithm.

Without loss of generality, we assume that after the repair crew visits all the nodes, it returns to the starting node (node 1) which is fixed beforehand. Scenarios where one is not interested in beginning from or returning to the starting node would be modeled slightly differently (the computational complexity remains the same).

Let a route be represented by $\pi : \{1, ..., M\} \mapsto \{1, ..., M\}$, ($\pi(i)$ is the ith node visited). Let the distances be such that a unit of distance is traversed in a unit of time. Given a route, the *latency* of a node $\pi(i)$ is the time (or equivalently distance) from the start at which node $\pi(i)$ is visited

$$L_\pi(\pi(i)) := \begin{cases} \sum_{k=1}^{M} d_{\pi(k)\pi(k+1)} \mathbf{1}_{[k<i]} & i = 2, ..., M \\ \sum_{k=1}^{M} d_{\pi(k)\pi(k+1)} & i = 1, \end{cases} \quad (4)$$

where we let $d_{\pi(M)\pi(M+1)} = d_{\pi(M)\pi(1)}$.

Cost 1. (*Cost is Proportional to Expected Number of Failures Before the Visit*). Up to the time that node $\pi(i)$ is visited, there is a probability $p(\tilde{x}_{\pi(i)})$ that a failure will occur in each unit time interval. This failure is determined by a Bernoulli random variable with parameter $p(\tilde{x}_{\pi(i)})$. Thus, in a time interval of length $L_\pi(\pi(i))$ units, the number of node failures follows a binomial distribution. For each node, we associate a cost proportional to the expected number of failures before the repair crew's visit:

$$\begin{aligned} \text{Cost of node } \pi(i) &\propto E(\text{number failures in } L_\pi(\pi(i)) \text{ time units}) \\ &= \text{mean of Bin}(L_\pi(\pi(i)), p(\tilde{x}_{\pi(i)})) = p(\tilde{x}_{\pi(i)})L_\pi(\pi(i)). \quad (5) \end{aligned}$$

If the failure probability for node $\pi(i)$ is small, we can afford to visit it later on in the route (the latency $L_\pi(\pi(i))$ is larger). If $p(\tilde{x}_{\pi(i)})$ is large, we visit node $\pi(i)$ earlier to keep our cost low. The failure cost for a route π is

$$\text{FailureCost}(\pi, f_\lambda, \{\tilde{x}_i\}_{i=1}^M, \{d_{i,j}\}_{i,j=1}^M) = \sum_{i=1}^M p(\tilde{x}_{\pi(i)})L_\pi(\pi(i)).$$

Substituting the definition of $L_\pi(\pi(i))$ from (4):

$$\text{FailureCost}(\pi, f_\lambda, \{\tilde{x}_i\}_{i=1}^M, \{d_{i,j}\}_{i,j=1}^M) =$$
$$\sum_{i=2}^M p(\tilde{x}_{\pi(i)}) \sum_{k=1}^M d_{\pi(k)\pi(k+1)} \mathbf{1}_{[k<i]} + p(\tilde{x}_{\pi(1)}) \sum_{k=1}^M d_{\pi(k)\pi(k+1)}, \qquad (6)$$

where $p(\tilde{x}_{\pi(i)})$ is given in (2). In a more general setting (explored in a longer version of this work [11]), we could relax the assumption of setting $p(\tilde{x}_{\pi(i)}) = 0$ after the visit as we have implicitly done here. Note that since the cost is a sum of M terms, it is invariant to ordering or indexing (caused by π) and we can rewrite it as

$$\text{FailureCost}(\pi, f_\lambda, \{\tilde{x}_i\}_{i=1}^M, \{d_{i,j}\}_{i,j=1}^M) = \sum_{i=1}^M p(\tilde{x}_i)L_\pi(i).$$

Cost 2. (*Cost is Proportional to Probability that First Failure is Before the Visit*). This cost reflects the penalty for not visiting a node before the first failure occurs there. The model is governed by the geometric distribution: the probability that the first failure for node $\pi(i)$ occurs at time $L_\pi(\pi(i))$ is $p(\tilde{x}_{\pi(i)})(1 - p(\tilde{x}_{\pi(i)}))^{L_\pi(\pi(i))-1}$, and the cost of visiting node $\pi(i)$ is proportional to:

$$P\Big(\text{first failure occurs before } L_\pi(\pi(i))\Big) = 1 - (1 - p(\tilde{x}_{\pi(i)}))^{L_\pi(\pi(i))}$$
$$= 1 - \left(1 - \frac{1}{1 + e^{-f_\lambda(\tilde{x}_{\pi(i)})}}\right)^{L_\pi(\pi(i))} = 1 - \left(1 + e^{f_\lambda(\tilde{x}_{\pi(i)})}\right)^{-L_\pi(\pi(i))}. \qquad (7)$$

Similarly to Cost 1, $L_\pi(\pi(i))$ influences the cost at each node. If we visit a node early in the route, then the cost incurred is small because the node is less likely to fail before we reach it. Similarly, if we schedule a visit later on in the tour, the cost is higher because the node has a higher chance of failing prior to the repair crew's visit. The total failure cost is

$$\sum_{i=1}^M \left(1 - \left(1 + e^{f_\lambda(\tilde{x}_{\pi(i)})}\right)^{-L_\pi(\pi(i))}\right). \qquad (8)$$

This cost is not directly related to a weighted TRP cost in its present form, but building on this, we will derive a cost that is the same as a weighted TRP. Before doing so in Section 3, we formulate the integer program for the simultaneous formulation for Cost 1.

3 Optimization

Mixed-Integer Optimization for Cost 1. For both the sequential and simultaneous formulations, we need to solve the TRP subproblem:

$$\begin{aligned}\pi^* &\in \text{argmin}_{\pi\in\Pi}\text{FailureCost}(\pi, f_\lambda^*, \{\tilde{x}_i\}_{i=1}^M, \{d_{i,j}\}_{i,j=1}^M),\\ &= \text{argmin}_{\pi\in\Pi}\sum_{i=2}^M p(\tilde{x}_{\pi(i)})\sum_{k=1}^M d_{\pi(k)\pi(k+1)}\mathbf{1}_{[k<i]} + p(\tilde{x}_{\pi(1)})\sum_{k=1}^M d_{\pi(k)\pi(k+1)}. \quad (9)\end{aligned}$$

The standard TRP objective is a special case of weighted TRP (9) when $\forall i = 1, ..., M$, $p(\tilde{x}_i) = p$. The TRP is different from the traveling salesman problem (TSP); the goal of the TSP is to minimize the total traversal time (in this case, this is the same as the distance traveled) needed to visit all nodes once, whereas the goal of the TRP is to minimize the sum of the waiting times to visit each node. Both the problems are known to be NP-complete in the general case [12].

We extend the integer linear program (ILP) of [6] to include "unequal flow values" in (9). For interpretation, consider the sum $\sum_{i=1}^M \bar{p}(\tilde{x}_i)$ as the total "flow" through a route where $\bar{p}(\tilde{x}_i)$ will be chosen later according to either Cost 1 or Cost 2. At the beginning of the tour, the repair crew has flow $\sum_{i=1}^M \bar{p}(\tilde{x}_i)$. Along the tour, flow of the amount $\bar{p}(\tilde{x}_i)$ is dropped when the repair crew visits node $\pi(i)$ at latency $L_\pi(\pi(i))$. We introduce two sets of variables $\{z_{i,j}\}_{i,j}$ and $\{y_{i,j}\}_{i,j}$ which can together represent a route (instead of the π notation). Let $z_{i,j}$ represent the flow on edge (i, j) and let a binary variable $y_{i,j}$ represent whether there exists a flow on edge (i, j). Then the mixed ILP is:

$$\min_{z,y}\sum_{i=1}^M\sum_{j=1}^M d_{i,j}z_{i,j} \quad \text{s.t.} \qquad (10)$$

$$\text{No flow from node } i \text{ to itself: } z_{i,i} = 0 \ \ \forall i = 1, ..., M \qquad (11)$$

$$\text{No edge from node } i \text{ to itself: } y_{i,i} = 0 \ \ \forall i = 1, ..., M \qquad (12)$$

$$\text{Exactly one edge into each node: } \sum_{i=1}^M y_{i,j} = 1 \ \ \forall j = 1, ..., M \qquad (13)$$

$$\text{Exactly one edge out from each node: } \sum_{j=1}^M y_{i,j} = 1 \ \ \forall i = 1, ..., M \qquad (14)$$

$$\text{Flow coming back at the end of the loop is } \bar{p}(\tilde{x}_1)\text{: } \sum_{i=1}^M z_{i,1} = \bar{p}(\tilde{x}_1) \qquad (15)$$

Change of flow after crossing node k is:

$$\sum_{i=1}^M z_{i,k} - \sum_{j=1}^M z_{k,j} = \begin{cases}\bar{p}(\tilde{x}_1) - \sum_{i=1}^M \bar{p}(\tilde{x}_i) & k = 1\\ \bar{p}(\tilde{x}_k) & k = 2, ..., M\end{cases} \qquad (16)$$

$$\text{Connects flows } z \text{ to indicators of edge } y\text{: } z_{i,j} \le r_{i,j}y_{i,j} \qquad (17)$$

$$\text{where } r_{i,j} = \begin{cases} \bar{p}(\tilde{x}_1) & j = 1 \\ \sum_{i=1}^{M} \bar{p}(\tilde{x}_i) & i = 1 \\ \sum_{i=2}^{M} \bar{p}(\tilde{x}_i) & \text{otherwise.} \end{cases}$$

Constraints (11) and (12) restrict self-loops from forming. Constraints (13) and (14) impose that every node should have exactly one edge coming in and one going out. Constraint (15) represents the flow on the last edge coming back to the starting node. Constraint (16) quantifies the flow change after traversing a node k. Constraint (17) represents an upper bound on $z_{i,j}$ relating it to the corresponding binary variable $y_{i,j}$.

Mixed-Integer Optimization for Cost 2. By applying the log function to the cost of each node (7) (and subtracting a constant), we can minimize a more tractable cost objective:

$$\text{FailureCost} = \min_{\pi} \sum_{i=1}^{M} L_{\pi}(\pi(i)) \log\left(1 + e^{f_{\lambda}(\tilde{x}_{\pi(i)})}\right).$$

This failure cost term is now a weighted sum of latencies where the weights are of the form $\log\left(1 + e^{f_{\lambda}(\tilde{x}_{\pi(i)})}\right)$. We can thus reuse the mixed ILP (10)-(17) where the weights are redefined as $\bar{p}(\tilde{x}_i) := \log\left(1 + e^{\lambda \cdot \tilde{x}_i}\right)$.

We have thus shown how to solve the weighted TRP subproblem, and we will now present ways to solve the full ML&TRP.

Mixed-Integer Nonlinear Programs (MINLPs) for Simultaneous Formulation. The full objective using Cost 1 is:

$$\min_{\lambda} \left(\sum_{i=1}^{m} \ln\left(1 + e^{-y_i f_{\lambda}(x_i)}\right) + C_2 ||\lambda||_2^2 + C_1 \min_{\{z_{i,j}, y_{i,j}\}} \sum_{i=1}^{M} \sum_{j=1}^{M} d_{i,j} z_{i,j} \right) \quad (18)$$

$$\text{such that constraints (11) to (17) hold, where } \bar{p}(\tilde{x}_i) = \frac{1}{1 + e^{-\lambda \cdot \tilde{x}_i}}.$$

The full objective using the modified version of Cost 2 is:

$$\min_{\lambda} \left(\sum_{i=1}^{m} \ln\left(1 + e^{-y_i f_{\lambda}(x_i)}\right) + C_2 ||\lambda||_2^2 + C_1 \min_{\{z_{i,j}, y_{i,j}\}} \sum_{i=1}^{M} \sum_{j=1}^{M} d_{i,j} z_{i,j} \right) \quad (19)$$

$$\text{such that constraints (11) to (17) hold, where } \bar{p}(\tilde{x}_i) = \log\left(1 + e^{\lambda \cdot \tilde{x}_i}\right).$$

If we have an algorithm for solving (18), then the same scheme can be used to solve (19). There are multiple ways of solving (or approximately solving) a mixed integer nonlinear optimization problem of the form (18) or (19). We consider three methods. The first method is to directly use a generic *mixed integer non-linear programming* (MINLP) solver. The second and third methods (called *Nelder-Mead* and *Alternating Minimization*, denoted NM and AM respectively) are iterative schemes over the λ parameter space. At every iteration of these algorithms, we will need to evaluate the objective function. This evaluation involves solving an instance of the weighted TRP subproblem. For the AM algorithm, define Obj as follows:

$$\begin{aligned}\text{Obj}(\lambda, \pi) = &\ \text{TrainingError}(f_\lambda, \{x_i, y_i\}_{i=1}^m) \\ &+ C_1 \text{FailureCost}\left(\pi, f_\lambda, \{\tilde{x}_i\}_{i=1}^M, \{d_{i,j}\}_{i,j=1}^M\right). \end{aligned} \quad (20)$$

Starting from an initial vector λ_0, Obj is minimized alternately with respect to λ and then with respect to π, as shown in Algorithm 1.1.

Inputs: $\{x_i, y_i\}_1^m, \{\tilde{x}_i\}_1^M, \{d_{ij}\}_{ij}, C_1, C_2, T$ and initial vector λ_0.
for t=1:T **do**
  Compute $\pi_t \in \text{argmin}_{\pi \in \Pi} \text{Obj}(\lambda_{t-1}, \pi)$ (mixed ILP).
  Compute $\lambda_t \in \text{argmin}_{\lambda \in \mathbb{R}^d} \text{Obj}(\lambda, \pi_t)$ (Gradient descent).
end for
Output: π_T.

Algorithm 1.1. AM: Alternating minimization algorithm

4 Experiments

We have now defined two formulations (sequential and simultaneous), each with two possible definitions for the failure cost (Cost 1 and Cost 2), and three algorithms for the simultaneous formulation (MINLP solver, NM, and AM). In what follows, we will highlight the advantage of the simultaneous method over the less general sequential method through two experiments. The first involves a very simple synthetic dataset, designed to show differences between the two methods. The second experiment involves a real dataset, designed as part of a collaboration with NYC's power company, Con Edison (see [10] for a more detailed description of these data). In each experiment, we solve the simultaneous formulation over a range of values of C_1 and compare the routes and failure estimates obtained over this range. Our goal for this section is to illustrate that incorporating the routing cost into the machine learning model can produce lower cost solutions in at least some scenarios, without harming prediction accuracy. For both experiments, we have a fixed training set and separate test set to evaluate predictions of the model, and the unlabeled set of nodes with distances. In both experiments, there is a lot of uncertainty in the estimates for the unlabeled set. In the toy example, the unlabeled set is in a low density region, so the probabilities could reasonably change without substantially affecting prediction ability. In the second experiment, the data are very imbalanced (the positive class is very rare), so there is a lot of uncertainty in the estimates, and further, there is a prior belief that a low-cost route exists. In particular, we have reason to believe that some of the probabilities are overestimated in this particular experiment using the particular unlabeled set we chose, and that knowing the repair route can help to determine these probabilities; this is because there are underground electrical cables traversing each linear stretch of the repair route.

Toy Example. We illustrate how the simultaneous formulation takes advantage of uncertainty; it is because a small change in the probabilities can give a completely different route and cost. Consider the graph G shown in Figure 1(a) and Figure 1(b). Figure 1(c) shows unlabeled points $\{\tilde{x}_i\}_{i=1}^4 \in \mathbb{R}^2$ along with the training instances (represented by two gray clusters). The sequential formulation produces a function f_λ^* whose 0.5-probability level set is shown as a black line here. The route corresponding to that solution is given in Figure 1(a), which is $\pi^* = 1 - 3 - 2 - 4 - 1$. If we were to move the 0.5-probability level set slightly, for instance to the dashed line in Figure 1(c) by using an appropriate tradeoff parameter C_1 in the simultaneous formulation, the probability estimates on the finite training set change only slightly, but the cost and the corresponding route change entirely (Figure 1(b)). The new route is $\pi^* = 1 - 3 - 4 - 2 - 1$, and yields a lower value of Cost 1 (a decrease of ~16.4%). In both cases, the probability estimators have very similar validation performance, but the solutions on the graph are different.

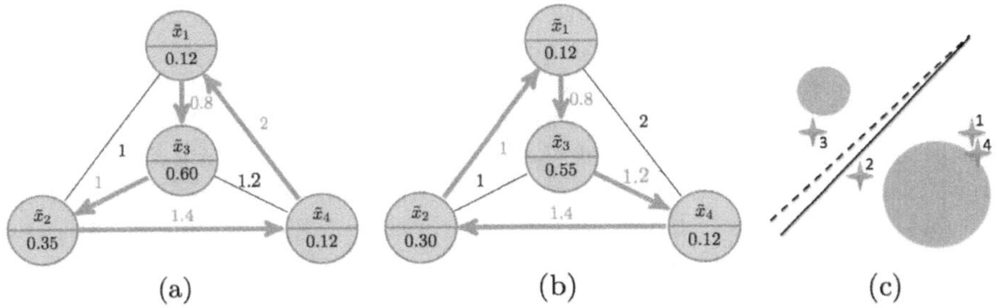

Fig. 1. For the above graphs, the numbers in the nodes indicate their probability of failures and the numbers on the edges indicate distances. (a) Route as determined by sequential formulation (*highlighted*). (b) Route determined by the simultaneous formulation. (c) The feature space.

The NYC Power Grid. We have information related to manholes from the Bronx, NYC (~23K manholes). Each manhole is represented by (4 dimensional) features that encode the number and type of electrical cables entering the manhole and the number and type of past events involving the manhole. The training features encode events prior to 2008, and the training labels are 1 if the manhole was the source of a serious event (fire, explosion, smoking manhole) during 2008. The prediction task is to predict events in 2009. The test set (for evaluating the performance of the predictive model) consists of features derived from the time period before 2009, and labels from 2009. Predicting manhole events can be a difficult task for machine learning, because one cannot necessarily predict an event using the available data. The operational task is to design a route for a repair crew that is fixing seven manholes in 2009 on which we want the cost of failures to be low. Because of the large class imbalance, the misclassification error is almost always the size of the whole positive class. Because of this, we evaluate the quality of the predictions from f_{λ^*} using the area under the ROC curve (AUC), for both training and test.

We solve (18) and (19) using an appropriate range of values for the regularization parameter C_1, with the goal of seeing whether for the same level of estimation performance, we can get a reduction in the cost of failures. Note that the uncertainty in the estimation of failure probabilities is due to the finite number of examples in the training set. The other regularization parameter C_2 is kept fixed throughout (in practice one might use cross-validation if C_2 is allowed to vary). The evaluation metric AUC is a measure of ranking quality; it is sensitive to the rank-ordering of the nodes in terms of their probability to fail, and it is not as sensitive to changes in the values of these probabilities. This means that as the parameter C_1 increases, the estimated probability values will tend to decrease, and thus the failure cost will decrease; it may be possible for this to happen without impacting the prediction quality as measured by the AUC, but this depends on the routes and it is not guaranteed. In our experiments, for both training and test we had a large sample (~23K examples). The test AUC values for the simultaneous method were all within 1% of the values obtained by the sequential method; this is true for both Cost 1 and Cost 2, for each of the AM, NM, and MINLP solvers, see Figures 3(a) and 3(b). The variation in TrainingError across the methods was also small, about 2%, see Figure 3(c). So, changing C_1 did not dramatically impact the prediction quality as measured by the AUC. On the other hand, the failure costs varied widely over the different methods and settings of C_1, as a result of the decrease in the probability estimates, as shown in Figure 3(d). As C_1 was increased from 0.05 to 0.5, Cost 1 went from 27.5 units to 3.2 units, which is over eight times smaller. This means that with a 1-2% variation in the predictive model's AUC, the failure cost can decrease a lot, potentially yielding a more cost-effective route for inspection and/or repair work. The reason for an order of magnitude change in the failure cost is because the probability estimates are reducing by an order of magnitude due to uncertainty; yet our model still maintained the same level of AUC performance on training and test sets. Figure 2(a) shows the route provided by the sequential formulation. For the simultaneous formulation, there are changes in the cost and

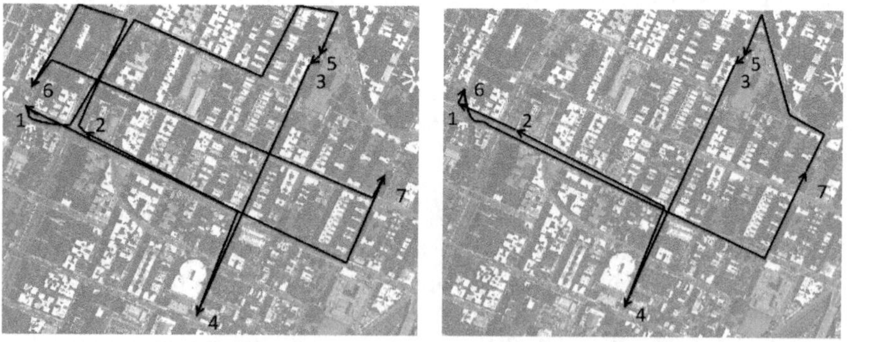

(a) (b)

Fig. 2. (a) Sequential formulation route: 1-5-3-4-2-6-7-1. (b) Simultaneous formulation route ($C_1 = 0.5$): 1-6-7-5-3-4-2-1.

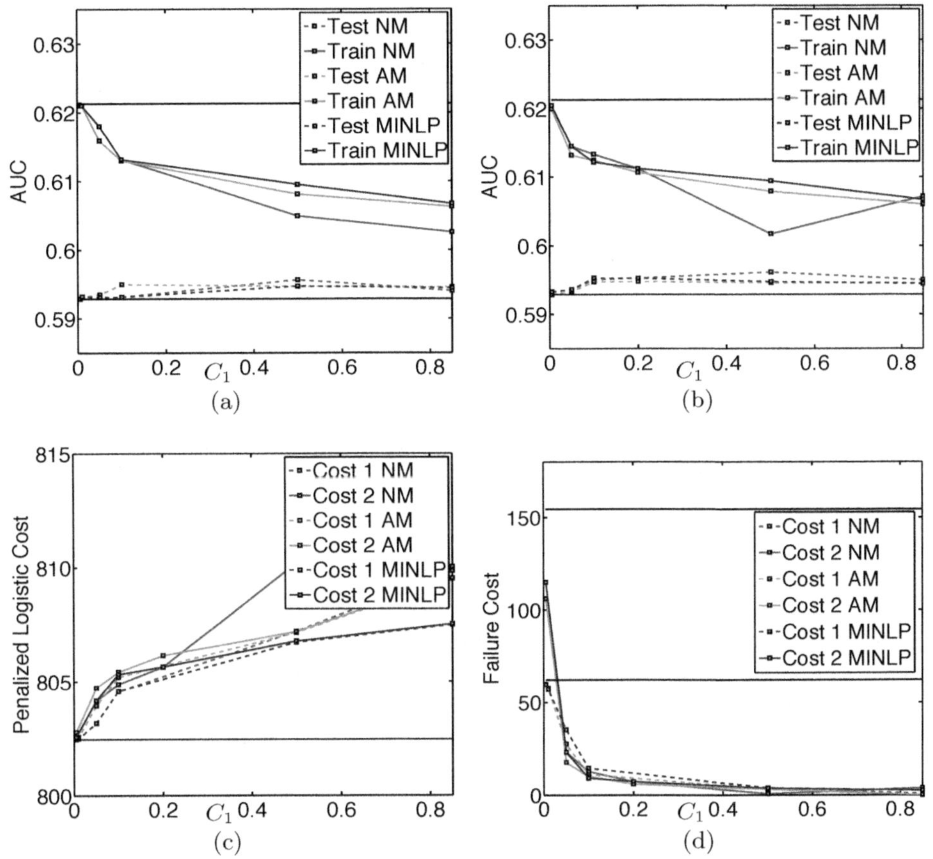

Fig. 3. For all the figures, horizontal lines represent baseline sequential formulation values for training or testing; x-axes represent values of C_1; the curves for the three algorithms (NM, AM and MINLP) are very similar to each other and the focus is on their trend with respect to C_1. (a) AUC values with Cost 1. (b) AUC values with Cost 2. (c) ℓ_2-regularized logistic loss. (d) Decreasing failure cost for both Cost 1 and 2.

the route as the coefficient C_1 increases. When the failure cost term starts influencing the optimal solution of the objective (18), we get a new route as shown in Figure 2(b). This demonstration on data from the Bronx illustrates that it is possible to take advantage of uncertainty in modeling, in order to create a much more cost-effective solution.

5 Generalization Bound

We initially introduced the failure cost regularization term in order to find scenarios where the data would support low-cost (more actionable) repair routes. From another point of view, incorporating regularization increases bias and reduces variance, and may thus allow us to obtain better prediction guarantees as we increase C_1. Any type of bias can either help or hurt the quality of the a statistical model, depending on whether the "prior belief" associated the bias is correct (this relates to "approximation error"). At the same time, incorporating bias helps to reduce the variance of the solution, reducing the difference between the training error we measure and the true error on the full population ("generalization error"). This difference is what we discuss in this section.

The "hypothesis space" is the set of models that an algorithm can choose from. When C_1 is large, it means we are only allowing models that yield low-cost solutions. This restriction on the hypothesis space (to the set of low-cost solutions) is a reduction in the size of this space. In statistical learning theory, the size of the hypothesis space is recognized as one of the most important quantities in the learning process, and this idea is formalized through probabilistic guarantees, *i.e.*, bounds on the generalization error. The bound we provide below shows how the TRP cost term (using Cost 1) reduces the size of the hypothesis space by removing a spherical cap, and how this could affect the generalization ability of the ML&TRP algorithms.

Define the true risk as the expectation of the logistic loss:

$$R(f_\lambda) := E_{(x,y)\sim\mathcal{X}\times\mathcal{Y}} l(f_\lambda(x), y) = \int \ln\left(1 + e^{-yf_\lambda(x)}\right) \partial\mu_{\mathcal{X}\times\mathcal{Y}}(x, y).$$

We bound $R(f_\lambda)$ by the empirical risk $R(f_\lambda, \{x_i, y_i\}_1^m) = \frac{1}{m}\sum_{i=1}^m \ln\left(1 + e^{-y_i f(x_i)}\right)$ plus a complexity term that depends on the geometry of where the nodes are located. Before we do this, we need to replace the Lagrange multiplier C_1 in (18) with an explicit constraint, so f_λ is subject to a specific limit on the failure cost:

$$\min_{\pi} \sum_{i=1}^{M} \frac{1}{1 + e^{-f_\lambda(\tilde{x}_{\pi(i)})}} L_\pi(\pi(i)) \leq C_g.$$

C_g is a constant (inversely related to C_1), and C_g will be a bias-variance tradeoff in the bound. Let $\sup_{x\in\mathcal{X}} ||x||_2 \leq M_2$, so $f_\lambda : \mathcal{X} \to [-M_1M_2, M_1M_2]$. Let us define the set of functions that are subject to a constraint on the failure cost as:

$$\mathcal{F}_0 := \left\{ f_\lambda : f_\lambda \in \mathcal{F}, \min_{\pi\in\Pi} \sum_{i=1}^{M} L_\pi(\pi(i)) \frac{1}{1 + e^{-f_\lambda(\tilde{x}_{\pi(i)})}} \leq C_g \right\}.$$

Now we incorporate the geometry. Let d_i to be the shortest distance from the starting node to node i and let d_1 be the length of the shortest tour that visits all the nodes and returns to node 1. Define a vector c element-wise by:

$$c^j = \frac{\tilde{c}^j}{C_g - \tilde{c}_0} \text{ where } \tilde{c}^j = \frac{e^{M_1M_2}}{(1 + e^{M_1M_2})^2} \left(\sum_i d_i \tilde{x}_i^j\right), \text{ where}$$

$$\tilde{c}_0 = \left(M_1M_2 \frac{e^{M_1M_2}}{(1 + e^{M_1M_2})^2} + \frac{1}{1 + e^{M_1M_2}}\right) \sum_i d_i.$$

This vector c incorporates both C_g and the d_i's that are the important ingredients in providing a generalization guarantee.

Theorem 1. *(**Generalization Bound**) Let $\mathcal{X} = \{x \in \mathbb{R}^d : ||x||_2 \leq M_2\}$, $\mathcal{Y} = \{-1, 1\}$. Let $\mathcal{F}_0$ be defined as above with respect to $\{\tilde{x}_i\}_{i=1}^M$, $\tilde{x}_i \in \mathcal{X}$ (not necessarily random). Let $\{x_i, y_i\}_{i=1}^m$ be a sequence of m examples drawn independently according to an unknown distribution $\mu_{\mathcal{X}\times\mathcal{Y}}$. Then for any $\epsilon > 0$,*

$$P\Big(\exists f_\lambda \in \mathcal{F}_0 : |R(f_\lambda, \{x_i, y_i\}_1^m) - R(f_\lambda)| > \epsilon\Big)$$
$$\leq 4\alpha(d, C_g, c)\left(\frac{32M_1M_2}{\epsilon} + 1\right)^d \exp\left(\frac{-m\epsilon^2}{128(e^{M_1M_2} - e^{-M_1M_2})^2}\right),$$

where

$$\alpha(d, C_g, c) := \frac{1}{2} + \frac{||c||_2^{-1} + \frac{\epsilon}{32M_2}}{M_1 + \frac{\epsilon}{32M_2}} \frac{\Gamma\left[1 + \frac{d}{2}\right]}{\sqrt{\pi}\Gamma\left[\frac{d+1}{2}\right]} {}_2F_1\left(\frac{1}{2}, \frac{1-d}{2}; \frac{3}{2}; \left(\frac{||c||_2^{-1} + \frac{\epsilon}{32M_2}}{M_1 + \frac{\epsilon}{32M_2}}\right)^2\right) \tag{21}$$

and where ${}_2F_1(a, b; c; d)$ is the hypergeometric function.

The term $\alpha(d, C_g, c)$ comes directly from formulae for the normalized volume of a spherical cap. Our goal was to establish that generalization can depend on C_g. As C_g decreases, the norm $||c||_2$ increases, and thus (21) decreases, and the whole bound decreases. Decreasing C_g may thus improve generalization. The proof is lengthy and is provided in a longer version [11].

6 Conclusion

In this work, we present a machine learning algorithm that takes into account the way its recommendations will be ultimately used. This algorithm takes advantage of uncertainty in the model in order to potentially find a much more practical solution. Including these operating costs is a new way of incorporating "structure" into machine learning algorithms, and we plan to explore this in other ways in ongoing work. We discussed the tradeoff between estimation error and operating cost for the specific application to the ML&TRP. In doing so, we showed a new way in which data dependent regularization can influence an algorithm's prediction ability, formalized through generalization bounds.

Acknowledgements. This work is supported by an International Fulbright Science and Technology Award, the MIT Energy Initiative, and the National Science Foundation under Grant IIS-1053407.

References

1. Picard, J.-C., Queyranne, M.: The time-dependent traveling salesman problem and its application to the tardiness problem in one-machine scheduling. Operations Research 26(1), 86–110 (1978)
2. Chapelle, O., Schölkopf, B., Zien, A. (eds.): Semi-Supervised Learning. MIT Press, Cambridge (2006)

3. Agarwal, S.: Ranking on graph data. In: Proceedings of the 23rd International Conference on Machine Learning (2006)
4. Belkin, M., Niyogi, P., Sindhwani, V.: Manifold regularization: A geometric framework for learning from labeled and unlabeled examples. Journal of Machine Learning Research 7, 2399–2434 (2006)
5. Zhou, D., Weston, J., Gretton, A., Bousquet, O., Schölkopf, B.: Ranking on data manifolds. In: Advances in Neural Information Processing Systems, vol. 16, pp. 169–176. MIT Press, Cambridge (2004)
6. Fischetti, M., Laporte, G., Martello, S.: The delivery man problem and cumulative matroids. Oper. Res. 41, 1055–1064 (1993)
7. van Eijl, C.A.: A polyhedral approach to the delivery man problem. Memorandum COSOR 95–19, Department of Mathematics and Computer Science, Eindhoven University of Technology, The Netherlands (1995)
8. Lechmann, M.: The traveling repairman problem - an overview, pp. 1–79, Diplomarbeit, Universitat Wein (2009)
9. Urbina, I.: Mandatory safety rules are proposed for electric utilities. New York Times. Late Edition, Sec B, Col 3, Metropolitan Desk, p. 2 (08-21-2004)
10. Rudin, C., Passonneau, R., Radeva, A., Dutta, H., Ierome, S., Isaac, D.: A process for predicting manhole events in Manhattan. Machine Learning 80, 1–31 (2010)
11. Tulabandhula, T., Rudin, C., Jaillet, P.: Machine Learning and the Traveling Repairman. arXiv:1104.5061 (2011)
12. Blum, A., Chalasani, P., Coppersmith, D., Pulleyblank, B., Raghavan, P., Sudan, M.: On the minimum latency problem. In: Proceedings of the 26th ACM Symposium on Theory of Computing, pp. 163–171 (September 1994)

Learning Complex Concepts Using Crowdsourcing: A Bayesian Approach

Paolo Viappiani[1], Sandra Zilles[2], Howard J. Hamilton[2], and Craig Boutilier[3]

[1] Department of Computer Science, Aalborg University, Denmark
[2] Department of Computer Science, University of Regina, Canada
[3] Department of Computer Science, University of Toronto, Canada

Abstract. We develop a Bayesian approach to concept learning for crowdsourcing applications. A probabilistic belief over possible concept definitions is maintained and updated according to (noisy) observations from experts, whose behaviors are modeled using discrete types. We propose recommendation techniques, inference methods, and query selection strategies to assist a user charged with choosing a configuration that satisfies some (partially known) concept. Our model is able to simultaneously learn the concept definition and the types of the experts. We evaluate our model with simulations, showing that our Bayesian strategies are effective even in large concept spaces with many uninformative experts.

1 Introduction

Crowdsourcing is the act of outsourcing a problem to a group or a community. It is often referred to as *human computation*, as human "experts" are used to solve problems that present difficulties for algorithmic methods; examples include Amazon's Mechanical Turk, the ESP game (for image labeling), and *reCaptcha* (for book digitization). Multiple human teachers, or experts, give feedback about (label) a particular problem instance. For instance, users refer to sites such as Yahoo! Answers to ask questions about everything from cooking recipes to bureaucratic instructions to health suggestions (e.g., which ingredients do I need to make tiramisu? how do I apply for a Chinese visa? how do I lose 20 pounds?).

As the information obtained with crowdsourcing is inherently noisy, effective strategies for aggregating multiple sources of information are critical. Aggregating noisy labels and controlling workflows are two problems in crowdsourcing that have recently been addressed with principled techniques [5,11,4]. In this work, we address the problem of generating recommendations for a user, where recommendation quality depends on some latent concept. The knowledge of the concept can only be refined by aggregating information from noisy information sources (e.g., human experts), and the user's objective is to maximize the quality of her choice as measured by satisfaction of the unknown latent concept. Achieving *complete* knowledge of the concept may be infeasible due to the quality of information provided by the experts. Fortunately, complete concept knowledge is generally unnecessary to select a satisfying instance of that concept. For instance, to successfully make tiramisu (a type of cake), certain ingredients might be necessary, while others may be optional. The concept c represents all possible "correct"

R.I. Brafman, F. Roberts, and A. Tsoukiàs (Eds.): ADT 2011, LNAI 6992, pp. 277–291, 2011.

recipes that are consistent with the abstract notion of the cake. A configuration or instance x is a candidate recipe, and it satisfies c iff it can be used to make the cake (i.e., is correct). By asking various, possibly noisy, experts about particular ingredients, the user may "learn" a recipe satisfying c without ever learning *all* recipes satisfying c.

Following [2], our aim is not to learn the concept definition *per se*; rather we want to learn just enough about it to make a (near-)optimal decision on the user's behalf. By exploiting the structure of the concept, a recommender system can adopt a strategy that queries only concept information that is relevant to the task at hand. For instance, if the system knows that an ingredient is extremely unlikely to be used in tiramisu, or is unlikely to be available, querying about this ingredient is unlikely to be helpful. Finally, the system needs to select the experts whose answers are (predicted to be) as informative as possible.

Our main contributions are 1) computational procedures to aggregate concept information (originating from noisy experts) into a probabilistic belief, 2) algorithms to generate recommendations that maximize the likelihood of concept satisfaction and 3) strategies to interactively select queries and experts to pose them to.

Our work is related to the model of Boutilier et al. [2,3], who present a regret-based framework for learning subjective features in the context of preference elicitation. Our approach can be seen both as a Bayesian counterpart of that model, and as an extension to the case of multiple experts.

2 Bayesian Concept Learning Approach

We consider the problem of learning a latent concept by aggregating information from several sources called *experts*. Each expert may have a partial and incorrect definition of the concept. As in traditional concept learning [10,9], we assume an abstract concept c is drawn from a concept class $\mathcal{C}$. However, instead of trying to identify the concept explicitly, we maintain a distribution over possible concept definitions, and update the distribution according to the information acquired from the experts, in order to recommend an instance that is highly likely to satisfy the concept.

2.1 Concepts

We consider the problem of learning an abstract boolean concept drawn from a fixed concept class. A boolean concept c is a function $\{0,1\}^n \to \{0,1\}$, where $\{\mathcal{X}_1, \ldots, \mathcal{X}_n\}$ is a set of n boolean features. A *solution* (goal of the learning problem) is any boolean vector (configuration) $(x_1, \ldots, x_n) \in \{0,1\}^n$ for which $c(x_1, \ldots, x_n) = 1$. We allow the solution space to be restricted by feasibility constraints; below we assume linear constraints of the type $A \cdot x \leq B$ (with matrix A and vector B of the right dimensions). For example, *budget constraints* associate a vector of costs $(a_1, \ldots, a_n)$ with each feature and require the total cost of a solution not to exceed the available budget b.

Throughout the paper, we restrict our focus to conjunctions [6] as the class of latent concepts, although our abstract model can be extended to boolean functions in general. A conjunctive concept c is a conjunction of literals over (some of) the atoms $\mathcal{X}_1, \ldots, \mathcal{X}_n$, e.g., $c = \mathcal{X}_2 \wedge \neg\mathcal{X}_4 \wedge \mathcal{X}_7$. A conjunction c can be equivalently represented as an assignment $(X_1^c, \ldots, X_n^c)$ of features $\mathcal{X}_1, \ldots, \mathcal{X}_n$ to the domain $\{T, F, DC\}$; in other words X_i^c can have one of the values T (true; the literal $\mathcal{X}_i$ occurs in c), F (false; the literal $\neg\mathcal{X}_i$ occurs in c), or DC (don't care; the atom $\mathcal{X}_i$ does not occur in c). In the above example, $X_2^c = X_7^c = T$, $X_4^c = F$, and $X_i^c = DC$ for $i \in \{1, 3, 5, 6\}$.

Since the latter representation is used throughout the text, we write $c = (X_1^c, \ldots, X_n^c)$ and, with a slight abuse of notation, we will sometime refer to X_i^c as the "value" of feature i in concept c; we will also drop the superscript when c is clear from context. A configuration $x = (x_1, \ldots, x_n)$ yields $c(x) = 1$ (we say x satisfies c) iff (i) $x_i = 1$ for each i such that the literal $\mathcal{X}_i$ occurs in c ($X_i^c = T$), and (ii) $x_i = 0$ for each i such that the literal $\neg\mathcal{X}_i$ occurs in c ($X_i^c = F$).

Because the concept is unknown, the system maintains a belief $P(c) = P(X_1^c, \ldots, X_n^c)$. We assume some prior distribution over concepts. It is sometimes convenient to reason with the marginal probabilities, $P(X_i)$, representing the distribution over feature i, i.e., $P(X_i = T)$, $P(X_i = F)$, and $P(X_i = DC)$; for convenience, we write these terms as $P(T_i)$, $P(F_i)$, and $P(DC_i)$, respectively.

2.2 Query Types

The system acquires information about the concept by posing queries to a set of experts. These concept queries can be of different forms (e.g., membership, equivalence, superset, or subset queries [1]) and their answers partition the hypothesis space. For instance, a membership query asks whether a given configuration x satisfies the concept (e.g., *"Is this a valid recipe for tiramisu?"*). Membership queries can be too cognitively demanding for a crowd-sourcing domain, as an expert would have to verify every problem feature to check whether the provided instance is satisfied. Thus, in this work we focus on *literal* queries, a special form of *superset queries*. A literal query q_i on feature i asks for the value of X_i; possible answers to the query are T, F, or DC.[1] Literal queries can be thought of as requests for a piece of information such as *"Are eggs needed for tiramisu?"*. Query strategies for selecting literal queries are discussed in Section 4.[2]

2.3 Expert Types

In practice, experts do not always provide correct answers. Hence we assume that experts belong to different populations or *types* from a set $\mathcal{T} = \{t_1, \ldots, t_k\}$. The type of an expert represents the expert's capacity and commitment to correctly answering queries about the concept (or aspects thereof). For instance, as in [4], types might discriminate "good" or knowledgeable experts, whose answers are likely to be correct, from "bad" experts, whose answers are drawn randomly. Our model generalizes to any number of types.

We indicate the assignments of types to experts with a vector $\theta = (\theta^1, \ldots, \theta^m)$, where $\theta^j \in \mathcal{T}$ is the type of expert j. A further natural assumption is that experts are

[1] Alternatively, one could ask queries such as "Is X_i positive in the concept definition?" Adapting our model to such queries is straightforward.

[2] Notice that literal queries cannot be answered unambiguously in general since dependencies may exist; but the value of a literal in a conjunctive concept is independent of the value of any other literal.

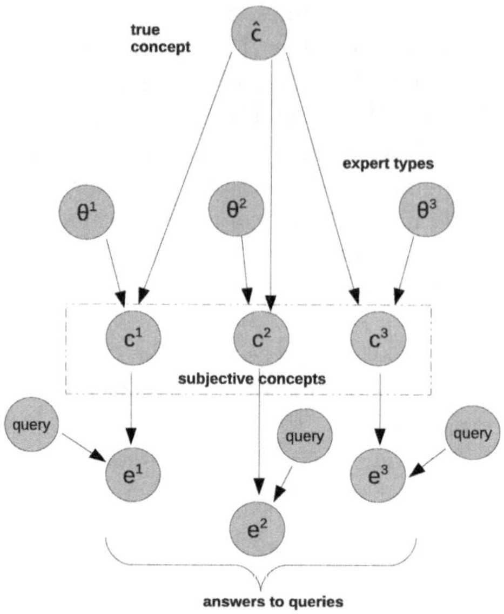

Fig. 1. Abstract model for learning an unknown concept from mutiple noisy experts

noisy and provide feedback with respect to their subjective definition of the concept. In other words, we assume that there exists one underlying (true) concept definition $\hat{c} = (X_1, \ldots, X_n)$, but each expert's response is based on its own subjective concept $c^j = (X_1^j, \ldots, X_n^j)$. When a query q_i^j on feature i is posed to expert j, the expert reveals its subjective value X_i^j for that feature (either T, F or DC). Subjective concepts are distributed, in turn, according to a generative model $P(c^j|\hat{c}, \theta^j)$, given expert type θ^j and true concept $\hat{c}$. For example, an "uninformed" expert may have a subjective concept that is probabilistically independent of $\hat{c}$, while an "informed" expert may have a concept that is much more closely aligned with $\hat{c}$ with high probability. In our experiments below, we assume a factored model $P(X_i^j|X_i, \theta^j)$. Moreover, since we always ask about a specific literal, we call this distribution the *response model*, as it specifies the probability of expert responses as a function of their type. This supports Bayesian inference about the concept given expert answers to queries (note that we do not assume expert types are themselves observed; inference is also used to estimate a distribution over types).

The graphical model for the general case is shown in Figure 1. In Figure 2 we show the model for conjunctions with 3 features and 2 experts; the subjective concept c^j of expert $j \in \{1, 2\}$ is composed of X_1^j, X_2^j and X_3^j.

As queries provide only "noisy" information about the true concept $\hat{c}$, the system cannot fully *eliminate* hypotheses from the version space given expert responses. To handle concept uncertainty, the system maintains a distribution or *belief* $P(c)$ over concept definitions, as well as a distribution over expert types $P(\theta)$. Both distributions are updated whenever queries are answered.

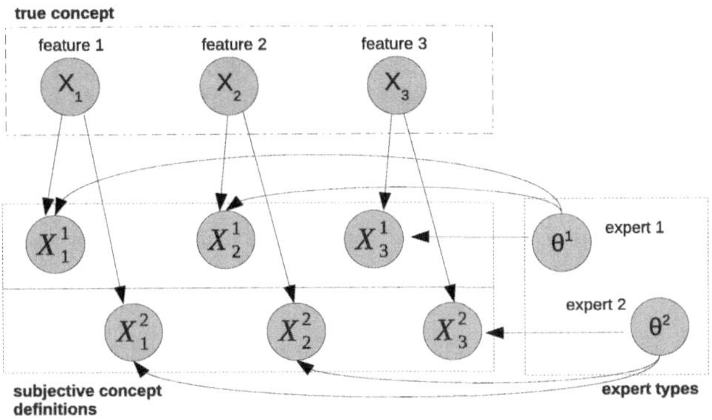

Fig. 2. Graphical model for Bayesian learning of conjunctions: 3 features, 2 experts

Beliefs about the true concept and expert subjective concepts will generally be correlated, as will beliefs about the types of different experts. Intuitively, if two experts consistently give similar answers, we expect them to be of the same type. When we acquire additional evidence about the type of one expert, this evidence affects our belief about the type of the other expert as well. Thus, when new evidence e is acquired, the joint posterior $P(c, \theta|e)$ cannot be decomposed into independent marginals over c and the θ^j, since c and θ are not generally independent. Similarly, new evidence about feature X_i might change one's beliefs about types, and therefore influence beliefs about another feature X_j. We discuss the impact of such dependence on inference below.

2.4 Decision Making

The system needs to recommend a configuration $x = (x_1, \ldots, x_n) \subseteq \{0,1\}^n$ that is likely to satisfy the concept (e.g., a recipe for tiramisu), based on the current belief $P(c)$. A natural approach is to choose a configuration x^* that maximizes the *a posteriori* probability of concept satisfaction (MAPSAT) according to the current belief: $x^* \in \arg\max_x P(c(x))$.

Exact maximization typically requires enumerating all possible configurations and concept definitions. Since this is not feasible, we consider the marginalized belief over concept features and optimize, as a surrogate, the product of probabilities of the individual features satisfying the configuration: $P(c(x)) \approx \tilde{P}(c(x)) = \prod_i P(c_i(x_i))$, where c_i is the restriction of concept c to feature i. In this way, optimization without feasibility or budget constraints can be easily handled. For each feature i, we choose $x_i = 1$ whenever $P(T_i) \geq P(F_i)$, and choose $x_i = 0$ otherwise.

However, in the presence of feasibility constraints, we cannot freely choose to set attributes in order to maximize the probability of concept satisfaction. We show how, using a simple reformulation, this can be solved as an integer program. Let $p_i^+ = P(T_i) + P(DC_i)$ be the probability that setting $x_i = 1$ is consistent with the concept definition for the i-th feature; similarly let $p_i^- = P(F_i) + P(DC_i)$ be the probability that setting $x_i = 0$ is consistent. Then the probability of satisfying the i-th feature is

$P(c_i(x_i)) = p_i^+ x_i + p_i^-(1 - x_i)$. The overall (approximated) probability of concept satisfaction can be written as:

$$P(c(x)) \approx \prod_{1 \leq i \leq n} p_i^+ x_i + p_i^-(1 - x_i) = \prod_{1 \leq i \leq n} (p_i^+)^{x_i} \prod_{1 \leq i \leq n} (p_i^-)^{(1-x_i)} \quad (1)$$

The latter form is convenient because we can linearize the expression by applying logarithms. To obtain the feasible configuration x^* maximizing the probability of satisfaction, we solve the following integer program (the known term has been simplified):

$$\max_{x_1,\ldots,x_n} \sum_{1 \leq i \leq n} [log(p_i^+) - log(p_i^-)] \cdot x_i \quad (2)$$

$$s.t. \quad A \cdot x \leq B \quad (3)$$

$$x \in \{0, 1\}^n \quad (4)$$

3 Inference

When a query is answered by some expert, the system needs to update its beliefs. Let e_i^j represent the evidence (query response) that expert j offers about feature i. Using Bayes' rule, we update the probability of the concept: $P(c|e_i^j) \propto P(e_i^j|c)P(c)$. Since the type θ^j of expert j is also uncertain, inference requires particular care. We consider below several strategies for inference. When discussing their complexity, we let n denote the number of features, m the number of experts, and k the number of types.

Exact Inference. Exact inference is intractable for all but the simplest concepts. A naive implementation of exact inference would be exponential in both the number of features and the number of experts. However, inference can be made more efficient by exploiting the independence in the graphical model. Expert types are mutually independent given concept c: $P(\theta|c) = \prod_{1 \leq j \leq m} P(\theta^j|c)$. This means that each concept can be "safely" associated with a vector of m probabilities $P(\theta^1|c), \ldots, P(\theta^m|c)$, one for each expert. For a concept space defined over n features, we explicitly represent the 3^n possible concept definitions, each associated with a matrix (of dimension m by k) representing $P(\theta|c)$. The probability of a concept is updated by multiplying the likelihood of the evidence and renormalizing: $P(c|e_i^j) \propto P(e_i^j|c)P(c)$. As the queries we consider are *local* (i.e., only refer to a single feature), the likelihood $P(e_i^j|c)$ of c is

$$P(e_i^j|c) = \sum_{t \in \mathcal{T}} P(e_i^j|\theta^j = t, X_i^c)P(\theta^j = t|c), \quad (5)$$

where X_i^c is the value of c for feature $\mathcal{X}_i$. The vector $(P(\theta^1|c, e_i^j), \ldots, P(\theta^m|c, e_i^j))$ is updated similarly. The overall complexity of this approach to exact inference is $O(3^n mk)$. Since the number of experts m is usually much larger than the number of features n, exact inference is feasible for small concept spaces, in practice, those with up to 5–10 features. In our implementation, exact inference with $n = 8$ and $m = 100$ requires 1–2 seconds per query.

Naive Bayes. This approach to inference makes the strong assumption that X_i and θ^j are mutually conditionally independent. This allows us to factor the concept distribution into marginals over features: $P(X_1), \ldots, P(X_n)$; similarly beliefs about experts are represented as $P(\theta^1), \ldots, P(\theta^m)$. The likelihood $P(e_i^j|X_i)$ of an answer to a query can be related to $P(e_i^j|\theta^j, X_i)$ (the response model) by marginalization over the possible types of expert j: $P(e_i^j|X_i) = \sum_{v \in \{t_1, t_2, \ldots\}} P(e_i^j|\theta^j = v, X_i) P(\theta^j = v|X_i)$. We write the expression for the updated belief about X_i given evidence as follows:[3]

$$P(X_i|e_i^j) = \frac{P(e_i^j|X_i)P(X_i)}{P(e_i^j)} \tag{6}$$

$$= \frac{\sum_{t \in \mathcal{T}} P(e_i^j|X_i, \theta^j{=}t) P(\theta^j, X_i)}{\sum_{z \in \{T,F,DC\}} \sum_{t \in \mathcal{T}} P(e_i^j|X_i{=}z, \theta^j{=}t) P(\theta^j, X_i)} \tag{7}$$

We update belief $P(X_i)$ using current type beliefs $P(\theta^1), \ldots, P(\theta^m)$. Our strong independence assumption allows simplification of Eq. 7:

$$P(X_i|e_i^j) = \frac{\sum_{t \in \mathcal{T}} P(e_i^j|X_i, \theta^j{=}t) P(\theta^j{=}t)}{\sum_z \sum_{t'} P(e_i^j|X_i{=}z, \theta^j = t')\, P(\theta^j{=}t')\, P(X_i{=}z)} P(X_i) \tag{8}$$

Similarly, for beliefs about types we have:

$$P(\theta^j|e_i^j) = \frac{\sum_z P(e_i^j|X_k = z, \theta^j)\, P(X_i = z)}{\sum_{z'} \sum_t P(e_i^j|X_i = z', \theta^j = t) P(\theta^j) P(X_i = z')} P(\theta^j) \tag{9}$$

This approximation is crude, but performs well in some settings. Moreover, with space complexity $O(n+m)$ and time complexity $O(nm)$, it is very efficient.

Monte Carlo. This approximate inference technique maintains a set of l particles, each representing a specific concept definition, using importance sampling. As with exact inference, we can factor beliefs about types. The marginal probability $P(X_i)$ that a given feature is true in the concept is approximated by the fraction of the particles in which X_i is true (marginalization over types is analogous). Whenever queries are answered, the set of particles is updated recursively with a resampling scheme. Each particle is weighted by the likelihood of the concept definition associated with that particle when evidence e_k^u is observed (the higher the likelihood, the higher the chance of resampling). Formally, the expression of the likelihood of a particle is analogous to the case of exact inference, but we only consider a limited number of possible concepts. Monte Carlo has $O(lmk)$ complexity; hence, it is more expensive than Naive Bayes but much less expensive than exact inference.

4 Query Strategies

We now present elicitation strategies for selecting queries. Each strategy is a combination of methods that, given the current beliefs about the concept and the types: i) selects

[3] Using Naive Bayes, we only update concept beliefs about X_i, the feature we asked about. Similarly, for types, we only update relative to θ^j, the expert that answered the query.

a feature to ask about; and ii) selects the expert to ask. Expert selection depends on the semantics of the types; here, as in [4], we assume experts are either "knowledgeable" (type t_1) or "ignorant" (type t_2). As a baseline, we consider two inefficient strategies for comparison purposes: (i) *broadcast* iterates over the features and, for each, asks the same query to a fixed number of experts; and (ii) *dummy* asks random queries of random experts; both baselines simply recommend solutions based on the most frequent answers received, without any optimization w.r.t. beliefs about concept satisfaction.

Feature Selection. We consider three strategies aimed at directly reducing concept uncertainty. The *maximum entropy* (or *maxent*) strategy selects the feature whose probability distribution over $\{T, F, DC\}$ has the greatest entropy. Unfortunately, this measure treats being uncertain between a T and F as the same as being uncertain between T and DC. The *minval* strategy selects the feature $\mathcal{X}_f$ with the lowest probability of "getting it right:" that is, $f = \arg\min_i\{\max(p_i^+, p_i^-)\}$ is viewed as the feature with the greatest potential for improvement. Each feature is "scored" using the probability, given our current beliefs, that the best guess for its feature value will match the true concept. The intention is to reduce the uncertainty that most hinders the chance of satisfying the concept. Finally, queries can be evaluated with respect to their capacity to improve decision quality using *value of information* [8]. Value of information can be optimized myopically or non myopically [7]. As there are m experts and n features, brute force maximization of myopic value of information would require considering mn queries, and for each performing the necessary Bayesian updates. We optimize *expected value of perfect information (EVPI)*; as shown below, this criterion can be computed using the current belief without expensive Bayesian updates. In this setting, EVPI measures the expected gain in the quality of a decision should we have access to perfect information about a particular feature. In other words, given an oracle able to provide the actual value (T, F or DC) of a feature, which should we ask about? The value of querying feature X_i is:[4]

$$EVPI_i = \sum_{z \in \{T,F,DC\}} P(X_i = z) \max_x P(c(x)|X_i = z). \tag{10}$$

Since we aim to select queries quickly, we also consider *Naive EVPI*, where $P(c(x)|X_i)$ is approximated by the product of the probabilities of satisfying each feature.

Observation 1. *In unconstrained problems, the feature selected with the minval heuristic strategy is identical to that selected by maximum Naive EVPI.*

A proof is provided in the appendix. It relies on the fact that, without feasibility constraints, one can optimize features independently. For the more general case, given feature i, we define $x^{+i} = \arg\max_{x \in X : x_i = 1} P(c(x))$ to be the optimal configuration among those where feature i is true; we define x^{-i} analogously. We write the approximate satisfaction probabilities as $\tilde{P}(c(x^{+i})) = p_i^+ \cdot p_{\neq i}^{+i}$, where $p_{\neq i}^{+i} = \prod_{j \neq i} P(c_j(x^{+i}))$, and $\tilde{P}(c(x^{-i})) = p_i^- \cdot p_{\neq i}^{-i}$.

[4] We consider each possible response (T, F or DC) by the oracle, the recommended configuration conditioned to the oracle's answer, and weight the results using the probability of the oracle's response.

Observation 2. *Naive EVPI can readily be computed using the current belief:*

$$EVPI_i = P(T_i)p^{+i}_{\neq i} + P(F_i)p^{-i}_{\neq i} + P(DC_i) \cdot \max\{p^{+i}_{\neq i}, p^{-i}_{\neq i}\}$$

A proof is provided in the appendix. From this observation it follows that if $P(DC_i) = 0$ (i.e., we know that a feature is either true or false in the concept definition), then $EVPI_i = \tilde{P}(c(x^{+i})) + \tilde{P}(c(x^{-i}))$. The most informative feature is the feature i that maximizes the sum of the probabilities of concept satisfaction of x^{+i} and x^{-i}. This, in particular, is true when one considers a concept space where "don't care" is not allowed.

Naive EVPI query maximization is generally very efficient. As the current best configuration x^* will coincide with either x^{+i} or x^{-i} for any feature i, it requires only $n+1$ MAPSAT-optimizations and n evaluations of EVPI using Observation 2. Its computational complexity is not affected by the number of experts m.

Expert Selection. For a given feature, the *greedy* strategy selects the expert with the highest probability of giving an informative answer (i.e., the one with the highest probability of having type t_1). It is restricted to never ask the the same expert about the same feature, which would be useless in our model. However, there can be value in posing a query to an expert other than that predicted to be most "knowledgeable" because we may learn more about the types of other experts. The *soft max* heuristic accomplishes this by selecting an expert j according to a Boltzmann distribution $\frac{e^{P(\theta^j = t_1)/\tau}}{\sum_r e^{P(\theta^r = t_1)/\tau}}$ with "temperature" τ, so that experts that are more likely to be of type t_1 are queried more often.

Combined Selection. There can be value in jointly choosing the feature and expert to ask as a pair. We consider strategies inspired by work on multi-armed bandit problems [12], which strives to resolve the tradeoff between *exploration* and *exploitation*. In our setting, we use the term exploitation to refer to strategies such as EVPI that try to directly learn more about the concept in question; in exploitation mode, we select experts greedily. We use the term exploration to refer to strategies such as soft max, whose goal is to learn more about expert types; in exploration mode, we select the feature we are most certain about because it will provide the most information about any given expert's type. The *explore-exploit* strategy embodies this tradeoff. We first generate the pair (i, j), where $\mathcal{X}_i$ is the feature that maximizes EVPI and j is the expert chosen greedily as above. We then use our current belief $P(\theta^j)$ about j's type to switch between exploitation and exploration: (a) we sample j's type using our belief $P(\theta^j)$; (b) if the sampled type is t_1 (knowledgeable), we pose query q^j_i (exploitation); (c) otherwise, we generate a new pair (i', j'), where i' is the index of the feature we are most certain about and expert j' is chosen using soft max, and pose query $q^{j'}_{i'}$ (exploration). In practice this method is more effective using a Boltzmann distribution over types; in the experiments below we "exploit" with probability $0.5 + 0.5\frac{e^{P(\theta^j = t_1)/\tau}}{e^{P(\theta^j = t_1)/\tau} + e^{P(\theta^j = t_2)/\tau}}$.

5 Experiments

We experimented with the query strategies described in Section 4 by comparing their effectiveness on randomly generated configuration problems and concepts. Queries posed

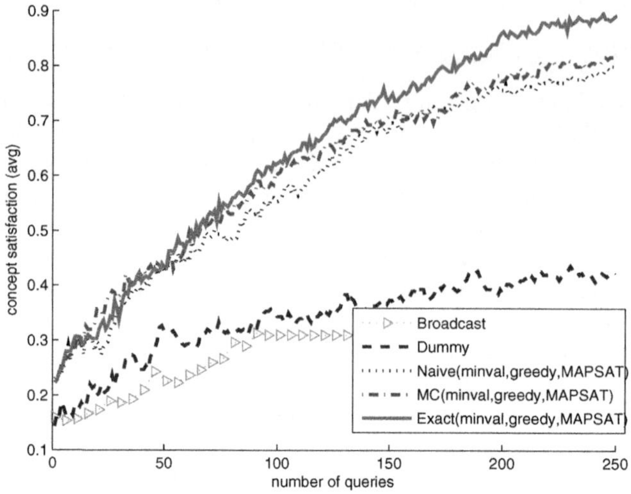

Fig. 3. Simulation with 5 features, 100 experts (20% knowledgeable experts); 300 runs

to simulated experts, each with a type and a subjective concept drawn from a prior distribution.[5] At any stage, each strategy recommends a configuration (decision) based on the current belief and selects the next query to ask; we record whether the current configuration satisfies the true concept.

The concept prior (which is available to the recommender system) is sampled using independent Dirichlet priors for each feature; this represents cases where prior knowledge is available about which features are most likely to be involved (either positively or negatively) in the concept. A *strategy* is a combination of: an inference method; a heuristic for selecting queries (feature and expert); and a method for making recommendations (either MAPSAT or *Most Popular*, the latter a heuristic that recommends each configuration feature based on the most common response from the experts).

Our results below show that good recommendations can be offered with very limited concept information. Furthermore, our decision-theoretic heuristics generate queries that allow a concept-satisfying recommendation to be found quickly (i.e., with relatively few expert queries). In the first experiment (see Figure 3), we consider a setting with 5 features and 100 experts, and compare all methods for Bayesian inference (Exact, Naive and Monte Carlo with 100 particles). All three methods generate queries using

[5] The type is either "knowledgeable" or "ignorant." We define probabilities for subjective concept definitions such that 70% of the time, knowledgeable experts reveal the true value of a particular feature (i.i.d. over different features), and a true T value is reported to be DC with higher probability than is F (0.2 and 0.1, respectively; and the values are symmetric when T and F are interchanged). Ignorant experts are uniformative (in expectation): each feature of the subjective concept is given a value T, F, and DC sampled i.i.d. from a random multinomial, the latter is drawn from a Dirichlet prior *Dir(4,4,4)* once for each run of the simulation. Since an expert's answers are consistent with its subjective concept, repeating a query to some expert has no value.

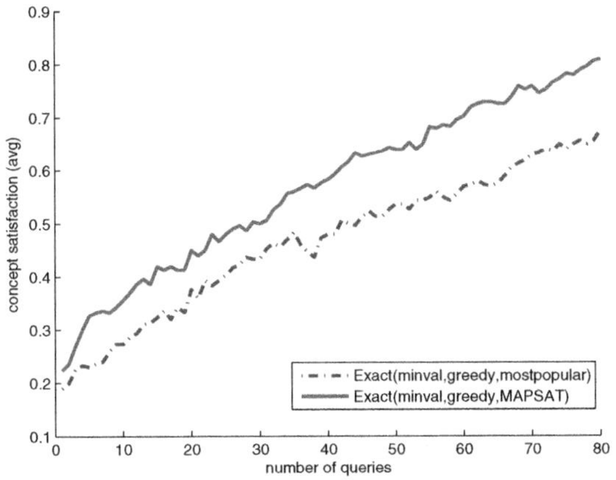

Fig. 4. MAPSAT vs Most Popular (5 features, 100 experts, 30% knowledgeable, 300 runs)

minval (to select features) and greedy (to select experts). We also include broadcast and dummy. Only 20% of the experts are knowledgeable, which makes the setting very challenging, but potentially realistic in certain crowdsourcing domains. Nonetheless our Bayesian methods identify a satisfactory configuration relatively quickly. While the exact method performs best, naive inference is roughly as effective as the more computationally demanding Monte Carlo strategy, and both provide good approximations to Exact in terms of recommendation quality. Dummy and broadcast perform poorly; one cannot expect to make good recommendations by using a simple "majority rule" based on answers to poorly selected queries. In a similar setting with a different proportion of informative experts, we show that MAPSAT outperforms Most Popular for choosing the current recommendation also when used with exact inference (Figure 4).[6]

In the next experiment, we consider a much larger concept space with 30 boolean variables (Figure 5). In this more challenging setting, exact inference is intractable; so we use naive Bayes for inference and compare heuristics for selecting features for queries. *Minval* is most effective, though maxent and random perform reasonably well.

Finally we evaluate heuristics for selecting experts (random, greedy and softmax) and the combined strategy (explore-exploit) in the presence of budget constraints. Each feature is associated with a cost a_i uniformly distributed between 1 and 10; this cost is only incurred when setting a feature as positive (e.g., when buying an ingredient); the available budget b is set to $0.8 \cdot \sum_i a_i$.

Figure 6 shows that the explore-exploit strategy is very effective, outperforming the other strategies. This suggests that our combined method balances exploration (asking queries in order to learn more about the types of experts) and exploitation (asking queries of the expert predicted to be most knowledgeable) in an appropriate fashion. Interestingly, *Naive(EVPI,greedy,MAPSAT)*, while using the same underlying heuristic for selecting features as *Naive(explore-exploit,MAPSAT)*, asks very useful queries initially, but after approximately 50-60 queries begins to underperform the explore-exploit

[6] As our heuristics only ask queries that are relevant, recommendations made by the Most Popular strategy are relatively good in this case.

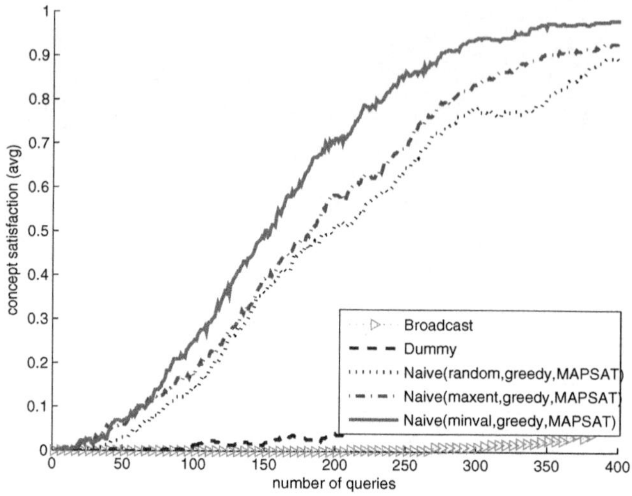

Fig. 5. Evaluation of feature selection methods in a larger concept space (30 features; 50% knowledgeable; 500 runs)

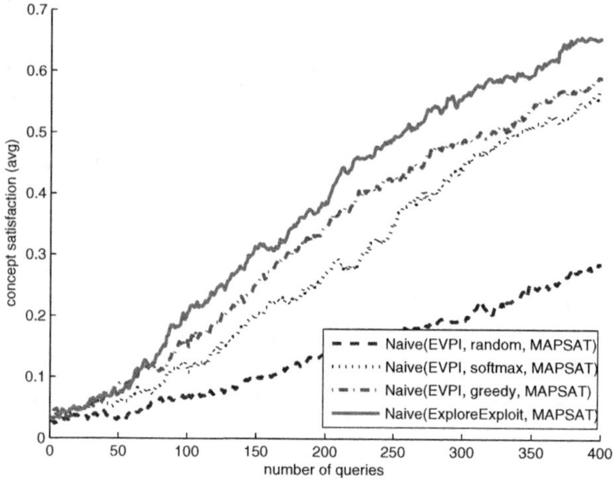

Fig. 6. Evaluation of expert selection methods (20 features; 20% of experts are knowledgeable; 500 runs)

method: it never explicitly asks queries aimed at improving its knowledge about the types of experts. Although the number of queries posed in these results may seem large, it is important to realize that they are posed of different experts: a single expert is asked at most n queries, with most experts asked only 1 or 2 queries. Figure 7 shows a histogram of the number of queries posed to each expert by the explore-exploit method in this last experiment. At the extremes we see that 34 experts are asked just a single query, while only 3 experts are asked 20 queries. Indeed, only 9 experts are asked more than 10 queries.

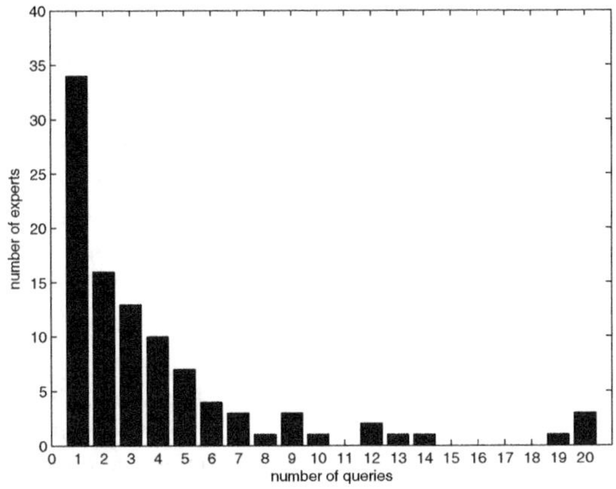

Fig. 7. Distribution of the number of queries posed to experts

6 Discussion and Future Work

We have presented a probabilistic framework for learning concepts from noisy experts in a crowdsourcing setting, with an emphasis on learning just enough about the concept to support the identification of a positive concept instance with high probability. We described methods for making recommendations given uncertain concept information and how to determine the most "relevant" queries. Since experts are noisy, our methods acquire indirect information about their reliability by aggregating their responses to form a distribution over expert types. Our experiments showed the effectiveness of our query strategies and our methods for inference and recommendations, even in large concept spaces, with many uninformative experts, and even when "good" experts are noisy.

The are many interesting future directions. Development of practical applications and validation with user studies is of critical importance. While we have focused on conjunctive concepts in this paper, we believe our model can be extended to more general concept classes. Special care, however, must be taken in developing several key aspects of such an extended model, including: the exact semantics of queries; the representation of the concept distribution; and inference over types and concepts. We are also interested in game-theoretic extensions of the model that allow (some or all) experts to provide responses that reflect their self-interest (e.g., by guiding a recommender system to specific products) and in adopting inference methods that can learn the hyperparameters without relying on the availability of informative priors.

Further investigation of query selection strategies is important. Our strategies incorporate notions from the multi-armed bandit literature, including means of addressing the exploration-exploitation tradeoff, a connection we would like to develop further. We are currently exploring a formulation of the query selection problem as a Markov decision process, which will allow *sequential* optimal query selection. Principled methods for query optimization in preference elicitation [13] could also provide valuable insights in this domain.

Our model values configurations based on their probability of satisfying the concept (i.e., assuming binary utility for concept satisfaction). Several other utility models can be considered. For instance, we might define utility as a sum of some concept-independent reward for a configuration—reflecting user preferences over features that are independent of the latent concept—plus an additional reward for concept satisfaction (as in [2,3]). One could also consider cases in which it is not known with certainty which features are available: the problem of generating recommendations under both concept and availability uncertainty would be of tremendous interest.

References

1. Angluin, D.: Queries and concept learning. Machine Learning 2, 319–342 (1988)
2. Boutilier, C., Regan, K., Viappiani, P.: Online feature elicitation in interactive optimization. In: Proceedings of the Twenty-sixth International Conference on Machine Learning (ICML 2009), Montreal, pp. 73–80 (2009)
3. Boutilier, C., Regan, K., Viappiani, P.: Simultaneous elicitation of preference features and utility. In: Proceedings of the Twenty-fourth AAAI Conference on Artificial Intelligence (AAAI 2010), Atlanta, pp. 1160–1167 (2010)
4. Chen, S., Zhang, J., Chen, G., Zhang, C.: What if the irresponsible teachers are dominating? In: Proceedings of the Twenty-Fourth AAAI Conference on Artificial Intelligence (AAAI 2010), Atlanta, pp. 419–424 (2010)
5. Dai, P., Mausam, Weld, D.S.: Decision-theoretic control of crowd-sourced workflows. In: Proceedings of the Twenty-Fourth AAAI Conference on Artificial Intelligence (AAAI 2010), Atlanta, pp. 1168–1174 (2010)
6. Haussler, D.: Learning conjunctive concepts in structural domains. Machine Learning 4, 7–40 (1989)
7. Heckerman, D., Horvitz, E., Middleton, B.: An approximate nonmyopic computation for value of information. IEEE Transactions on Pattern Analysis and Machine Intelligence 15(3), 292–298 (1993)
8. Howard, R.: Information value theory. IEEE Transactions on Systems Science and Cybernetics 2(1), 22–26 (1966)
9. Kearns, M.J., Li, M.: Learning in the presence of malicious errors. SIAM Journal on Computing 22, 807–837 (1993)
10. Mitchell, T.M.: Version spaces: A candidate elimination approach to rule learning. In: Proceedings of the Fifth International Joint Conference on Artificial Intelligence (IJCAI 1977), Cambridge, pp. 305–310 (1977)
11. Shahaf, D., Horvitz, E.: Generalized task markets for human and machine computation. In: Proceedings of the Twenty-Fourth AAAI Conference on Artificial Intelligence (AAAI 2010), Atlanta, pp. 986–993 (2010)
12. Sutton, R.S., Barto, A.G.: Reinforcement Learning: An Introduction. MIT Press, Cambridge (1998)
13. Viappiani, P., Boutilier, C.: Optimal Bayesian recommendation sets and myopically optimal choice query sets. In: Advances in Neural Information Processing Systems (NIPS), Vancouver, vol. 23, pp. 2352–2360 (2010)

7 Appendix

Proof of Observation 1: Assume we ask the oracle about feature i. Let $p^*_j = \max(p^+_j, p^-_j)$ for any feature j. The optimal configuration x^* *in the updated belief* given the oracle's response is such that $x^* = \arg\max_x \prod_i P(c_i(x)|X_i = v)$, where v (either T,F or DC) is the oracle's response. Since there are no constraints, it can be optimized independently for the different features. Feature i of the optimal configuration x_i^* will necessarily be set to 1 or 0 in a way consistent with v (in case of DC, either is equivalent) and we are sure that x_i^* satisfies feature i; all other features j will be set according to p_j^*. The (approximated) probability of concept satisfaction is:

$$\max_x \prod_j P(c_j(x)|X_i = v) = \prod_{j \neq i} \max(p_j^+, p_j^-) = \prod_{j \neq i} p_j^* = p_{\neq i}^*. \tag{11}$$

Therefore, $EVPI_i = \sum_{v=T,F,DC} P(X_i = v) \cdot p_{\neq i}^* = p_{\neq i}^*$. The argument follows from observing that $i = \arg\max p_{\neq i}^*$ iff $i = \arg\min p_i^*$. ∎

Proof of Observation 2: Note that x^{+i} and x^{-i} are the optimal configurations in the posterior beliefs $P(c|X_i = T)$ and $P(c|X_i = F)$ respectively. In the case that the oracle's answer is DC ("don't care") then the optimal configuration is either x^{+i} or x^{-i} depending on which of the two gives higher probability of satisfying all features beside i. The argument follows from Equation 10. ∎

Online Cake Cutting

Toby Walsh

NICTA and UNSW Sydney, Australia
toby.walsh@nicta.com.au

Abstract. We propose an online form of the cake cutting problem. This models situations where agents arrive and depart during the process of dividing a resource. We show that well known fair division procedures like cut-and-choose and the Dubins-Spanier moving knife procedure can be adapted to apply to such online problems. We propose some fairness properties that online cake cutting procedures can possess like online forms of proportionality and envy-freeness. We also consider the impact of collusion between agents. Finally, we study theoretically and empirically the competitive ratio of these online cake cutting procedures. Based on its resistance to collusion, and its good performance in practice, our results favour the online version of the cut-and-choose procedure over the online version of the moving knife procedure.

1 Introduction

Congratulations. Today is your birthday so you take a cake into the office to share with your colleagues. At tea time, people slowly start to arrive. However, as some people have to leave early, you cannot wait for everyone to arrive before you start sharing the cake. How do you proceed fairly?

This is an example of an *online* cake cutting problem. Most previous studies of fair division assume that all agents are available at the time of the division [Brams and Taylor, 1996]. Here, agents arrive and depart as the cake is being divided. Online cake cutting provides an abstract model for a range of practical problems besides birthday parties. Consider, for instance, allocating time on a large telescope. Astronomers will have different preferences for when to use the telescope depending on what objects are visible, the position of the sun, etc. How do we design a web-based reservation system so that astronomers can asynchronously choose observation times that is fair to all? As a second example, consider allocating space at an exhibition. Exhibitors will have different preferences for space depending on the size, location, cost, etc. How do allocate space when not all exhibitors arrive at the same time but those who have arrived want to start setting up immediately?

Online cake cutting poses some interesting new challenges. On the one hand, the online aspect of such problems makes fair division more difficult than in the offline case. How can we ensure that agents do not envy cake already given to other agents? On the other hand, the online aspect of such problems may make fair division easier than in the offline case. Perhaps agents do not envy cake that has already been eaten before they arrive?

R.I. Brafman, F. Roberts, and A. Tsoukiàs (Eds.): ADT 2011, LNAI 6992, pp. 292–305, 2011.

2 Online Cake Cutting

We assume that agents are risk averse. That is, they declare valuations of the cake that maximizes the minimum value of the cake that they receive, regardless of what the other agents do. This is a common assumption in cake cutting. For instance, Brams, Jones and Klamler (2006) argue:

> "...*As is usual in the cake-cutting literature, we postulate that the goal of each person is to maximize the value of the minimum-size piece (maximin piece) that he or she can guarantee, regardless of what the other person does. Thus, we assume that each person is risk-averse: He or she will never choose a strategy that may yield a more valuable piece of cake if it entails the possibility of getting less than a maximin piece* ..."

We will formulate cake cutting as dividing the unit interval, $[0, 1]$ between n agents.

Definition 1 (Cutting). *A* cutting *of a set of intervals S is a set of intervals recursively defined as follows: S is a cutting, and if S is a cutting and $[a, b] \in S$ then $S \setminus \{[a, b]\} \cup \{[a, c], [c, b]\}$ is a cutting where $a < c < b$.*

A partition of a set S is a set of subsets of S whose union equals the original set which have an empty pairwise intersection. That is, $\{S_i \mid 1 \leq i \leq n\}$ is a partition of S iff $S = \bigcup_{1 \leq i \leq n} S_i$ and $S_i \cap S_j = \{\}$ for $1 \leq i < j \leq n$.

Definition 2 (Division). *A* division *of the cake amongst n agents is a partition of some cutting of $\{[0, 1]\}$ into n subsets.*

A special type of division is where each agent receives a single continuous interval. That is, the cutting contains n intervals, and each agent receives a subset containing just one interval. Note that we suppose there is no waste and that all cake is allocated. We can either relax this assumption, or introduce an additional "dummy" agent who is allocated any remaining cake.

Agents may value parts of the cake differently. For instance, one may prefer the iced part, whilst another prefers the candied part. As a second example, as we argued before, astronomers may prefer different observation times. We capture these differences by means of valuation functions on intervals.

Definition 3 (Valuation). *Each agent i has an additive (but possibly different) valuation function with $v_i([0, 1]) = 1$, $v_i([a, b]) = v_i([a, c]) + v_i([c, b])$ for any $a \leq c \leq b$, and for some set of intervals S, $v_i(S) = \sum_{[a,b] \in S} v_i([a, b])$.*

In an online cake cutting problem, the agents are assumed to arrive in a fixed order. We assume without loss of generality that the arrival order is agent 1 to agent n. Once agents are allocated all their cake, they depart. The order in which agents are allocated cake and depart depends on the cake cutting procedure. For example, the agent present who most values the next slice of cake could be the next to be allocated cake and to depart. We can now formally define the online cake cutting problem.

Definition 4 (Online cake cutting). *An online cake cutting procedure is a procedure that given the total number of agents yet to arrive, a set of agents currently present, and a set of intervals R, either returns "wait" (indicating that we will wait for the next agent to arrive) or returns an agent from amongst those present and two sets of intervals S and T such that $S \cup T$ is a cutting of R. The agent returned by the procedure is allocated S, and T is then left to be divided amongst the agents not yet allocated cake. When no agents are left to arrive and there is only one agent present, the procedure must return $S = R, T = \{\}$. That is, the last agent is allocated whatever is left of the cake. When no agents are left to arrive and there is more than one agent present, the procedure cannot return "wait" but must cut the cake and assign it to one agent.*

Our definition of online cake cutting does not assume that all agents receive cake. Any agent can be allocated an empty set containing no intervals. However, our definition does assume the whole cake is eventually allocated, and that each agent receives all their cake at one time. We assume that at least one agent is allocated some cake before the last arrives otherwise the problem is not online. A special type of online cake cutting procedure is when the departure order is fixed in advance. For instance, if the procedure waits for the first agent to arrive, and whenever a new agent arrives, allocates cake to the longest waiting agent then the departure order is the same as the arrival order. Another special type of cake online cake cutting procedure is one in which the cake is only cut from one or other of the ends of the cake. There are many interesting possible generalisations of this problem. For example, there may only be a bound on the total number of agents to arrive (e.g. you've invited 20 work colleagues to share your birthday cake but not all of them might turn up). Another generalisation is when an agent is not allocated cake all at one time but at several moments during the process of division.

3 Fairness Properties

What properties do we want from an online cake cutting procedure? The literature on cake cutting studies various notions of fairness like envy freeness, as well as various forms of strategy proofness [Brams and Taylor, 1996; Robertson and Web, 1998; Chen *et al.*, 2010]. These are all properties that we might want from an online cake cutting procedure.

Proportionality: A cake cutting procedure is *proportional* iff each of the n agents assigns at least $\frac{1}{n}$ of the total value to their piece(s). We call such an allocation *proportional*.

Envy Freeness: This is a stronger notion of fairness. A cake cutting procedure is *envy free* iff no agent values another agent's pieces more than their own. Note that envy freeness implies proportionality but not vice versa.

Equitability: A cake cutting procedure is *equitable* iff agents assign the same value to the cake which they are allocated (and so no agent envies the valuation that another agent gives to their cake). For 3 or more agent, equitability and envy freeness can be incompatible [Brams and Taylor, 1996].

Efficiency: This is also called Pareto optimality. A cake cutting procedure is *Pareto optimal* iff there is no other allocation to the one returned that is more valuable for one agent and at least as valuable for the others. Note that Pareto optimality does not in itself ensure fairness since allocating all the cake to one agent is Pareto optimal. A cake cutting procedure is *weakly Pareto optimal* iff there is no other allocation to the one returned that is more valuable for all agents. A cake cutting procedure that is Pareto optimal is weakly Pareto optimal but not vice versa.

Truthfulness: Another consideration is whether agents can profit by being untruthful about their valuations. As in [Chen *et al.*, 2010], we say that a cake cutting procedure is *weakly truthful* iff there exists some valuations of the other agents such that an agent will do at least as well by telling the truth. A stronger notion (often called strategy proofness in social choice) is that agents must not be able to profit even when they know how others value the cake. As in [Chen *et al.*, 2010], we say that a cake cutting procedure is *truthful* iff there are no valuations where an agent will do better by lying.

The fact that some agents may depart before others arrive places some fundamental limitations on the fairness of online cake cutting procedures. In particular, unlike the offline case, we can prove a strong impossibility result.

Proposition 1. *No online cake cutting procedure is proportional, envy free or equitable.*

Proof: Consider any cake cutting procedure. As the procedure is online, at least one agent i departs before the final agent n arrives. Since the valuation function of agent n, v_n is not revealed before agent i departs, the set of intervals S_i allocated to agent i cannot depend on v_n. Similarly, v_n cannot change who is first to depart. Suppose agent n has a valuation function with $v_n(S_i) = 1$. As v_n is additive and $v_n([0,1]) = 1$, agent n only assigns value to the intervals assigned to agent i. Hence, any interval outside S_i that is allocated to agent n is of no value to agent n. Hence the procedure is not proportional. Since envy-freeness implies proportionality, by modus tollens, the procedure is also not envy-free.

To demonstrate that no cake cutting procedure is equitable, we restrict ourselves to problems in which all agents assign non-zero value to any non-empty interval. Suppose that the procedure is equitable. As all the cake is allocated, at least one agent must receive cake. Since the procedure is equitable, it follows that all agents must receive some cake. Now, the first agent i to depart and the set of intervals S_i allocated to agent i cannot depend on v_n, the valuation function of the last agent to arrive. Suppose $v_i(S_i) = a$. Now we have argued that S_i is non-empty. Hence, by assumption, $a > 0$. We now modify the valuation function of agent n so that $v_n(S_i) = 1 - \frac{a}{2}$. Then $v_n(S_n) \leq \frac{a}{2} < a = v_i(S_i)$. Hence the procedure is not equitable. □

By comparison, the other properties of Pareto optimality and truthfulness are achievable in the online setting.

Proposition 2. *There exist online cake cutting procedures that are Pareto optimal and truthful.*

Proof: Consider the online cake cutting procedure which allocates all cake to the first agent to arrive. This is Pareto optimal as any other allocation will be less desirable for this agent. It is also truthful as no agent can profit by lying about their valuations. □

Of course, allocating all cake to the first agent to arrive is not a very "fair" procedure. Therefore we need to consider other weaker properties of fairness that online procedures can possess. We introduce such properties in the next section

4 Online Properties

We define some fairness properties that are specific to online procedures.

Proportionality: We weaken the definition of proportionality to test whether agents receive a fair proportion of the cake that *remains* when they arrive. A cake cutting procedure is *weakly proportional* iff each agent assigns at least $\frac{r}{k}$ of the total value of the cake to their pieces where r is the fraction of the total value assigned by the agent to the (remaining) cake when they arrive and k is the number of agents yet to be allocated cake at this point.

Envy Freeness: We can weaken the definition of envy freeness to consider just agents allocated cake after the arrival of a given agent. A cake cutting procedure is *weakly envy free* iff agents do not value cake allocated to agents after their arrival more than their own. Note that weak envy freeness implies weak proportionality but not vice versa. Similarly, envy freeness implies weak envy freeness but not vice versa. An even weaker form of envy freeness is when an agent only envies cake allocated to other agents whilst they are present. A cake cutting procedure is *immediately envy free* iff agents do not value cake allocated to any agent after their arrival and before their departure more than their own. Weak envy freeness implies immediate envy freeness but not vice versa.

Order Monotonicity: An agent's allocation of cake typically depends on when they arrive. We say that a cake cutting procedure is *order monotonic* iff an agent's valuation of their cake does not decrease when they are moved earlier in the arrival ordering and all other agents are left in the same relative positions. Note that as the moved agent can receive cake of greater value, other agents may receive cake of less value. A positive interpretation of order monotonicity is that agents are encouraged to participate as early as possible. On the other hand, order monotonicity also means that agents who have to arrive late due to reasons beyond their control may receive less value.

The online versions of the proportional and envy free properties are weaker than their corresponding offline properties. We consider next two well known offline procedures that naturally adapt to the online setting and demonstrate that they have many of the online properties introduced here.

5 Online Cut-and-Choose

The cut-and-choose procedure for two agents dates back to antiquity. It appears nearly three thousand years ago in Hesiod's poem *Theogeny* where Prometheus divides a cow and Zeus selects the part he prefers. Cut-and-choose is enshrined in the UN's 1982 Convention of the Law of the Sea where it is used to divide the seabed for mining. In cut-and-choose, one agent cuts the cake and the other takes the "half" that they most prefer. We can extend cut-and-choose to more than two agents by having one agent cut a "proportional" slice and giving this slice to the agent who values it most. We then repeat with one fewer agent. The two person cut-and-choose procedure is proportional, envy free, Pareto optimal and weakly truthful. However, it is not equitable nor truthful.

We can use cut-and-choose as the basis of an online cake cutting procedure. The first agent to arrive cuts off a slice of cake and waits for the next agent to arrive. Either the next agent to arrive chooses this slice and departs, or the next agent to arrive declines this slice and the waiting agent takes this slice and departs. If more agents are to arrive, the remaining agent cuts the cake and we repeat the process. Otherwise, the remaining agent is the last agent to be allocated cake and departs with whatever is left. We assume that all agents know how many agents will arrive. A natural extension (which we do not consider further) is when multiple agents arrive and can choose or reject the cut cake. By insisting that an agent cuts the cake before the next agent is allowed to arrive, we will make the procedure more resistant to collusion. We discuss this in more detail later.

Example 1. Suppose there are three agents, the first values only $[\frac{1}{2}, 1]$, the second values only $[\frac{1}{3}, 1]$, and the third values only $[0, \frac{3}{4}]$. We suppose that they uniformly value slices within these intervals. If we operate the online cut-and-choose procedure, the first agent arrives and cuts off the slice $[0, \frac{2}{3}]$ as they assign this slice $\frac{1}{3}$ the total value of the cake. The second agent then arrives. As they assign this slice with $\frac{1}{2}$ the total value of the cake and they are only expecting $\frac{1}{3}$ of the total, the second agent is happy to take this slice and depart. The first agent then cuts off the slice $[\frac{2}{3}, \frac{5}{6}]$ as they assign this $\frac{1}{3}$ of the total value of the cake (and $\frac{1}{2}$ of the value remaining after the second agent departed with their slice). The third agent then arrives. As they assign the slice $[\frac{2}{3}, \frac{5}{6}]$ with all of the total value of the remaining cake and they are only expecting $\frac{1}{2}$ of whatever remains, the third agent is happy to take this slice and depart. The first agent now takes what remains, the slice $[\frac{5}{6}, 1]$. We can argue that everyone is happy as the first agent received a "fair" proportion of the cake, whilst the other two agents received slices that were of even greater proportional value to them.

The online cut-and-choose procedure has almost all of the online fairness properties just introduced.

Proposition 3. *The online cut-and-choose procedure is weakly proportional, immediately envy free, and weakly truthful. However, it is not proportional, (weakly) envy free, equitable, (weakly) Pareto optimal, truthful or order monotonic.*

Proof: Suppose agent i cuts the slice c_i. As agent i is risk averse, and as there is a chance that agent i is allocated c_i, agent i will cut c_i to ensure that $v_i(c_i) \geq \frac{r}{k}$ where k is the number of agents still to be allocated cake and r is the fraction of cake remaining when agent i arrived. Similarly as there is a chance that agent i is not allocated c_i, but will have to take a share of what remains, they will cut c_i so that $v_i(c_i) \leq \frac{r}{k}$. Hence, $v_i(c_i) = \frac{r}{k}$, and the procedure is both weakly proportional and weakly truthful. It is also immediately envy free since each slice that agent i cuts (and sees allocated) has the same value, $\frac{r}{k}$.

To show that this procedure is not proportional, (weakly) envy free, equitable, (weakly) Pareto optimal, truthful or order monotonic consider four agents who value the cake as follows: $v_1([0, \frac{1}{4}]) = \frac{1}{4}$, $v_1([\frac{1}{4}, \frac{3}{4}]) = \frac{1}{12}$, $v_1([\frac{3}{4}, 1]) = \frac{2}{3}$, $v_2([\frac{1}{4}, \frac{1}{2}]) = \frac{1}{3}$, $v_2([\frac{1}{2}, \frac{5}{8}]) = \frac{2}{3}$, $v_3([0, \frac{1}{4}]) = \frac{1}{2}$, $v_3([\frac{1}{2}, \frac{5}{8}]) = \frac{1}{12}$, $v_3([\frac{5}{8}, \frac{3}{4}]) = \frac{1}{6}$, $v_3([\frac{3}{4}, 1]) = \frac{1}{4}$, $v_4([\frac{1}{4}, \frac{1}{2}]) = \frac{3}{4}$, $v_4([\frac{1}{2}, \frac{3}{4}]) = \frac{1}{12}$, and $v_4([\frac{3}{4}, 1]) = \frac{1}{6}$. All other slices have zero value. For instance, $v_2([0, \frac{1}{4}]) = v_3([\frac{1}{4}, \frac{1}{2}]) = 0$.

If we apply the online cut-and-choose procedure, agent 1 cuts off the slice $[0, \frac{1}{4}]$ as $v_1([0, \frac{1}{4}]) = \frac{1}{4}$ and 4 agents are to be allocated cake. Agent 2 places no value on this slice so agent 1 takes it. Agent 2 then cuts off the slice $[\frac{1}{4}, \frac{1}{2}]$ as $v_2([\frac{1}{4}, \frac{1}{2}]) = \frac{1}{3}v_2([\frac{1}{4}, 1])$ and 3 agents remain to be allocated cake. Agent 3 places no value on this slice so agent 2 takes it. Agent 3 then cuts the cake into two pieces of equal value: $[\frac{1}{2}, \frac{3}{4}]$ and $[\frac{3}{4}, 1]$. Agent 4 takes the slice $[\frac{3}{4}, 1]$ as it has greater value, leaving agent 3 with the slice $[\frac{1}{2}, \frac{3}{4}]$

The procedure is not proportional as agent 4 receives the slice $[\frac{3}{4}, 1]$ but $v_4([\frac{3}{4}, 1]) = \frac{1}{6}$. The procedure is not (weakly) envy free as agent 1 receives the slice $[0, \frac{1}{4}]$ and agent 4 receives the slice $[\frac{3}{4}, 1]$, but $v_1([0, \frac{1}{4}]) = \frac{1}{4}$ and $v_1([\frac{3}{4}, 1]) = \frac{2}{3}$. Hence agent 1 envies the slice allocated to agent 4. The procedure is not equitable as agents receive cake of different value. The procedure is not (weakly) Pareto optimal as allocating agent 1 with $[\frac{3}{4}, 1]$, agent 2 with $[\frac{1}{2}, \frac{3}{4}]$, agent 3 with $[0, \frac{1}{4}]$, and agent 4 with $[\frac{1}{4}, \frac{1}{2}]$ gives all agents greater value.

The procedure is not truthful as agent 2 can get a more valuable slice by misrepresenting their preferences and cutting off the larger slice $[\frac{1}{4}, \frac{5}{8}]$. This slice contains all the cake of any value to agent 2. Agent 3 has $v_3([\frac{1}{4}, \frac{5}{8}]) = \frac{1}{12}$ so lets agent 2 take this larger slice. Finally, the procedure is not order monotonic as the value of the cake allocated to agent 4 decreases from $\frac{1}{6}$ to $\frac{1}{8}$ when they arrive before agent 3. □

6 Online Moving Knife

Another class of cake cutting procedures uses one or more moving knives. For example, in the Dubins-Spanier procedure for n agents [Dubins and Spanier, 1961], a knife is moved across the cake from left to right. When an agent shouts "stop", the cake is cut and this agent takes the piece to the left of the knife. The procedure continues with the remaining agents until one agent is left (who takes whatever remains). This procedure is proportional but is not envy-free. However, only the first $n-2$ agents allocated slices of cake can be envious.

We can use the Dubins-Spanier procedure as the basis of an online moving knife procedure. The first k agents ($k \geq 2$) to arrive perform one round of a moving knife procedure to select a slice of the cake. Whoever chooses this slice, departs. At this point, if all agents have arrived, we continue the moving knife procedure with $k-1$ agents. Alternatively the next agent arrives and we start again a moving knife procedure with k agents.

Example 2. Consider again the example in which there are three agents, the first values only $[\frac{1}{2}, 1]$, the second values only $[\frac{1}{3}, 1]$, and the third values only $[0, \frac{3}{4}]$. If we operate the online moving knife procedure, the first two agents arrive and perform one round of the moving knife procedure. The second agent is the first to call "cut" and departs with the slice $[0, \frac{5}{9}]$ (as this has $\frac{1}{3}$ of the total value of the cake for them). The third agent then arrives and performs a round of the moving knife procedure with the first agent using the remaining cake. The third agent is the first to call "cut" and departs with the slice $[\frac{5}{9}, \frac{47}{72}]$ (as this has $\frac{1}{2}$ the total value of the remaining cake for them). The first agent takes what remains, the slice $[\frac{47}{72}, 1]$. We can argue that everyone is happy as the second and third agents received a "fair"' proportion of the cake that was left when they arrived, whilst the first agent received an even greater proportional value.

The online moving knife procedure has similar fairness properties as the online cut-and-choose procedure. However, as we shall show in the following sections, it is neither as resistant to collusion nor as fair in practice.

Proposition 4. *The online moving knife procedure is weakly proportional, immediately envy free and weakly truthful. However, it is not proportional, (weakly) envy free, equitable, (weakly) Pareto optimal, truthful or order monotonic.*

Proof: Suppose j agents ($j > 1$) have still to be allocated cake. Consider any agent who has arrived. They call "cut" as soon as the knife reaches $\frac{1}{j}$ of the value of the cake left for fear that they will receive cake of less value at a later stage. Hence, the procedure is weakly truthful and weakly proportional. The procedure is also immediately envy free as they will assign less value to any slice that is allocated after their arrival and before their departure.

To show that this procedure is not proportional, (weakly) envy free, equitable, (weakly) Pareto optimal, or truthful consider again the example with four agents used in the last proof. Suppose $k = 2$ so that two agents perform each round of the moving knife procedure. Agent 1 and 2 arrive and run a round of the moving knife procedure. Agent 1 calls "cut" and departs with the slice $[0, \frac{1}{4}]$. Agent 3 then arrives and agent 2 and 3 perform a second round of the moving knife procedure. Agent 2 calls "cut" and departs with the slice $[\frac{1}{4}, \frac{1}{2}]$. Agent 4 then arrives and agent 3 and 4 perform the third and final round of the moving knife procedure. Agent 3 calls "cut" and departs with the slice $[\frac{1}{2}, \frac{3}{4}]$, leaving agent 4 with the slice $[\frac{3}{4}, 1]$. This is the same allocation as the online cut-and-choose procedure. Hence, for the same reasons as before, the online moving knife procedure is not proportional, (weakly) envy free, (weakly) Pareto optimal or truthful.

Finally, to show that the online moving knife procedure is not order monotonic consider again $k = 2$, and three agents with valuation functions: $v_1([0, \frac{1}{3}]) = v_1([\frac{1}{3}, \frac{2}{3}]) = v_1([\frac{2}{3}, 1]) = \frac{1}{3}$, $v_2([0, \frac{1}{3}]) = 0$, $v_2([\frac{1}{3}, \frac{2}{3}]) = v_2([\frac{2}{3}, 1]) = \frac{1}{2}$, $v_3([0, \frac{1}{6}]) = \frac{1}{3}$, $v_3[\frac{1}{6}, \frac{1}{3}]) = v_3([\frac{1}{3}, \frac{2}{3}]) = 0$, and $v_3([\frac{2}{3}, 1]) = \frac{2}{3}$. Agent 1 and 2 arrive and run a round of the moving knife procedure. Agent 1 calls "cut" and departs with the slice $[0, \frac{1}{3}]$. Agent 3 then arrives and agent 2 and 3 perform a second and final round of the moving knife procedure. Agent 2 calls "cut" and departs with the slice $[\frac{1}{3}, \frac{2}{3}]$, leaving agent 3 with the slice $[\frac{2}{3}, 1]$. On the other hand, if agent 3 arrives ahead of agent 2 then the value of the interval allocated to agent 3 drops from $\frac{2}{3}$ to $\frac{1}{3}$. Hence the procedure is not order monotonic. □

7 Online Collusion

An important consideration in online cake cutting procedures is whether agents present together in the room can collude together to increase the amount of cake they receive.

We shall show that this is a property that favours the online cut-and-choose procedure over the online moving knife procedure. We say that a cake cutting procedure is vulnerable (resistant) to *online collusion* iff there exists (does not exist) a protocol to which the colluding agents can agree which increases or keeps constant the value of the cake that each receives. We suppose that agents do not meet in advance so can only agree to a collusion when they meet during cake cutting. We also suppose that other agents can be present when agents are colluding. Note that colluding agents cannot change their arrival order and can only indirectly influence their departure order. The arrival order is fixed in advance, and the departure order is fixed by the online cake cutting procedure.

7.1 Online Cut-and-Choose

The online cut-and-choose procedure is resistant to online collusion. Consider, for instance, the first two agents to participate. The first agent cuts the cake before the second agent is present (and has agreed to any colluding protocol). As the first agent is risk averse, they will cut the cake proportionally for fear that the second agent will decline to collude. Suppose the second agent does not assign a proportional value to this slice. It would be risky for the second agent to agree to any protocol in which they accept this slice as they might assign less value to any cake which the first agent later offers in compensation. Similarly, suppose the second agent assigns a proportional or greater value to this slice. It would be risky for the second agent to agree to any protocol in which they reject this slice as they might assign less total value to the slice that they are later allocated and any cake which the first agent offers them in compensation. Hence, assuming that the second agent is risk averse, the second agent will follow the usual protocol of accepting the slice iff it is at least proportional. A similar argument can be given for the other agents.

7.2 Online Moving Knife

On the other hand, the online moving knife procedure is vulnerable to online collusion. Suppose four or more agents are cutting a cake using the online moving knife procedure, but the first two agents agree to the following protocol:

1. Each agent will (silently) indicate when the knife is over a slice worth $\frac{3}{4}$ of the total.
2. Each will only call "stop" once the knife is over a slice worth $\frac{3}{4}$ of the total and the other colluding agent has given their (silent) indication that the cake is also worth as much to them;
3. Away from the eyes of the other agents, the two colluding agents will share this slice of cake using a moving knife procedure.

Under this protocol, both agents will receive slices that they value more than $\frac{1}{4}$ of the total. This is better than not colluding. Note that it is advantageous for the agents to agree to a protocol in which they call "stop" later than this. For example, they could agree to call stop at $\frac{(p-1)}{p}$ of the total value for some $p > 3$. In this way, they would receive more than $\frac{(p-1)}{2p}$ of the total value of the cake (which tends to half the total value as $p \rightsquigarrow \infty$).

8 Competitive Analysis

An important tool to study online algorithms is competitive analysis. We say that an online algorithm is *competitive* iff the ratio between its performance and the performance of the corresponding offline algorithm is bounded. But how do we measure the performance of a cake cutting algorithm?

8.1 Egalitarian Measure

An egalitarian measure of performance would be the reciprocal of the smallest value assigned by any agent to their slice of cake. We take the reciprocal so that the performance measure increases as agent gets less valuable slices of cake. Using such a measure of performance, neither the online cut-and-choose nor the online moving knife procedures are competitive. There exist examples with just 3 agents where the competitive ratio of either online procedure is unbounded. The problem is that the cake left to share between the late arriving agents may be of very little value to these agents.

8.2 Utilitarian Measure

An utilitarian measure of performance would be the reciprocal of the sum of the values assigned by the agents to their slices of cake (or equivalently the reciprocal of the mean value). With such a measure of performance, the online cut-and-choose and moving knife procedures are competitive provided the total number of agents, n is bounded. By construction, the first agent in the online cut-and-choose or moving knife procedure must receive cake of value at least $\frac{1}{n}$ of the total. Hence, the sum of the valuations is at least $\frac{1}{n}$. On the other hand, the sum of the valuations of the corresponding offline algorithm cannot be more than n. Hence the competitive ratio cannot be more than n^2. In fact, there exist examples where the ratio is $O(n^2)$. Thus the utilitarian competitive ratio is bounded iff n itself is bounded.

9 Experimental Results

To test the performance of these procedures in practice, we ran some experiments in which we computed the competitive ratio of the online moving knife and cut-and-choose procedures compared to their offline counterparts. We generated piecewise linear valuations for each agent by dividing the cake into k random segments, and assigning a random value to each segment, normalizing the total value of the cake. It is an interesting research question whether random valuations are more challenging than valuations which are more correlated. For instance, if all agents have the same valuation function (that is, if we have perfect correlation) then the online moving knife procedure performs identically to the offline. On the other hand, if the valuation functions are not correlated, online cake cutting procedures can struggle to be fair especially when late arriving agents more greatly value the slices of cake allocated to early departing agents. Results obtained uncorrelated instances need to be interpreted with some care as there are

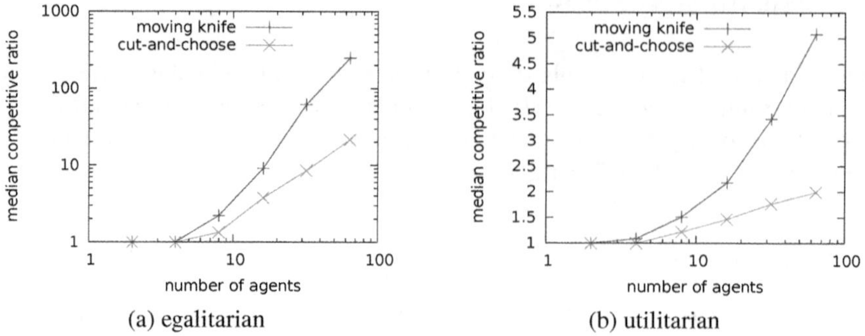

Fig. 1. Competitive ratio between online and offline cake cutting procedures for (a) the egalitarian and (b) utilitarian performance measures. Note different scales to y-axes.

many pitfalls to using instances that are generated entirely at random [Gent *et al.*, 1997; MacIntyre *et al.*, 1998; Gent *et al.*, 2001].

We generated cake cutting problems with between 2 and 64 agents, where each agent's valuation function divides the cake into 8 random segments. At each problem size, we ran the online and offline moving knife and cut-and-choose procedures on the same 10,000 random problems. Overall, the online cut-and-choose procedure performed much better than the online moving knife procedure according to both the egalitarian and utilitarian performance measures. By comparison, the offline moving knife procedure performed slightly better than the offline cut-and-choose procedure according to both measures. See Figure 1 for plots of the competitive ratios between the performance of the online and offline procedures. Perhaps unsurprisingly, the egalitarian performance is rather disappointing when there are many agents since there is a high probability that one of the late arriving agents gets cake of little value. However, the utilitarian performance is reasonable, especially for the online cut-and-choose procedure. With 8 agents, the average value of cake assigned to an agent by the online cut-and-choose procedure is within about 20% of that assigned by the offline procedure. Even with 64 agents, the average value is within a factor of 2 of that assigned by the offline procedure.

10 Online Mark-and-Choose

A possible drawback of both of the online cake cutting procedures proposed so far is that the first agent to arrive can be the last to depart. What if we want a procedure in which agents can depart soon after they arrive? The next procedure has this property. Agents depart as soon as the next agent arrives (except for the last agent to arrive who takes whatever cake remains). However, the new procedure may not allocate cake from one end. In addition, the new procedure does not necessarily allocate continuous slices of cake.

In the online mark-and-choose procedure, the first agent to arrive marks the cake into n pieces. The second agent to arrive selects one piece to give to the first agent who then departs. The second agent then marks the remaining cake into $n-1$ pieces and waits for the third agent to arrive. The procedure repeats in this way until the last agent arrives. The last agent to arrive selects which of the two halves marked by the penultimate agent should be allocated to the penultimate agent, and takes whatever remains.

Example 3. Consider again the example in which there are three agents, the first values only $[\frac{1}{2}, 1]$, the second values only $[\frac{1}{3}, 1]$, and the third values only $[0, \frac{3}{4}]$. If we operate the online mark-and-choose procedure, the first agent arrives and marks the cake into 3 equally valued pieces: $[0, \frac{2}{3}]$, $[\frac{2}{3}, \frac{5}{6}]$, and $[\frac{5}{6}, 1]$. The second agent then arrives and selects the least valuable piece for the first agent to take. In fact, both $[\frac{2}{3}, \frac{5}{6}]$ and $[\frac{5}{6}, 1]$ are each worth $\frac{1}{4}$ of the total value of the cake to the second agent. The second agent therefore chooses between them arbitrarily. Suppose the second agent decides to give the slice $[\frac{2}{3}, \frac{5}{6}]$ to the first agent. Note that the first agent assigns this slice with $\frac{1}{3}$ of the total value of the cake. This leaves behind two sections of cake: $[0, \frac{2}{3}]$ and $[\frac{5}{6}, 1]$. The second agent then marks what remains into two equally valuable pieces: the first is the interval $[0, \frac{7}{12}]$ and the second contains the two intervals $[\frac{7}{12}, \frac{2}{3}]$ and $[\frac{5}{6}, 1]$. The third agent then arrives and selects the least valuable piece for the second agent to take. The first piece is worth $\frac{7}{12}$ of the total value of the cake to the third agent. As this is over half the total value, the other piece must be worth less. In fact, the second piece is worth $\frac{1}{4}$ of the total value. The third agent therefore gives the second piece to the second agent. This leaves the third agent with the remaining slice $[0, \frac{7}{12}]$. It can again be claimed that everyone is happy as the first agents received a "fair"" proportion of the cake that was left when they arrived, whilst both the second and third agent received an even greater proportional value.

This procedure again has the same fairness properties as the online cut-and-choose and moving knife procedures.

Proposition 5. *The online mark-and-choose procedure is weakly proportional, immediately envy free and weakly truthful. However, it is not proportional, (weakly) envy free, equitable, (weakly) Pareto optimal, truthful, or order monotonic.*

Proof: Any agent marking the cake divides it into slices of equal value (for fear that they will be allocated one of the less valuable slices). Similarly, an agent selecting a slice for another agent selects the slice of least value to them (to maximize the value that they receive). Hence, the procedure is weakly truthful and weakly proportional. The procedure is also immediately envy free as they will assign less value to the slice that they select for the departing agent than the value of the slices that they mark.

To show that this procedure is not proportional, (weakly) envy free, equitable, (weakly) Pareto optimal or truthful consider again the example with four agents used in earlier proofs. The first agent marks and is assigned the slice $[0, \frac{1}{4}]$ by the second agent. The second agent then marks and is assigned the slice $[\frac{1}{4}, \frac{1}{2}]$. The third agent then marks and is assigned the slice $[\frac{1}{2}, \frac{3}{4}]$, leaving the fourth agent with the slice $[\frac{3}{4}, 1]$. The procedure is not proportional as the fourth agent only receives $\frac{1}{6}$ of the total value, not (weakly) envy free as the first agent envies the fourth agent, and not equitable as agents receive cake of different value. The procedure is not (weakly) Pareto optimal as allocating the first agent with $[\frac{3}{4}, 1]$, the second with $[\frac{1}{2}, \frac{3}{4}]$, the third with $[0, \frac{1}{4}]$, and the fourth with $[\frac{1}{4}, \frac{1}{2}]$ gives all agents greater value.

The procedure is not truthful as the second agent can get a larger and more valuable slice by misrepresenting their preferences and marking the cake into the slices $[\frac{1}{4}, \frac{5}{8}]$, $[\frac{5}{8}, \frac{3}{4}]$, and $[\frac{3}{4}, 1]$. In this situation, the third agent allocates the second agent with the slice $[\frac{1}{4}, \frac{5}{8}]$ which is of greater value to the second agent.

Finally, to show that the procedure is not order monotonic consider three agents and a cake in which the first agent places equal value on each of $[0, \frac{1}{3}]$, $[\frac{1}{3}, \frac{2}{3}]$ and $[\frac{2}{3}, 1]$, the second places no value on $[0, \frac{1}{3}]$, half the total value on $[\frac{1}{3}, \frac{2}{3}]$, and one quarter on each of $[\frac{2}{3}, \frac{5}{6}]$, and $[\frac{5}{6}, 1]$, and the third places a value of one sixth the total value on $[0, \frac{1}{6}]$, no value on $[\frac{1}{6}, \frac{1}{3}]$ and $[\frac{1}{3}, \frac{2}{3}]$, and half the remaining value on $[\frac{2}{3}, \frac{5}{6}]$ and $[\frac{5}{6}, 1]$. The first agent marks and is allocated the slice $[0, \frac{1}{3}]$. The second agent marks and is allocated the slice $[\frac{1}{3}, \frac{2}{3}]$, leaving the third agent with the slice $[\frac{2}{3}, 1]$. On the other hand, suppose the third agent arrives ahead of the second agent. In this case, the third agent marks the cake into two slice, $[\frac{1}{3}, \frac{5}{6}]$ and $[\frac{5}{6}, 1]$. The second agent allocates the third agent the slice $[\frac{5}{6}, 1]$. Hence, the value of the interval allocated to the third agent halves when they go second in the arrival order. Hence the procedure is not order monotonic. □

11 Related Work

There is an extensive literature on fair division and cake cutting procedures. See, for instance, [Brams and Taylor, 1996]. There has, however, been considerably less work on fair division problems similar to those considered here. Thomson considers a generalization where the number of agents may increase [Thomson, 1983]. He explores whether it is possible to have a procedure in which agents' allocations are monotonic (i.e. their values do not increase as the number of agents increase) combined with other common properties like weak Pareto optimality. Cloutier *et al.* consider a different generalization of the cake cutting problem in which the number of agents is fixed but there are multiple cakes [Cloutier *et al.*, 2010]. This models situations where, for example, agents wish to choose shifts across multiple days. This problem cannot be reduced to multiple single cake cutting problems if the agents' valuations across cakes are linked (e.g. you prefer the same shift each day). A number of authors have studied distributed mechanisms for fair division (see, for example, [Chevaleyre *et al.*, 2009]). In such mechanisms, agents typically agree locally on deals to exchange goods. The usual goal is to identify conditions under which the system converges to a fair or envy free allocation.

12 Conclusions

We have proposed an online form of the cake cutting problem. This permits us to explore the concept of fair division when agents arrive and depart during the process of dividing a resource. It can be used to model situations, such as on the internet, when we need to divide resources asynchronously. There are many possible future directions for this work. One extension would be to undesirable goods (like chores) where we want as little of them as possible. It would also be interesting to consider the variation of the problem where agents have partial information about the valuation functions of the other agents. For voting and other forms of preference aggregation, there has been considerable interest of late in reasoning about preferences that are incomplete or partially known [Pini *et al.*, 2007; Walsh, 2007; Pini *et al.*, 2008]. With cake cutting, agents can act more strategically when they have such partial knowledge.

Acknowledgments. Toby Walsh is supported by the Australian Department of Broadband, Communications and the Digital Economy, the ARC, and the Asian Office of Aerospace Research and Development (AOARD-104123).

References

[Brams and Taylor, 1996] Brams, S.J., Taylor, A.D.: Fair Division: From cake-cutting to dispute resolution. Cambridge University Press, Cambridge (1996)

[Brams *et al.*, 2006] Brams, S.J., Jones, M.A., Klamler, C.: Better ways to cut a cake. Notices of the AMS 53(11), 1314–1321 (2006)

[Chen *et al.*, 2010] Chen, Y., Lai, J.K., Parkes, D.C., Procaccia, A.D.: Truth, justice, and cake cutting. In: Proceedings of the 24th National Conference on AI. Association for Advancement of Artificial Intelligence (2010)

[Chevaleyre *et al.*, 2009] Chevaleyre, Y., Endriss, U., Maudet, N.: Distributed fair allocation of indivisible goods. Working paper, ILLC, University of Amsterdam (2009)

[Cloutier *et al.*, 2010] Cloutier, J., Nyman, K.L., Su, F.E.: Two-player envy-free multi-cake division. Mathematical Social Sciences 59(1), 26–37 (2010)

[Dubins and Spanier, 1961] Dubins, L.E., Spanier, E.H.: How to cut a cake fairly. The American Mathematical Monthly 68(5), 1–17 (1961)

[Gent *et al.*, 1997] Gent, I.P., Grant, S.A., MacIntyre, E., Prosser, P., Shaw, P., Smith, B.M., Walsh, T.: How Not to Do it. Research Report 97.27, School of Computer Studies, University of Leeds, 1997. An earlier and shorter version of this report by the first and last authors appears In: Proceedings of the AAAI 1994 Workshop on Experimental Evaluation of Reasoning and Search Methods and as Research Paper No 714, Dept. of Artificial Intelligence, Edinburgh (1994)

[Gent *et al.*, 2001] Gent, I.P., MacIntyre, E., Prosser, P., Smith, B.M., Walsh, T.: Random constraint satisfaction: Flaws and structure. Constraints 6(4), 345–372 (2001)

[MacIntyre *et al.*, 1998] MacIntyre, E., Prosser, P., Smith, B.M., Walsh, T.: Random constraint satisfaction: Theory meets practice. In: Maher, M.J., Puget, J.-F. (eds.) CP 1998. LNCS, vol. 1520, pp. 325–339. Springer, Heidelberg (1998)

[Pini *et al.*, 2007] Pini, M., Rossi, F., Venable, B., Walsh, T.: Incompleteness and incomparability in preference aggregation. In: Proceedings of 20th IJCAI. International Joint Conference on Artificial Intelligence (2007)

[Pini *et al.*, 2008] Pini, M.S., Rossi, F., Venable, K.B., Walsh, T.: Dealing with incomplete agents' preferences and an uncertain agenda in group decision making via sequential majority voting. In: Brewka, G., Lang, J. (eds.) Principles of Knowledge Representation and Reasoning: Proceedings of the Eleventh International Conference (KR 2008), pp. 571–578. AAAI Press, Menlo Park (2008)

[Robertson and Web, 1998] Robertson, J., Web, W.: Cake-Cutting Algorithms: Be Fair If You Can. A K Peters/CRC Press (1998)

[Thomson, 1983] Thomson, W.: The fair division of a fixed supply among a growing population. Mathematics of Operations Research 8(3), 319–326 (1983)

[Walsh, 2007] Walsh, T.: Uncertainty in preference elicitation and aggregation. In: Proceedings of the 22nd National Conference on AI. Association for Advancement of Artificial Intelligence (2007)

Influence Diagrams with Memory States: Representation and Algorithms

Xiaojian Wu, Akshat Kumar, and Shlomo Zilberstein

Computer Science Department
University of Massachusetts
Amherst, MA 01003
{xiaojian,akshat,shlomo}@cs.umass.edu

Abstract. Influence diagrams (IDs) offer a powerful framework for decision making under uncertainty, but their applicability has been hindered by the exponential growth of runtime and memory usage—largely due to the *no-forgetting* assumption. We present a novel way to maintain a limited amount of memory to inform each decision and still obtain near-optimal policies. The approach is based on augmenting the graphical model with *memory states* that represent key aspects of previous observations—a method that has proved useful in POMDP solvers. We also derive an efficient EM-based message-passing algorithm to compute the policy. Experimental results show that this approach produces high-quality approximate polices and offers better scalability than existing methods.

1 Introduction

Influence diagrams (IDs) present a compact graphical representation of decision problems under uncertainty [8]. Since the mid 1980's, numerous algorithms have been proposed to find optimal decision policies for IDs [4,15,9,14,5,11,12]. However, most of these algorithms suffer from limited scalability due to the exponential growth in computation time and memory usage with the input size. The main reason for algorithm intractability is the no-forgetting assumption [15], which states that each decision is conditionally dependent on all previous observations and decisions. This assumption is widely used because it is necessary to guarantee a policy that achieves the highest expected utility. Intuitively, the more information is used for the policy, the better it will be. However, as the number of decision variables increases, the number of possible observations grows exponentially, requiring a prohibitive amount of memory and a large amount of time to compute policies for the final decision variable, which depends on all the previous observations.

This drawback can be overcome by pruning irrelevant and non-informative variables without sacrificing the expected utility [16,17]. However, the analysis necessary to establish irrelevant variables is usually nontrivial. More importantly, this irrelevance or independence analysis is based on the graphical representation of the influence diagram. In some cases the actual probability distribution implies

R.I. Brafman, F. Roberts, and A. Tsoukiàs (Eds.): ADT 2011, LNAI 6992, pp. 306–319, 2011.

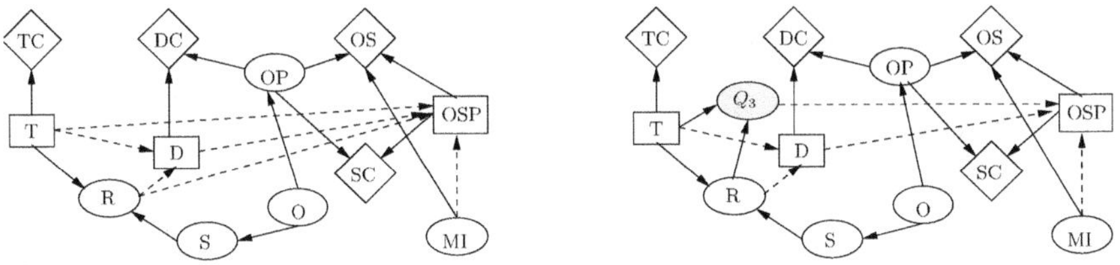

Fig. 1. a) Influence diagram of the oil wildcatter problem (left); b) with a shaded memory node (right). Dotted arrows denote informational arcs.

additional independence relationships among variables that cannot be inferred from the graphical structure. This is usually the case when variables have a large number of successors. Therefore it is beneficial to extract additional (exact or approximate) independence relations in a principled way, thereby decreasing the number of variables that each decision must memorize. In this work, we address this issue by introducing the notion of memory nodes.

Finite-state controllers have been proved very effective in solving infinite-horizon POMDPs [7]. Instead of memorizing long sequences of observations, the idea is to maintain a relatively small number of internal memory states and to choose actions based on this bounded memory. Computing a policy in that case involves determining the action selection function as well as the controller transition function, both of which could be either deterministic or stochastic. With bounded memory, the resulting policy may not be optimal, but with an increasing controller size ϵ-optimality can be guaranteed [2]. A number of search and optimization methods have been used to derive good POMDP policies represented as controllers [1]. More recently, efficient probabilistic inference methods have been proposed as well [19].

Our goal in this paper is to leverage these methods in order to develop more scalable algorithms for the evaluation of IDs. To achieve that, first we introduce a technique to augment IDs with memory nodes. Then, we derive an expectation-maximization (EM) based algorithm for approximate policy iteration for the augmented ID. In the evaluation section, we examine the performance of our algorithm against standard existing techniques.

2 Influence Diagram

An influence diagram (ID) is defined by a directed acyclic graph $G = \{N, A\}$, where N is a set of nodes and A is a set of arcs. The set of nodes, N, is divided into three disjoint groups $\langle X, D, R\rangle$. The set $X = \{X_1, X_2, ..., X_n\}$ is a set of n *chance* nodes, the set $D = \{D_1, D_2, ..., D_m\}$ is a set of m *decision* nodes and $R = \{R_1, R_2, ..., R_T\}$ is a set of T *reward* nodes. Fig. 1(a) shows the influence diagram of the oil wildcatter problem [21], in which decision nodes are illustrated by squares, chance nodes by ellipses and reward nodes by diamonds.

Let $\pi(\cdot)$ and $\Omega(\cdot)$ denote the parents and domain of a node respectively. The domain of a set $Z = \{Z_1, Z_2, ...Z_k\} : Z \subseteq N$, is defined to be the Cartesian product $\times_{Z_i \in Z} \Omega(Z_i)$ of its individual members' domains. Associated with each chance node is a conditional probability table $P(X_i|\pi(X_i))$. The domain of each decision node is a discrete set of actions. The parents $\pi(D_i)$ of a decision node D_i are called observations, denoted by $O(D_i)$. In other words, decisions are conditioned on the value of their parents [15]. Each reward node R_i defines a *utility function* $g_i(\pi(R_i))$ which maps every joint setting of its parents to a real valued utility.

A stochastic decision rule for a decision node D_i is denoted by δ_i and models the CPT $P(D_i|\pi(D_i); \delta_i) = \delta_i(D_i, \pi(D_i))$. A policy Δ for the ID is a set of decision rules $\{\delta_1, \delta_2, ..., \delta_m\}$, containing one rule for each decision node. Given a complete assignment $\{\boldsymbol{x}, \boldsymbol{d}\}$ of chance nodes X and decision nodes D, the total utility is:

$$U(\boldsymbol{x}, \boldsymbol{d}) = \sum_{i=1}^{T} g_i(\{\boldsymbol{x}, \boldsymbol{d}\}_{\pi(R_i)}) \tag{1}$$

where $\{\boldsymbol{x}, \boldsymbol{d}\}_{\pi(R_i)}$ is the value of $\pi(R_i)$ assigned according to $\{\boldsymbol{x}, \boldsymbol{d}\}$. The expected utility (EU) of a given policy Δ is equal to

$$\sum_{\boldsymbol{x} \in \Omega(X), \boldsymbol{d} \in \Omega(D)} P(\boldsymbol{x}, \boldsymbol{d}) U(\boldsymbol{x}, \boldsymbol{d})$$

The probability of a complete assignment $\{\boldsymbol{x}, \boldsymbol{d}\}$ is calculated using the chain rule as follows: $P(\boldsymbol{x}, \boldsymbol{d}) = \prod_{i=1}^{n} P(x_i|\pi(X_i)) \prod_{j=1}^{m} \delta_j(d_j, \pi(D_j); \Delta)$. Therefore, the expected utility is:

$$EU(\Delta; G) = \sum_{\boldsymbol{x} \in \Omega(X), \boldsymbol{d} \in \Omega(D)} \prod_{i=1}^{n} P(x_i|\pi(X_i)) \prod_{j=1}^{m} \delta_j(d_j, \pi(D_j); \Delta) U(\boldsymbol{x}, \boldsymbol{d}) \tag{2}$$

The goal is to find the optimal policy $\Delta^\star$ for a given ID that maximizes the expected utility.

A standard ID is typically required to satisfy two constraints [8,15]:

- **Regularity:** The decision nodes are executed sequentially according to some specified total order. In the oil wildcatter problem of Fig. 1(a), the order is $T \prec D \prec OSP$. With this constraint, the ID models the decision making process of a single agent as no decisions can be made concurrently.
- **No-forgetting:** This assumption requires an agent to remember the entire observation and decision history. This implies $\pi(D_i) \subseteq \pi(D_{i+1})$ where $D_i \prec D_{i+1}$. With the no-forgetting assumption, each decision is made based on all the previous information.

3 Influence Diagram with Memory States

The *no-forgetting* assumption makes the policy optimization computationally challenging. In this work, we introduce the notion of influence diagrams with

Algorithm 1. IDMS representation of an influence diagram

input : An ID $G = (N, A)$, k as the number of memory states

1 Create a copy $G_{ms} \leftarrow G$

2 **foreach** *decision node* $i \geq 2$ **do**

3   Add a memory node Q_i to G_{ms} with $|Q_i| = k$

4   Add incoming arcs into Q_i s.t.

5     $\pi(Q_i; G_{ms}) \leftarrow \begin{cases} \pi(D_1; G) & (i = 2) \\ \big(\pi(D_{i-1}; G) \cup Q_{i-1}\big) \backslash \pi(D_{i-2}; G) & (i > 2) \end{cases}$

6   If $\pi(Q_i; G_{ms}) \equiv \phi$, then delete Q_i

7 **foreach** *decision node* $i \geq 2$ **do**

8   **if** $\exists Q_i$ **then**

9   Delete all incoming arcs to D_i in G_{ms}

10   Set the parent of D_i s.t.

11         $\pi(D_i; G_{ms}) \leftarrow \big(Q_i \cup \pi(D_i; G)\big) \backslash \pi(D_{i-1}; G)$

return: the memory bounded ID G_{ms}

memory states (IDMS). The key idea is to approximate the no-forgetting assumption by using limited memory in the form of memory nodes. We start with an intuitive definition and then describe the exact steps to convert an ID into its memory bounded IDMS counterpart.

Definition 1. *Given an influence diagram (ID), the corresponding influence diagram with memory states (IDMS) generated by Alg. 1 approximates the no-forgetting assumption by using new memory states for each decision node, which summarize the past information and provide the basis for current and future decisions.*

The set of memory states for a decision node is represented by a *memory node.* Memory nodes fall into the category of chance nodes in the augmented ID. Such memory nodes have been quite popular in the context of sequential decision making problems, particularly for solving single and multiagent partially observable MDPs [7,13,2]. In these contexts, they are also known as finite-state controllers and are often used to represent policies compactly. Such bounded memory representation provides a flexible framework to easily tradeoff accuracy with the computational complexity of optimizing the policy. In fact, we will show that given sufficient memory states, the optimal policy of an IDMS is equivalent to the optimal policy of the corresponding original ID.

Alg. 1 shows the procedure for converting a given ID, G, into the corresponding memory states based representation G_{ms} using k memory states per memory node. We add one memory node Q_i for each decision node D_i, except for the first decision. The memory nodes are added according to the decision node ordering dictated by the regularity constraint (see line 1). Intuitively, the memory node Q_i summarizes all the information observed up to (not including) the decision node D_{i-1}. Therefore the parents of Q_i include the information summary until the decision D_{i-2} represented by the node Q_{i-1} and the *new* information obtained after (and including) the decision D_{i-2} and before the decision D_{i-1} (see line 1). Once all such memory nodes are added, we base each decision D_i upon the memory node Q_i and the new information obtained after (and including) the decision D_{i-1} (see line 1). The rest of the incoming arcs to the decision nodes are deleted.

The IDMS approach is quite different from another bounded-memory representation called limited memory influence diagrams (LIMIDs) [11]. A LIMID also approximates the no-forgetting assumption by assuming that each decision depends only upon the variables that can be directly observed while taking the decision. In general, it is quite non-trivial to convert a given ID into LIMID as domain knowledge may be required to decide which information arcs must be deleted and the resulting LIMID representation is not unique. In contrast, our approach requires no domain knowledge and it augments the graph with new nodes. The automatic conversion produces a *unique* IDMS for a given ID using the Alg. 1, parameterized by the number of memory states.

Fig. 1(b) shows an IDMS created by applying Alg. 1 to the ID of the oil wildcatter problem. In the original ID, the order of the decisions is $T \prec D \prec OSP$, namely $D_1 = T$, $D_2 = D$ and $D_3 = OSP$. In the first iteration (see lines 2-6), Q_2 is created as a parent of the node D. However, since T has no parents in the original ID, no parents are added for Q_2 and Q_2 is deleted (see line 6). In the second iteration, Q_3 is created as a parent of OSP, and T, R are linked to Q_3 as its parents because both T and R are parents of D (see line 4 with condition "$i > 2$"). Then, the parents of OSP are reset to be $Q3, D$ and MI (see line 11 with "$i = 3$") because the additional parent of OSP other than D in the original ID is MI.

The CPT of memory nodes, which represents stochastic transitions between memory states, is parameterized by λ: $P(Q_i|\pi(Q_i);\lambda_i) = \lambda_i(Q_i,\pi(Q_i))$. The decision rules δ for an IDMS are modified according to the new parents. The policy for the IDMS is defined as $\Delta_{ms} = \{\lambda_2,\ldots,\lambda_m,\delta_1,\ldots,\delta_m\}$. The expected utility for an IDMS with policy Δ_{ms}, denoted $EU(\Delta_{ms};G_{ms})$, is:

$$\sum_{\boldsymbol{x},\boldsymbol{q},\boldsymbol{d}}\prod_{i=1}^{n}P(x_i|\pi(X_i))\prod_{j=2}^{m}\lambda_j(q_j,\pi(Q_j);\Delta_{ms})\prod_{l=1}^{m}\delta_l(d_l,\pi(D_l);\Delta_{ms})U(\boldsymbol{x},\boldsymbol{d}) \quad (3)$$

The goal is to find an optimal policy $\Delta^{\star}_{ms}$ for the IDMS G_{ms}. As the IDMS approximates the no-forgetting assumption and the value of information is non-negative, it follows that $EU(\Delta^{\star}_{ms};G_{ms}) \leq EU(\Delta^{\star};G)$. As stated by the following proposition, an IDMS has far fewer parameters than the corresponding ID. Therefore optimizing the policy for the IDMS will be computationally simpler than for the ID.

Proposition 1. *The number of policy parameters in the IDMS increases quadratically with the number of memory states and remains asymptotically fixed w.r.t. the number of decisions. In contrast, the number of parameters in an ID increases exponentially w.r.t. the number of decisions.*

Proof. The no-forgetting assumption implies that $\pi(D_{i-1}; G) \subseteq \pi(D_i; G)$ in the ID G. Therefore the number of parameters $P(D_i|\pi(D_i); G)$ increases exponentially with the number of decisions. In the IDMS G_{ms}, the size of the parent set of a decision node D_i is $|\pi(D_i; G_{ms})| = |\pi(D_i; G) \backslash \pi(D_{i-1}; G)| + 1$. In many IDs, one can often bound the amount of new information available after each decision by some constant $\mathcal{I} \geq |\pi(D_i; G) \backslash \pi(D_{i-1}; G)|$ for every i. If there are k memory states and the maximum domain size of any node is d, then the number of parameters is $O(d^{\mathcal{I}+1} \cdot k)$ for each decision rule. We can use the same reasoning to show that there are at most $\mathcal{I} + 1$ parents for a controller node Q_i. Therefore the total number of parameters for a controller node is $O(d^{\mathcal{I}} \cdot k^2)$. This shows that overall, parameters increase quadratically w.r.t. the memory states.

Proposition 2. *With a sufficiently large number of memory states, the best policy of an IDMS has the same utility as the best policy of the corresponding ID. Specifically, when* $|\Omega(Q_i; G_{ms})| = |\Omega(\pi(Q_i; G_{ms}))|$ *for all* i, $EU(\Delta^\star_{ms}; G_{ms}) = EU(\Delta^\star; G)$.

Proof. Let O_i be the set of nodes observed up to (not including) D_i in an IDMS. First, we prove the statement that if in the IDMS, $|\Omega(Q_i)| = |\Omega(\pi(Q_i))|$, then a one-to-one mapping can be built from $\Omega(O_{i-1})$ to $\Omega(Q_i)$. For Q_2, the first memory node, $\pi(Q_2) = O_1$ and the size of Q_2 is equal to $|\Omega(O_1)|$. Thus the mapping can be easily built. Now suppose that the statement is correct for Q_{i-1}. Then for Q_i, since $\pi(Q_i) = \{Q_{i-1}\} \bigcup (O_{i-1}/O_{i-2})$ and a one-to-one mapping from $\Omega(O_{i-2})$ to $\Omega(Q_{i-1})$ already exists, then a one-to-one mapping from O_{i-1} to Q_i can be built similarly in which Q_{i-1} provides all the information of O_{i-2}. Thus, the statement is true for all i. As a result, for each D_i, a one-to-one mapping from O_i to $\pi(D_i; G_{ms})$ can be created such that the no-forgetting condition is satisfied. Therefore, we have $EU(\Delta^\star_{ms}; G_{ms}) = EU(\Delta^\star; G)$.

4 Approximate Policy Iteration for IDMS

In this section, we present an approximate policy iteration algorithm based on the well known expectation-maximization (EM) framework [6]. The key idea is to transform the policy optimization problem in the IDMS to that of probabilistic inference in an appropriately constructed Bayes net. Such planning-by-inference approach has been shown to be quite successful in Markovian planning problems [20,18,10]; we extend it to influence diagrams. To construct the Bayes net BN_{ms} for a given IDMS, we transform all the reward nodes R_t in the IDMS into binary chance nodes $\hat{R}_t$ with the domain $\Omega(\hat{R}_t) = \{0, 1\}$. The rest of the model is the same as the given IDMS. The CPT of $\hat{R}_t$ is set as follows:

$$P(\hat{R}_t = 1|\pi(R_t)) \propto g_t(\pi(R_t); G_{ms}) \quad (4)$$

This can be easily done in several ways such as setting $P(\hat{R}_t = 1|\pi(R_t)) = (g_t(\pi(R_t); G_{ms}) - g_{min})/(g_{max} - g_{min})$, where g_{max}, g_{min} denote the maximum and minimum values of the reward.

Proposition 3. *The expected utility of an IDMS is directly proportional to the sum of expectation of binary reward nodes in the corresponding Bayes net:* $EU(\Delta_{ms}; G_{ms}) \propto E(\sum_{t=1}^{T} \hat{R}_t) + \langle \text{Ind. terms} \rangle$.

Proof. By the linearity of the expectation, we have:

$$
\begin{aligned}
E(\sum_{t=1}^{T} \hat{R}_t; \Delta_{ms}) &= \sum_{t=1}^{T} E(\hat{R}_t; \Delta_{ms}) && (5)\\
&= \sum_{t=1}^{T} P(\hat{R}_t = 1; \Delta_{ms}) \cdot 1 + P(\hat{R}_t = 0; \Delta_{ms}) \cdot 0\\
&= \sum_{t=1}^{T} \sum_{\pi(R_t)} P(\pi(R_t); \Delta_{ms}) P(\hat{R}_t = 1 | \pi(R_t))\\
&= \sum_{t=1}^{T} \sum_{\pi(R_t)} \frac{1}{g_{max} - g_{min}} P(\pi(R_t); \Delta_{ms}) g_t(\pi(R_t); G_{ms}) - \frac{T \cdot g_{min}}{g_{max} - g_{min}}\\
&\propto \sum_{t=1}^{T} \sum_{\pi(R_t)} P(\pi(R_t); \Delta_{ms}) g_t(\pi(R_t)) + \langle \text{Ind. terms} \rangle\\
&= EU(\Delta_{ms}; G_{ms}) + \langle \text{Ind. terms} \rangle && (6)
\end{aligned}
$$

where $\langle$Ind. terms$\rangle$ is a constant with respect to different policies.

4.1 Bayes Net Mixture for IDMS

Intuitively, Proposition 3 and Eq. (5) suggest an obvious method for IDMS policy optimization: if we maximize the likelihood of observing each reward node $\hat{R}_t = 1$, then the IDMS policy will also be optimized. We now formalize this concept using a Bayes net mixture. In this mixture, there is one Bayes net for *each reward node* R_t. This Bayes net is similar to the Bayes net BN_{ms} of the given IDMS, except that it includes only one reward node $\hat{R}$ corresponding to a reward node $\hat{R}_t$ of BN_{ms}; all other binary reward nodes and their incoming arcs are deleted. The parents and the CPT of $\hat{R}$ are the same as that of $\hat{R}_t$. Fig. 2(a) shows this mixture for the oil wildcatter IDMS of Fig. 1(b). The first BN corresponds to the reward node TC, all other reward nodes (DC, OS, SC) are deleted; the second BN is for the node DC. The variable $\mathcal{T}$ is the mixture variable, which can take values from 1 to T, the total number of reward nodes. It has a fixed uniform distribution: $P(\mathcal{T} = i) = 1/T$. The overall approach is based on the following theorem.

Theorem 1. *Maximizing the likelihood* $L(\hat{R}; \Delta_{ms})$ *of observing the variable* $\hat{R} = 1$ *in the Bayes net mixture (Fig. 2(a)) is equivalent to optimizing the IDMS policy.*

Proof. The likelihood for each individual BN in the BN mixture is $L_t^{\Delta_{ms}} = P(\hat{R} = 1 | \mathcal{T}; \Delta_{ms})$, which is equivalent to $P(\hat{R}_t = 1; \Delta_{ms})$ in the Bayes net

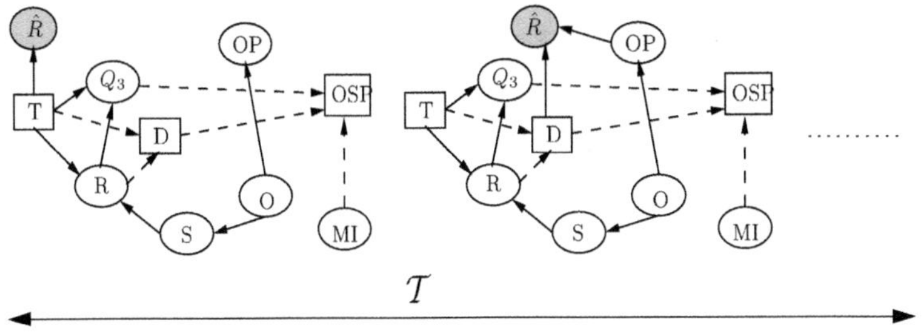

Fig. 2. Bayes net mixture for the oil wildcatter problem

BN_{ms}. Note that the deleted binary reward nodes in each individual BN of the mixture do not affect this probability. Therefore the likelihood for the complete mixture is:

$$L(\hat{R};\Delta_{ms}) = \sum_{t=1}^{T} P(\mathcal{T}=t)L_t^{\Delta_{ms}} = \frac{1}{T}\sum_{t=1}^{T} P(\hat{R}_t=1;\Delta_{ms}) \tag{7}$$

From Proposition 3, we now have $L(\hat{R};\Delta_{ms}) \propto EU(\Delta_{ms};G_{ms})$. Therefore maximizing the likelihood for the mixture would optimize the IDMS policy.

Note that for the implementation, we do not explicitly create the mixture; all the computations on this mixture can be directly performed on the single Bayes net BN_{ms}.

4.2 The Expectation Maximization (EM) Algorithm

We now derive the E and M-step of the expectation-maximzation framework that can be used to maximize the above likelihood [6]. In the EM framework, the observed data is $\hat{R}=1$; the rest of the variables are hidden. The parameters to optimize are the policy parameters for the IDMS: the λ's for the memory nodes and δ's for the decision nodes. The full joint $P(\hat{R},X,D,Q,\mathcal{T};\Delta_{ms})$ for the BN mixture is given by:

$$P(\hat{R}|\pi(\hat{R}),\mathcal{T})\prod_{i=1}^{n} P(X_i|\pi(X_i))\prod_{j=1}^{m}\delta_j(D_j,\pi(D_j))\prod_{l=2}^{m}\lambda_l(Q_l,\pi(Q_l)) \tag{8}$$

We will omit specifying Δ_{ms} as long as it is unambiguous. As EM maximizes the log-likelihood, we take the log of the above to get:

$$\log P(\hat{R},X,D,Q,\mathcal{T};\Delta_{ms}) = \sum_{j=1}^{m}\delta_j(D_j,\pi(D_j)) + \sum_{l=2}^{m}\lambda_l(Q_l,\pi(Q_l)) + \langle\text{Ind. terms}\rangle \tag{9}$$

where ⟨Ind. terms⟩ denote terms independent of the parameters λ and δ. EM maximizes the expected log-likelihood $Q(\Delta_{ms}, \Delta^{\star}_{ms})$ to be equal to:

$$\sum_{\mathcal{T}=1}^{T} \sum_{X,D,Q} P(\hat{R}=1, X, D, Q, \mathcal{T}; \Delta_{ms}) \log P(\hat{R}=1, X, D, Q, \mathcal{T}; \Delta^{\star}_{ms}) \tag{10}$$

where Δ_{ms} is the current policy and $\Delta^{\star}_{ms}$ is the policy to be computed for the next iteration. We first show the update rule for decision node parameters δ.

$$Q(\Delta_{ms}, \Delta^{\star}_{ms}) = \sum_{\mathcal{T}=1}^{T} \sum_{X,D,Q} P(\hat{R}=1, X, D, Q, \mathcal{T}; \Delta_{ms}) \sum_{j=1}^{m} \log \delta_j(D_j, \pi(D_j); \Delta^{\star}_{ms})$$

$$= \sum_{j=1}^{m} 1/T \sum_{D_j, \pi(D_j)} \sum_{\mathcal{T}=1}^{T} P(\hat{R}=1, D_j, \pi(D_j) | \mathcal{T}; \Delta_{ms}) \log \delta_j(D_j, \pi(D_j); \Delta^{\star}_{ms})$$

The above expression can be easily maximized for each parameter δ_j using the Lagrange multiplier for the normalization constraint:

$$\forall \pi(D_j) : \sum_{D_j} \delta_j(D_j | \pi(D_j)) = 1.$$

The final updated policy is:

$$\delta_j(D_j, \pi(D_j); \Delta^{\star}_{ms}) = \frac{\sum_{\mathcal{T}=1}^{T} P(\hat{R}=1, D_j, \pi(D_j) | \mathcal{T}; \Delta_{ms})}{C_{\pi(D_j)}} \tag{11}$$

where $C_{\pi(D_j)}$ is the normalization constant. The memory node parameter (λ) update equation is analogous to the above with the node D_i replaced by Q_l. The above equation describes the M-step. We next describe the E-step that involves computing the probabilities $P(\hat{R}=1, (\cdot), \pi(\cdot) | \mathcal{T}; \Delta_{ms})$ where $(\cdot)$ ranges over the decision and memory nodes.

4.3 Probabilities Computation

The join-tree algorithm is an efficient algorithm for computing marginal probabilities [3]. The algorithm performs inference on the Bayesian network by transforming it into a join-tree. The tree satisfies the running intersection property. Each tree node represents a clique containing a set of nodes in the BN_{ms}. An advantage of this algorithm is that any node and its parents are included in at least one clique. Therefore by performing a global message passing, the joint probabilities of nodes and its parents with a given evidence can be obtained from cliques implementing the E-step.

Alg. 2 describes the procedure to update the decision rules $\delta_i(D_i, \pi(D_i))$. In each iteration, one of the variables R_t is set to 1 and the corresponding probabilities are calculated. New parameters are computed using Eq. (11).

Algorithm 2. Procedure for updating $\delta_j(D_j, \pi(D_j))$

input : BN_{ms} – the transformed Bayesian network
Build the join-tree for BN_{ms}
Initialize parameters δ_i randomly $\forall i = 1 : m$
repeat
  Initialize $V(D_i, \pi(D_i)) \leftarrow 0$
  for $t = 1 : T$ **do**
    Set evidence $R_t = 1$ to every clique containing R_t
    Conduct a global message passing on the join-tree
    Compute $P(R_t = 1, D_i, \pi(D_i))$ by marginalization $\forall i = 1 : m$
    $V(D_i, \pi(D_i)) \leftarrow V(D_i, \pi(D_i)) + P(R_t = 1, D_i, \pi(D_i))$
    Recover potentials and clear evidence
  $\delta_i^\star(D_i, \pi(D_i)) = V(D_i, \pi(D_i))/C_{\pi(D_i)}$ (C $\equiv$ normalization constant)
  Set $\delta_i^\star$ into BN_{ms}
until *the convergence criterion is satisfied*
return: the BN_{ms} with updated policy parameters

Fig. 3 shows the join-tree of the oil wildcatter problem. The performance of Alg. 2 is mainly determined by the size of the largest clique or tree-width of the join-tree. The size of the cliques is influenced largely by the number of parents of each node because each node and its parent are contained in at least one clique (family preserving property). Therefore this algorithm will be more efficient for the IDMS as the number of parents of each node is much smaller that in the ID.

5 Experiments

In this section, we compare the EM algorithm against Cooper's algorithm [4], implemented in SMILE, a library created by the Decision Systems Lab at U. Pitt. We test the algorithms on two datasets: randomly generated IDs and Bayesian networks converted into IDs. The Cooper's algorithm provides optimal solutions.

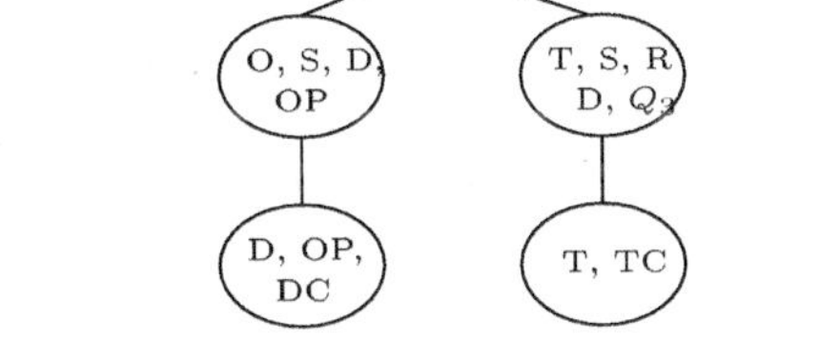

Fig. 3. Join Tree of Oil wildcatter problem

5.1 Randomly Generated IDs

We randomly generated IDs with different settings and fixed the number of parents of chance nodes and reward nodes to be 2. Each decision node has two more parents than the previous decision node (the no-forgetting assumption is forced). With 0.1 probability, a chance node degenerates into a deterministic

Table 1. 'C40' and 'C60' denote the number of chance nodes (40 and 60 respectively). All the networks have 6 reward nodes. 'D' is the number of decision nodes. '-' means that Cooper's algorithm ran out of memory before terminating. T denotes time in seconds. M denotes memory required in MB. Loss is equal to $(EU(Cooper) - EU(EM))/EU(Cooper)$.

C40	Cooper		EM			C60	Coopers		EM		
D	T	M	T	M	Loss	D	T	M	T	M	Loss
4	1.1	5.3	0.2	7.0	<1.0%	4	1.2	5.3	0.8	7.0	<1.0%
5	7.2	8.1	0.2	8.0	1.2%	5	7.1	8.1	1.6	8.0	<1.0%
6	24.2	11.5	0.4	10.7	1.0%	6	25.4	12.0	5.6	10.7	<1.0%
7	106.7	48.6	0.6	16.5	<1.0%	7	112.6	48.2	1.1	16.5	<1.0%
8	264.0	227.0	1.3	31.1	1.6%	8	256.8	227.0	6.8	31.1	<1.0%
9	-	>764	2.4	111.0	-	9	-	>763.8	2.7	111.0	-
10	-	-	3.1	111.0	-	10	-	-	2.0	111.0	-
11	-	-	5.1	150.0	-	11	-	-	16.7	150.0	-
12	-	-	6.7	207.0	-	12	-	-	18.9	207.0	-
13	-	-	5.6	207.0	-	13	-	-	37.2	207.0	-

node. In order to increase bias, for each reward node, the reward value is in range $[0, 20]$ with 40% probability, in [20,70] with 20% probability and in $[70, 100]$ with 40% probability. For each network setting, 10 instances are tested and the average is reported. The results are shown in Table 1. In these experiments, Cooper's algorithm ran on the original ID (no-forgetting) and EM on an IDMS with 2 states per memory node. As the number of decision nodes increases, the running time and memory usage of Cooper 's algorithm grows much faster than EM's. When the ID has 9 decision nodes, Cooper's algorithm fails to terminate, but EM can still solve the problem in less than 3 seconds using only 111 MB of memory. Furthermore, EM provides good solution quality. The value loss against Cooper's algorithm (which is optimal) is about 1%.

5.2 Bayesian Networks Transformed into IDs

Since real world decision problems are likely to have more structure and nodes are usually not randomly connected, we also experimented with the Bayesian network samples available on the GENIE website. We built IDs by transforming a portion of chance nodes into decision nodes and also adding a certain number of reward nodes. Two Bayesian network datasets were used. The average results are reported in Table 2. In both of these benchmarks, EM again performs much better w.r.t. runtime and the solution quality loss remains small, around 1%. On these benchmarks, both EM and Cooper's algorithm are faster than on the random graphs as many of these Bayes nets are tree-structured.

5.3 The Effect of Memory States

In this section, we examine how well memory states approximate the no-forgetting assumption and the effect of the number of memory states on the overall quality

Table 2. Results for the Hepar II and Win95pts Bayesian network datasets. D, C, R represent the number of decision nodes, chance nodes and reward nodes respectively. T is the running time in seconds. M is the amount of memory in MB.

Hepar II			Coopers		EM			Win95pts			Coopers		EM		
D	C	R	T	M	T	M	Loss	D	C	R	T	M	T	M	Loss
14	61	5	47.27	759.8	0.22	3.5	1.5%	13	63	5	45.05	759.8	0.26	3.4	1.5%
15	60	5	15.17	760.1	0.21	3.7	1.1%	14	62	5	14.81	760.7	0.23	3.6	< 1%
16	59	5	10.98	760.3	0.26	3.7	< 1%	15	61	5	10.66	761.1	0.21	3.6	< 1%
17	58	5	22.02	761.7	0.24	4.0	< 1%	16	60	5	21.29	762.6	0.22	3.9	< 1%
18	57	5	14.20	762.3	0.21	4.3	< 1%	17	59	5	13.86	763.2	0.22	4.3	< 1%
18	57	5	15.35	762.6	0.22	4.6	< 1%	18	58	5	14.71	763.4	0.21	4.6	< 1%

achieved by EM. For simplicity, we use a small ID containing only three nodes: a chance node, a decision node, and a reward node as their child. We let both the chance node and the decision node have 50 states and the value of the chance node is distributed uniformly. This simple ID can be easily made to represent more complex situations. For example, we can replace the chance node with a complex Bayes net and similarly replace the reward node by a Bayes net with additional reward nodes.

In this simple ID, we assume that the chance node models some events that occurred much earlier such that the current decision node does not observe them directly. However, the nodes have some effect on the reward obtained, so a memory node is provided that could record the value of the chance node so that the right decision can be made. I would like to change as: In order to test the effect of increasing the size of the memory node on the expected utility, we assign value for the reward node such that for each value of the chance node, only one action (selected randomly) of the decision node produces the reward "1" and all the other actions produce "0". In this way, it is crucial to know the value of the chance node in order to maximize the expected utility.

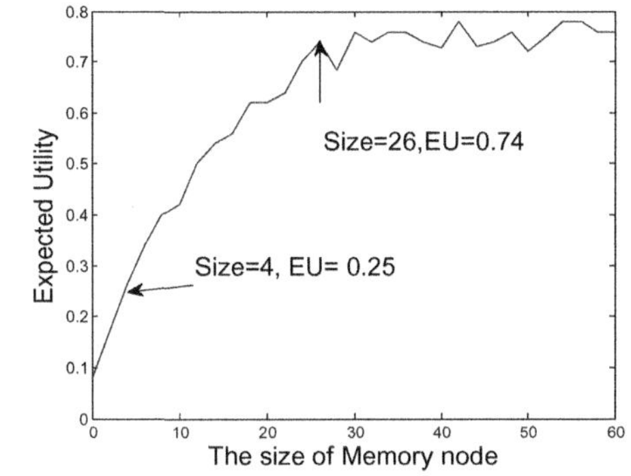

Fig. 4. The effects of memory states w.r.t. expected utility

When the size of the memory node is 50, then according to Proposition 2, the maximum expected utility that can be obtained by an *optimal* policy is 1. In these experiments, we tested the EM algorithm with different sizes of the memory node. The results, shown in Fig. 4, confirm that the EU increases quickly at the beginning and then remains almost constant at about 26 memory states.

Note that the EU does not reach 1 with 50 memory states because the EM algorithm converges to local optima. This example illustrates a case in which a large size of the memory node is needed in order to obtain good solutions. We also note that this experiment is deliberately designed to test the impact of violating the no-forgetting assumption in an extreme situation. In practice, we anticipate that smaller memory nodes will suffice because reward nodes are not as tightly coupled with chance nodes as in these experiments.

6 Conclusion

In this paper, we introduce a technique to transform an influence diagram into an influence diagram with memory states by relaxing the no-forgetting assumption. We also develop the EM algorithm to solve the resulting IDMS efficiently. We show that there exist problems that require large memory states to obtain good quality solutions. However, experiments with both randomly generated and standard benchmark IDs yield near-optimal policies using a small number of memory states. This work unifies techniques for solving (decentralized) POMDPs using finite-state controllers and solving large influence diagrams. The connections we establish in this work between various memory-bounded approximations will facilitate greater sharing of results between researchers working on these problems.

Acknowledgements. This work was funded in part by the National Science Foundation under grant IIS-0812149 and the Air Force Office of Scientific Research under grant FA9550-08-1-0181.

References

1. Amato, C., Bernstein, D.S., Zilberstein, S.: Optimizing fixed-size stochastic controllers for POMDPs and decentralized POMDPs. Autonomous Agents and Multi-Agent Systems 21, 293–320 (2010)
2. Bernstein, D.S., Amato, C., Hansen, E.A., Zilberstein, S.: Policy iteration for decentralized control of Markov decision processes. Journal of Artificial Intelligence Research 34, 89–132 (2009)
3. Cecil Huang, A.D.: Inference in belief networks: A procedural guide. International Journal of Approximate Reasoning 15, 225–263 (1994)
4. Cooper, G.: A method for using belief networks as influence diagrams. In: Proc. of the Conference on Uncertainty in Artificial Intelligence, pp. 55–63 (1988)
5. Dechter, R.: A new perspective on algorithims for optimizing policies under uncertainty. In: Proc. of the International Conference on Artificial Intelligence Planning Systems, pp. 72–81 (2000)
6. Dempster, A.P., Laird, N.M., Rubin, D.B.: Maximum likelihood from incomplete data via the EM algorithm. Journal of the Royal Statistical society, Series B 39(1), 1–38 (1977)
7. Hansen, E.A.: An improved policy iteration algorithm for partially observable MDPs. In: Proc. of Neural Information processing Systems, pp. 1015–1021 (1997)

8. Howard, R.A., Matheson, J.E.: Infuence diagrams. In: Readings on the Principles and Applications of Decision Analysis, vol. II, pp. 719–762. Strategic Decisions Group (1984)
9. Jensen, F., Jensen, F.V., Dittmer, S.L.: From influence diagrams to junction trees. In: Proc. of the Conference on Uncertainty in Artificial Intelligence, pp. 367–373 (1994)
10. Kumar, A., Zilberstein, S.: Anytime planning for decentralized POMDPs using expectation maximization. In: Proc. of the Conference on Uncertainty in Artificial Intelligence, pp. 294–301 (2010)
11. Nilsson, D., Lauritzen, S.: Representing and solving decision problems with limited information. Management Science 47(9), 1235–1251 (2001)
12. Marinescu, R.: A new approach to influence diagram evaluation. In: Proc. of the 29th SGAI International Conference on Innovative Techniques and Applications of Artificial Intelligence (2009)
13. Poupart, P., Boutilier, C.: Bounded finite state controllers. In: Proc. of Neural Information processing Systems, pp. 823–830 (2003)
14. Qi, R., Poole, D.: A new method for influence diagram evaluation. Computational Intelligence 11, 498–528 (1995)
15. Shachter, R.: Evaluating influence diagrams. Operations Research 34, 871–882 (1986)
16. Shachter, R.: Probabilistic inference and influence diagrams. Operations Research 36, 589–605 (1988)
17. Shachter, R.: An ordered examination of influence diagrams. Networks 20, 535–563 (1990)
18. Toussaint, M., Charlin, L., Poupart, P.: Hierarchical POMDP controller optimization by likelihood maximization. In: Proc. of the Conference on Uncertainty in Artificial Intelligence, pp. 562–570 (2008)
19. Toussaint, M., Harmeling, S., Storkey, A.: Probabilistic inference for solving (PO)MDPs. Technical Report EDI-INF-RR-0934, School of Informatics, University of Edinburgh (2006)
20. Toussaint, M., Storkey, A.J.: Probabilistic inference for solving discrete and continuous state Markov decision processes. In: Proc. of International Conference on Machine Learning, pp. 945–952 (2006)
21. Zhang, N.L., Qi, R., Poole, D.: A computational theory of decision networks. International Journal of Approximate Reasoning 11, 83–158 (1994)

Game Theory and Human Behavior: Challenges in Security and Sustainability

Rong Yang, Milind Tambe, Manish Jain,
Jun-young Kwak, James Pita, and Zhengyu Yin

University of Southern California, Los Angeles, CA, 90089
{yangrong,tambe,manishja,junyounk,jpita,zhengyuy}@usc.edu

Abstract. Security and sustainability are two critical global challenges that involve the interaction of many intelligent actors. Game theory provides a sound mathematical framework to model such interactions, and computational game theory in particular has a promising role to play in helping to address key aspects of these challenges. Indeed, in the domain of security, we have already taken some encouraging steps by successfully applying game-theoretic algorithms to real-world security problems: our algorithms are in use by agencies such as the US coast guard, the Federal Air Marshals Service, the LAX police and the Transportation Security Administration. While these applications of game-theoretic algorithms have advanced the state of the art, this paper lays out some key challenges as we continue to expand the use of these algorithms in real-world domains. One such challenge in particular is that classical game theory makes a set of assumptions of the players, which may not be consistent with real-world scenarios, especially when humans are involved. To actually model human behavior within game-theoretic framework, it is important to address the new challenges that arise due to the presence of human players: (i) human bounded rationality; (ii) limited observations and imperfect strategy execution; (iii) large action spaces. We present initial solutions to these challenges in context of security games. For sustainability, we lay out our initial efforts and plans, and key challenges related to human behavior in the loop.

Keywords: Decision-making, Human Behavior, Game Theory, Security, Sustainability.

1 Introduction

Many of today's critical national and global challenges involve interactions of large numbers of different agents (individuals, large and small corporations, government agencies). A key challenge in solving these problems is to model and analyze the strategic interactions among these multiple intelligent agents, with their different goals, strategies and capabilities. Game theory provides a fundamental tool to understand and analyze such challenges.

The goal of this paper is to point to some research issues in computational game theory as they relate to two such global challenges: security and sustainability. These are massive world challenges, and as such the paper only focuses on

R.I. Brafman, F. Roberts, and A. Tsoukiàs (Eds.): ADT 2011, LNAI 6992, pp. 320–330, 2011.

limited aspects of these challenges. The key thrust of the research issues we focus on is in the fusion of computational game theory and models of human behavior. More specifically, classical game theory makes assumptions on human behavior – such as perfect and infallible rationality, the ability to perfectly observe and perfectly execute strategies – that are not consistent with real-world scenarios. Indeed, it is well understood that humans are bounded in their computational abilities or may reach "irrational" decisions due to other reasons [16,2]. In both security and sustainability domains, many of the agents are humans, and it is therefore critical to integrate human behavior in the game-theoretic algorithms for these domains.

In security, many scenarios are naturally modeled as a game; much of our own research has focused on the use of Bayesian Stackelberg games for security resource allocation[13,18,17]. These games typically involve a defender who acts first by setting up a security policy and an adversary who may conduct surveillance and then react; in our work, given particular restrictions on payoffs in these games, they are often labeled as *security games*[17]. The research in this area mainly focuses on improving the allocation of security resources by more accurately modeling the human adversary's behavior [20,14,19]. We briefly discuss four key research challenges in this context. The first challenge comes from the basic assumption of classical game theory that all players are perfectly rational, which may not hold when dealing with human adversaries. It is therefore crucial to integrate more realistic models of human decision-making in security games to more accurately predict adversaries' response to defender's strategies. The second challenge is caused by uncertainties in security games that arise in particular due to human players: specifically, adversaries may not perfectly observe defender strategies, defenders may not perfectly execute their strategies, etc. Therefore, it is important to ensure a robust solution in designing defender's resource allocation strategies. The third challenge is modeling, particularly given that we face human adversaries with the capability to generate a very large number of potential threats in real security games. We create a new game-theoretic framework that allows for compact modeling of such threats. Finally, scalability is important as a result of growth in the number of defender strategies, the attacker strategies, and the attacker types. We need to develop efficient algorithms for computing optimal defender strategy of allocating defender resources.

In sustainability, we focus on energy as a key resource, and for providing a concrete scenario, outline our initial efforts using a multi-agent system to lower energy usage in an office building. This research once again requires that we not only model complex strategic interactions between individuals (humans and agents) and design successful mechanisms to influence the humans' behavior, but also ensure that our theoretical models are augmented by more realistic models of human behavior. While we outline just our initial steps, sustainability research in general will require further integration of game theory and human behavior as we consider the complex strategic interactions in the future of large and small energy producers and consumers, individuals, governments, utility companies and others.

In the following, we first discuss the challenges we face in applying game theory to real-world security scenarios, and outline our approaches to address these challenges. For sustainability, we describe a multi-agent system highlighting the challenges of applying game-theoretic framework to the domain.

2 Security

Stackelberg games are often used to model the interaction between defenders and adversaries (attackers) in security settings [13,17,18]. In such games, there is a defender, who plays the role of leader, taking action first, and a follower (attacker) who responds to the leader's actions. In particular, in Stackelberg security games, the defender decides on how to allocate their security resources taking into consideration the response of the adversary; the attacker conducts surveillance to learn defender's strategy and then launches an attack. The optimal defender strategy hence emphasizes randomized security allocation to maintain unpredictability in its actions. I In a Bayesian Stackelberg game, the defender faces multiple types of adversaries, who might have different preference and objectives. Computing the optimal defender strategy for Bayesian Stackelberg games, so as to reach a "strong Stackelberg Equlibrium" is known to be a NP-hard problem[1].

In this section, we first give a brief introduction to the actual deployed applications that we have developed for different security agencies based on fast algorithms for obtaining optimal defender strategies in Bayesian Stackelberg games. While these algorithms have significantly advanced the state of the art, new challenges arise as we continue to expand the role of these game-theoretic algorithms; we discuss these challenges next.

2.1 Background

Armor. (Assistant for Randomized Monitoring Over Routes) was our first application of security games [13]. It is deployed at the Los Angeles International Airport (LAX) since 2007. ARMOR helps LAX police officers to randomize deployment of their limited security resources. For example, they have eight terminals but not enough explosive-detecting canine units to patrol all terminals at all times of the day. Given that LAX may be under surveillance by adversaries, the question is where and when to have the canine units patrol the different terminals. The foundation of ARMOR are algorithms for solving Bayesian Stackelberg games [12,13]; they recommend a randomized pattern for setting up checkpoints and canine patrols so as to maintain unpredictability.

Iris. (Intelligent Randomization In Scheduling) was designed to help the Federal Air Marchals Service (FAMS) to randomize allocations of air marshals to flights to avoid predictability by adversaries conducting surveillance, yet provide adequate protection to more important flights [18]. The challenge is that there are a very large number of flights over a month, and not enough air marshals to cover all the flights. At its backend, IRIS casts the problem it solves as a Stackelberg game and in particular as a security game with a special payoff structure. IRIS uses the Aspen algorithm [3], and is in use by FAMS since 2009.

Guards. (Game-theoretic Unpredictable and Randomly Deployed Security) was developed in collaboration with the United States Transportation Security Administration (TSA) to assist in resource allocation tasks for airport protection at over four hundred United States airports [15]. In contrast with ARMOR and IRIS, which focus on one installation/applications and one security activity (e.g. canine patrol or checkpoints) per application, GUARDS reasons with multiple security activities, diverse potential threats and also hundreds of end users. The goal for GUARDS is to allocate TSA personnel to security activities conducted to protect the airport infrastructure. GUARDS again utilizes a Stackelberg game, but generalizes beyond security games and develops a novel solution algorithm for these games. GUARDS has been delivered to the TSA and is currently under evaluation and testing for scheduling practices at an undisclosed airport.

Protect. (Port Resilience Operational/Tactical Enforcement to Combat Terrorism) is a pilot project we recently started in collaboration with the United State Coast Guard. PROTECT aims to recommend randomized patrolling strategies for the coast guard while taking into account (i) weights of different targets protected in their area of operation; (ii) adversary reaction to any patrolling strategy. We have begun with a demonstration and evaluation in the port of Boston and depending on our results there, we may proceed to other ports.

2.2 Challenges in Integrating Human Behavior Models

The first-generation security game applications mentioned above have been a significant step forward over previous methods in allocating security resources. However, to continue expanding the use of game-theoretic methods in security settings, we must address the human behavior within game-theoretic frameworks. While classical game theory makes assumption of perfect rationality, flawless observations and perfect executions, human decision makers may have bounded rationality, suffer from limited observation power and introduce error in execution of strategies [16]. They may also create new strategies that are not originally defined in the game model. To address such challenges, we must integrate realistic models of human bahavior into game-theoretic algorithms. To that end, in this section, we outline our initial research effort to tackle these challenges and point out key future challenges.

2.3 Human Decision Making

In order to address the assumption of perfect rationality of human adversaries, we focus on integrating more realistic models of human decision-making into the computational analysis of security problems. In this context, we have developed COBRA (Combined Observability and Rationality Assumption) [14] to address (i) the anchoring bias of human while interpreting the probabilities of several

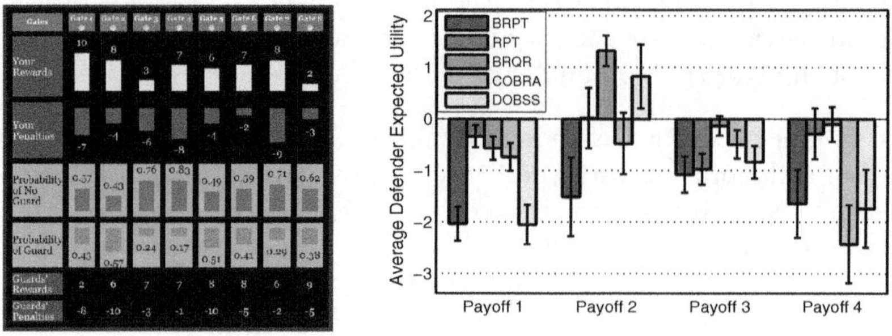

(a) Game Interface (b) Average Defender Expected Utility

Fig. 1. Experiments with human subjects

events; (ii) the bounded rationality they have in computing best response. Our most recent work in addressing human decision making [19] develops two new methods for generating defender strategies in security games based on using two well-known models of human behavior to model the attacker's decisions. The first is Prospect Theory (PT) [7], which provides a descriptive framework for decision-making under uncertainty that accounts for both, risk preferences (e.g. loss aversion) and variations in how humans interpret probabilities through a weighting function. The second model is Quantal Response Equilibrium (QRE) [11], which assumes that humans will choose better actions more frequently, but with some noise in the decision-making process that leads to stochastic choice probabilities. In this work, we develop new techniques to compute optimal defender strategies in Stackelberg security games under the assumption that the attacker will make choices according to either the PT or QRE model. More specifically, we present

- BRPT (Best Response to Prospect Theory), a mixed integer programming formulation, for computing the optimal leader strategy against players whose response follows a PT model;
- RPT (Robust-PT), modifying BRPT method to account for uncertainty about the adversaries choice, caused by imprecise computation [16].
- BRQR (Best Response to Quantal Response), to compute the optimal defender strategy assuming that the adversary's response is based on quantal response model.

In order to validate the performance of different models, we conducted the intensive empirical evaluation of different models against human subjects in security games. An online game called "The Guard and the Treasure" was designed to simulate a security scenario similar to the ARMOR program for the Los Angeles International (LAX) airport [13]. Figure 1(a) shows the interface of the game. Subjects played the role of followers and were able to observe the leader's mixed strategy. In the game, subjects were asked to choose one of the eight gates to open (attack). We conducted experiment with college students at USC to compare our five models: COBRA, BRPT, RPT, BRQR and the perfect rationality baseline (DOBSS) in the experiment.

Fig. 1(b) displays average performance for the different strategies in each payoff structure. Overall, BRQR performs best, RPT outperforms COBRA, and BRPT and DOBSS perform the worst. BRPT and DOBSS suffer from adversary's deviation from the optimal strategy. In comparison, BRQR, RPT and COBRA all try to be address such deviations. BRQR considers some (possibly very small) probability of adversary attacking any target. In contrast, COBRA and RPT separate the targets into two groups, the ϵ-optimal set and the non-ϵ-optimal set, using a hard threshold. They then try to maximize the worst case for the defender assuming the response will be in the ϵ-optimal set, but assign less resources to other targets. When the non-ϵ-optimal targets have high defender penalties, COBRA and RPT become vulnerable when the targets that are identified as non-ϵ-optimal are actually preferred by the subjects.

2.4 Robustness to Uncertainties in Attacker's Observation and Defender's Strategy Execution

As mentioned earlier, attacker-defender Stackelberg games have become a popular game-theoretic approach for security with deployments for the LAX Police, the FAMS and the TSA. Unfortunately, most of the existing solution approaches do not model two key uncertainties of the real-world: there may be noise in the defender's execution of the suggested mixed strategy and/or the observations made by an attacker can be noisy. In our recent work [20], we provide a framework to model these uncertainties, and demonstrate that previous strategies perform poorly in such uncertain settings. This work provides three key contributions: (i) RECON, a mixed-integer linear program that computes the risk-averse strategy for the defender given a fixed maximum execution and observation noise, $\boldsymbol{\alpha}$ and $\boldsymbol{\beta}$ respectively. RECON assumes that nature chooses noise to maximally reduce defenders utility, and RECON maximizes against this worst case; (ii) two novel heuristics that speed up the computation of RECON by orders of magnitude; (iii) experimental results that demonstrate the superiority of RECON in uncertain domains where existing algorithms perform poorly.

We compare the solution quality of RECON, ERASER, and COBRA under uncertainty: ERASER [6] is used to compute the SSE solution, and COBRA [14] is one of the latest algorithms that addresses attacker's observational error. Figure 2(a) and 2(b) present the comparisons of the worst-case utilities among RECON, ERASER and COBRA under two uncertainty settings: low uncertainty ($\boldsymbol{\alpha}=\boldsymbol{\beta}=\mathbf{0.01}$) and high uncertainty ($\boldsymbol{\alpha}=\boldsymbol{\beta}=\mathbf{0.1}$). MAXIMIN utility is provided as a benchmark. Here x-axis shows the number of targets and y-axis shows the defender's worst-case utility. RECON significantly outperforms MAXIMIN, ERASER and COBRA in both uncertainty settings. For example, in high uncertainty setting, for 80 targets, RECON on average provides a worst-case utility of -0.7, significantly better than MAXIMIN (-4.1), ERASER (-8.0) and COBRA (-8.4).

While RECON provides the best performance when we compare worst-case utilities, a key challenge that remains open is to compare its performance with BRQR mentioned in the previous section, and perform such comparison against human subjects. These are key topics for future work.

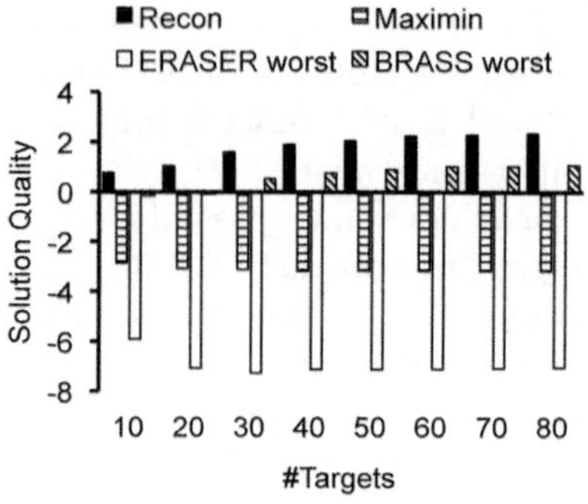

(a) Low uncertainty case ($\alpha = \beta = 0.01$)

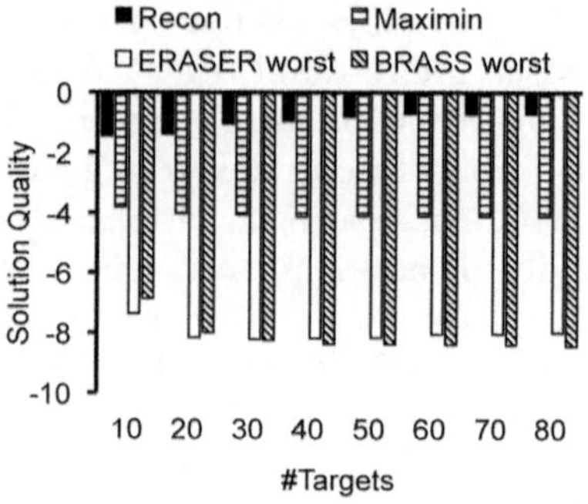

(b) High uncertainty case ($\alpha = \beta = 0.1$)

Fig. 2. Worst Case Defender Utility Comparison

2.5 Modeling Challenge

Security Circumvention Games (SCGs) are a modeling approach to address an adversary's potentially innumerable action space in security settings. While SCGs are motivated by an existing model of security games [21], SCGs make three new critical contributions to the security game model: (i) SCGs allow for defensive actions that consider heterogeneous security activities for each target, (ii) SCGs allow for multiple resources to be allocated to a target (i.e. targets are no longer covered or uncovered), and (iii) SCGs allow for heterogeneous threats on each target. For example, examining a security problem faced by the U.S. Transportation Security Administration (TSA), airports have ticketing areas, waiting areas, and cargo-holding areas. Within each of these areas, TSA has a number of security activities to choose from such as running perimeter patrols, screening cargo, and screening employees. TSA must both choose how many resources to assign to each area and which security activities to run. After observing the TSA's security policy, the attacker will choose which area to attack and what potential threat to execute. The key challenge is how to optimally allocate limited security resources between targets to specific activities, taking into account an attacker's response. SCGs provide the following additional key contributions: (i) a compact representation of the defender actions for efficiency; and (ii) an alternative approach for modeling attacker actions that avoids enumerating all possible threat scenarios. This attempt to avoid exhaustive enumeration of all possible threats is key in SCGs.

More specifically, SCGs create a list of potential threats that circumvent different combinations of specific security activities. By basing threats on circumventing particular combinations of security activities, we avoid the issue of enumerating all the possible potential threats. However, we also incorporate a cost to the attacker for circumventing more activities to capture the idea of causing maximal damage at minimal cost. Each individual security activity has a specific circumvention cost associated with it and more activities circumvented leads to a higher circumvention cost. This cost reflects the additional difficulty of executing an attack against increased security. This difficulty could be due to the need for additional resources, time, and other factors in executing an attack. Since attackers can now actively circumvent specific security activities, randomization becomes a key factor in the solutions leading to significant unpredictability in defender actions.

2.6 Addressing the Scalability Challenge

Real-world problems, like the FAMS security resource allocation problem, present trillions of action choices for the defender in security games. Such large problem instances cannot even be represented in modern computers, let alone solved using previous techniques. We provide new models and algorithms that compute optimal defender strategies for massive real-world security domains. In particular, we developed: (i) ASPEN and RUGGED, algorithms that compute the optimal defender strategy with a very large number of pure strategies for both the defender and the attacker [3,5]; (ii) a new hierarchical framework for Bayesian games that can scale-up to large number of attacker types and is applicable to all Stackelberg solvers [4]. Moreover, these algorithms have not only been experimentally validated, but ASPEN has also been deployed in the real-world [6].

Scaling Up in Pure Strategies: ASPEN and RUGGED provide scale-ups in real-world domains by efficiently analyzing the strategy space of the players. Both algorithms use strategy generation: the algorithms start by considering a minimal set of pure strategies for both the players (defender and attacker). Pure strategies are then generated iteratively, and a strategy is added to the set only if it would help increase the payoff of the corresponding player (a defender's pure strategy is added if it helps increase the defender's payoff). This process is repeated until the optimal solution is obtained.

Scaling Up with Attacker Types: The overarching idea of our approach to scale up attacker types is to improve the performance of branch-and-bound while searching for the solution of a Bayesian Stackelberg game. We decompose the Bayesian Stackelberg game into many hierarchically-organized smaller games, where Each smaller game considers only a few attacker types. The solutions obtained for the restricted games at the child nodes of the hierarchical game tree are used to provide: 1. pruning rules, 2. tighter bounds, and 3. efficient branching heuristics to solve the bigger game at the parent node faster. Additionally, these algorithms are naturally designed for obtaining quality bounded approximations since they are based on branch-and-bound, and provide a further order of magnitude scale-up without any significant loss in quality if approximate solutions are allowed.

3 Sustainability

To illustrate the challenges in applying game theoretic framework to the field of sustainability, we very briefly discuss a multi-agent system that affects both occupant behaviors and the operation of devices related to energy use. We also consider occupants as active participants in the energy reduction strategy by enabling them to engage in negotiations with intelligent agents that attempt to implement more energy conscious occupant planning. This occupant planning is carried out using multi-objective optimization methods to model the uncertainty of agent decisions, interactions and even general human behavior models. In these negotiations, minimizing energy and minimizing occupant discomfort resulting from various conditions in the space as well as from the negotiation process itself are the considered objectives.

In such energy domains, multi-agent interaction in the context of coordination presents novel challenges to optimize the energy consumption while satisfying the comfort level of occupants in the buildings. First, we should explicitly consider uncertainty while reasoning about coordination in a distributed manner. In particular, we suggest Bounded-parameters Multi-objective Markov Decision Problems (BM-MDPs) to model agent interactions/negotiations and optimize multiple competing objectives for human comfort and energy savings. Second, human behaviors and their occupancy preferences should be incorporated into planning and modeled as part of the system. As human occupants get involved in the negotiation process, it also becomes crucial to consider practical noise in human behavior models during the negotiation process. As a result, our goal is to eventually allow our system to be capable of generating an optimal and robust plan not only for building usage but also for occupants.

In our initial implementation, we compare four different energy control strategies: (i) manual control that simulates the current building control strategy maintained by USC facility managers; (ii) reactive control that building device agents reactively respond to the behaviors of human agents; (iii) proactive control that building agents predict human agent's occupancy and behavioral pattern given the schedules of human agents; and (iv) proactive control with a simple MDP that explicitly models agent negotiations [9,10]. As shown in [9,10], the simulation results indicate that our suggested control strategies could potentially achieve significant improvements in energy consumption while maintaining a desired occupant comfort level. However, this initial implementation is just a first step and significant additional research challenges need to be addressed for the intelligent energy-aware system to increase occupants' motivation to reduce their consumption as practicals by providing building occupants with feedback, especially understanding how their own or other neighbors' behavior influences energy consumption and long-term changes during negotiations.

4 Conclusion

Game theory provides a fundamental mathematical framework to model many real world problems involving strategic interactions among multiple intelligent actors; furthermore, computational game theory allows to scale-up the problems we can handle within this framework. However, classical game theory makes a set of assumptions of the rationality of the players which may not hold when dealing with real human players, thus requiring us to address new challenges in incorporating realistic models of human behavior in our game-theoretic algorithms. In this paper, we discussed our research in addressing these challenges in the context of security and sustainability. In security, we explained key challenges we face in addressing real world security problems, and presented the initial solutions for these challenges. In sustainability, the main concerns is the usage of energy and how to efficiently exploit the available reserves. The goal is to optimize the negotiation between minimizing energy use and minimizing occupant discomfort. Overall, this fusion of computational game theory and realistic models of human behaviors not only is critical in addressing real-world domains, but also leads to a whole new set of exciting research challenges.

References

1. Conitzer, V., Sandholm, T.: Computing the optimal strategy to commit to (2006)
2. Hastie, R., Dawes, R.M.: Rational Choice in an Uncertain World: the Psychology of Judgement and Decision Making. Sage Publications, Thounds Oaks (2001)
3. Jain, M., Kardes, E., Kiekintveld, C., Ordóñez, F., Tambe, M.: Security games with arbitrary schedules: A branch and price approach. In: AAAI (2010)
4. Jain, M., Kiekintveld, C., Tambe, M.: Quality-bounded solutions for finite bayesian stackelberg games: Scaling up. In: AAMAS (to appear, 2011)
5. Jain, M., Korzhyk, D., Vanek, O., Conitzer, V., Pechoucek, M., Tambe, M.: A double oracle algorithm for zero-sum security games on graphs. In: AAMAS (2011)
6. Jain, M., Tsai, J., Pita, J., Kiekintveld, C., Rathi, S., Tambe, M., Ordóñez, F.: Software Assistants for Randomized Patrol Planning for the LAX Airport Police and the Federal Air Marshals Service. Interfaces 40, 267–290 (2010)
7. Kahneman, D., Tvesky, A.: Prospect theory: An analysis of decision under risk. Econometrica 47(2), 263–292 (1979)
8. Kiekintveld, C., Jain, M., Tsai, J., Pita, J., Ordonez, F., Tambe, M.: Computing optimal randomized resource allocations for massive security games. In: AAMAS (2009)
9. Klein, L., Kavulya, G., Jazizadeh, F., Kwak, J., Becerik-Gerber, B., Varakantham, P., Tambe, M.: Towards optimization of building energy and occupant comfort using multi-agent simulation. In: The 28th International Symposium on Automation and Robotics in Construction (ISARC) (June 2011)
10. Kwak, J., Varakantham, P., Tambe, M., Klein, L., Jazizadeh, F., Kavulya, G., Gerber, B.B., Gerber, D.J.: Towards optimal planning for distributed coordination under uncertainty in energy domains. In: Workshop on Agent Technologies for Energy Systems (ATES) at AAMAS (2011)
11. McKelvey, R.D., Palfrey, T.R.: Quantal response equilibria for normal form games. Games and Economic Behavior 2, 6–38 (1995)
12. Paruchuri, P., Pearce, J.P., Marecki, J., Tambe, M., Ordonez, F., Kraus, S.: Playing games for security: An efficient exact algorithm for solving bayesian stackelberg games. In: AAMAS (2008)
13. Pita, J., Jain, M., Ordonez, F., Portway, C., Tambe, M., Western, C., Paruchuri, P., Kraus, S.: Deployed armor protection: The application of a game theoretic model for security at the los angeles international airport. In: AAMAS (2008)

14. Pita, J., Jain, M., Ordonez, F., Tambe, M., Kraus, S.: Solving stackelberg games in the real-world: Addressing bounded rationality and limited observations in human preference models. Artificial Intelligence Journal 174(15), 1142–1171 (2010)
15. Pita, J., Tambe, M., Kiekintveld, C., Cullen, S., Steigerwald, E.: Guards - game theoretic security allocation on a national scale. In: AAMAS (2011)
16. Simon, H.: Rational choice and the structure of the environment. Psychological Review 63(2), 129–138 (1956)
17. Tambe, M.: Security and Game Theory: Algorithms, Deployed Systems, Lessons Learned. Cambridge University Press, Cambridge (2011)
18. Tsai, J., Rathi, S., Kiekintveld, C., Ordonez, F., Tambe, M.: Iris - a tool for strategic security allocation in transportation networks. In: AAMAS (2009)
19. Yang, R., Kiekintveld, C., Ordonez, F., Tambe, M., John, R.: Improving resource allocation strategy against human adversaries in security games. In: IJCAI (2011)
20. Yin, Z., Jain, M., Tambe, M., Ordonez, F.: Risk-averse strategies for security games with execution and observational uncertainty. In: AAAI (2011)
21. Yin, Z., Korzhyk, D., Kiekintveld, C., Conitzer, V., Tambe, M.: Stackelberg vs. Nash in security games: interchangeability, equivalence, and uniqueness. In: AAMAS (2010)

Constrained Multicriteria Sorting Method Applied to Portfolio Selection

Jun Zheng, Olivier Cailloux, and Vincent Mousseau

Industrial Engineering Laboratory, Ecole Centrale Paris
Grande Voie des Vignes 92295 Châtenay Malabry Cedex, France
{jun.zheng,olivier.cailloux,vincent.mousseau}@ecp.fr

Abstract. The paper focuses on portfolio selection problems which aim at selecting a subset of alternatives considering not only the performance of the alternatives evaluated on multiple criteria, but also the performance of portfolio as a whole, on which balance over alternatives on specific attributes is required by the Decision Makers (DMs).

We propose a two-level method to handle such decision situation. First, at the individual level, the alternatives are evaluated by the sorting model ELECTRE TRI which assigns alternatives to predefined ordered categories by comparing alternatives to profiles separating the categories. The DMs' preferences on alternatives are expressed by some assignment examples they can provide, which reduces the DMs' cognitive efforts. Second, at the portfolio level, the DMs' preferences express requirements on the composition of portfolio and are modeled as constraints on category size. The method proceeds through the resolution of a Mixed Integer Program (MIP) and selects a satisfactory portfolio as close as possible to the DMs' preference.

The usefulness of the proposed method is illustrated by an example which integrates a sorting model with assignment examples and constraints on the portfolio definition. The method can be used widely in portfolio selection situation where the decision should be made taking into account the performances of individual alternatives and portfolio simultaneously.

Keywords: Multicriteria decision aiding, Portfolio selection, Preference elicitation.

1 Introduction

Let us consider the student enrollment in universities every year. Universities want to select students with good performances on several criteria (such as GPAs, motivation, maturity, ...). At the same time, the selected students should satisfy some specific requirements at a collective level. For instance, the number of students in each department should be more or less balanced. Each department tries to achieve a gender (nationality, etc.) diversity. Moreover, the positions available are limited. Therefore, the universities face a decision which consists of selecting a certain number of students, designing a waiting list and rejecting

R.I. Brafman, F. Roberts, and A. Tsoukiàs (Eds.): ADT 2011, LNAI 6992, pp. 331–343, 2011.

the other students (see similar example in universities [7], other examples are available in the book [28]). Another example of such portfolio selection problems concerns allocating grants to research proposals. The committee evaluates the merit of the proposal, including originality, novelty, rigor and the ability of the researchers to carry out the research individually. On a whole level, they try to balance the funding among disciplines, institutions and even regions. Therefore, a decision is to be made to select certain research proposals within limited budget.

The two problems above share some characteristics. Firstly, they involve evaluating individual alternatives according to their performances on multiple criteria. Secondly, a portfolio is to be selected based not only on individual alternative's performance, but also on the performance of the whole portfolio. Such situation typically corresponds to a portfolio selection problem.

There is a large number of methods in literature for evaluating and selecting portfolios [15,25,16,1,8]. Cost-benefit analysis [24], multiattribute utility theory [13], weighted scoring [9] are widely used. Some researchers combine preference programming with portfolio selection considering incomplete preference information [17,18]. However, to our knowledge, Multiple Criteria Decision Aiding (MCDA) outranking methods have not been applied to portfolio selection problem. Furthermore, the ability of the methods to express sophisticated preference on portfolios has little been explored. A balance model [14] is developed which measures the distribution of specific attributes by dispersion and uses such measurement to select subsets of multiattribute items. [15] uses constraints to eliminate the ones which do not fit in the requirement on whole portfolio.

We propose a two-level method for such portfolio selection problems. At individual level, the paper uses Electre Tri method [26,27] to evaluate the alternatives on multiple criteria, which assigns alternatives to predefined ordered categories by comparing an alternative with several profiles. The DMs' preference on individual evaluation can be taken into account by some assignment examples. At portfolio level, a wide class of preferences on portfolios (resource limitation, balance of the selected items over an attribute...) are represented using general category size constraints. An optimization procedure is performed by solving a MIP to infer the values of preference parameters and to identify a satisfactory portfolio.

The paper is organized as follows. Section 2 formulates portfolio selection problem as a constrained multicriteria sorting problems. Section 3 presents a mathematical program which computes the portfolio that best matches the DMs' preferences. Section 4 illustrates the proposed method with an example. The last section groups conclusions.

2 Problem Formulation

2.1 Evaluating Alternatives with Electre Tri Method

Alternatives to be included on a portfolio are evaluated by an outranking method Electre Tri [26,27]. This method assigns alternatives to predefined ordered categories by comparing the alternatives to profiles which define the frontiers of two successive categories. For example, for enrollment problem described in Section 1, the DMs want to sort the students into three categories: *accepted, waiting list* or *rejected* according to students' performances on multiple criteria. Thus the two profiles could be two frontiers which separate these three categories.

Formally, ELECTRE TRI assigns each alternative of a set $A = \{a_1, a_2, ..., a_n\}$ to k pre-defined ordered categories $\text{Cat}_1 \ll \text{Cat}_2 \ll ... \ll \text{Cat}_k$. K denotes the set of indices of the k categories ($K = \{1, 2, ..., k\}$). The alternatives are evaluated on m criteria. Let J denote the set of the indices of the criteria $g_1, g_2, ..., g_m$ ($J = \{1, 2, ..., m\}$). For all $j \in J$, $a \in A$, $g_j(a)$ represents the evaluation of a with respect to the jth criteria. In what follows we assume, without loss of generality, that preference increases with the value on each criterion. b_h is the upper limit of category h and the lower limit of category $h+1$, $h = 1, 2, ..., k-1$. In other words, the frontier separating two categories is represented by the evaluations on the set of criteria of the profile b_h: $g_j(b_h)$, $\forall j \in J$. The assignment of an alternative a results from the comparison of a to the profiles $b_1, b_2, ..., b_{k-1}$.

ELECTRE TRI uses an outranking relation $\succcurlyeq$ which represents assertions $a \succcurlyeq b_h$ whose meaning is "a is at least as good as b_h". In order to validate the assertion $a \succcurlyeq b_h$, a "sufficient" majority of criteria should be in favor of this assertion. A set of weights coefficients $(w_1, w_2, ..., w_m)$ which sum to 1 representing relative importance of the criteria is additively used in the concordance test when computing the strength of the coalitions of criteria being in favor of the assertion $a \succcurlyeq b_h$.

ELECTRE TRI builds a concordance index $C(a, b_h) \in [0, 1], \forall a \in A, \forall h \in K$, defined as $C(a, b_h) = \sum_{j \in J | g_j(a) \geq g_j(b_h)} w_j$. It represents the degree of credibility of the assertion $a \succcurlyeq b_h$. The assertion $a \succcurlyeq b_h$ is considered to be valid if $C(a, b_h) \geq \lambda$, λ being the "majority level" such that $\lambda \in [0.5, 1]$.

We consider a simplified ELECTRE TRI method which ignores discrimination thresholds (preference and indifference threshold) and veto threshold involved in the standard non-discordance condition [23]. Such simplification is in line with the axiomatic study of Bouyssou and Marchant [2,3].

Given the outranking relation $\succcurlyeq$, alternatives are assigned to categories on the basis of the way they compare to profiles b_h. ELECTRE TRI proposes two assignment procedures (so-called pessimistic and optimistic rule). In this paper we consider only the pessimistic rule which assigns alternative a to the highest category Cat_h for which $a \succcurlyeq b_{h-1}$ and not $a \succcurlyeq b_h$.

So as to implement ELECTRE TRI, an elicitation process is necessary to determine the values of preference parameters (profiles b_h, weights w_j and majority level λ). In a portfolio selection perspective, we consider DMs' preference at two levels. At alternative level, the DMs express preferences on alternatives individually. At a portfolio level, they express preferences on portfolios as a whole (resource limitation, balance of the selected items over an attribute, ...). These two preference levels are distinguished, as they are elicited in different ways, and could be provided by different DMs who have expertise and understanding of the portfolio selection at different levels.

2.2 DMs' Preference on Alternatives

The DMs have little understanding of the precise semantics of the preference parameters involved in Electre Tri. On the contrary, they can easily express their expertise on which category an alternative should be assigned to. Therefore, we propose to elicit the DMs' preference in an indirect way, in accordance with the disaggregation-aggregation paradigm. Instead of providing precise values for the parameters, the DMs provide assignment examples, i.e. alternatives which they are able to assign confidently to a category. For instance, in a student selection problem, the DMs may state that one particular student should be assigned to the best category (the set of accepted students). Inference procedure can thus be used to compute values for the preference parameters that best match the assignment examples. Several authors have proposed disaggregation methodologies from assignment examples expressed by the DMs. Mousseau and Słowiński use non-linear programming to infer all the parameters simultaneously [22], and some suggest to infer weights only assuming the profiles are fixed [21]. Researchers also proposed to compute robust assignment categories to which an alternative is possible to be assigned, considering all combinations of values compatible with the DMs' preference statements [11] and developed corresponding software [10]. Recently, an evolutionary approach has been presented to infer all parameters of Electre Tri model [12]. In this paper, we assume all the preference parameters are variables and infer them by solving a MIP.

2.3 DMs' Preference Information on Portfolios

The DMs' preferences can also be expressed at the portfolio level (resource limitation, balance on the composition of categories w.r.t. an attribute, ...). We formalize such preferences as general constraints on category size. For example, in the student enrollment case, let us denote the category of rejected students Cat_1, the category of waiting list Cat_2 and the category of admitted students Cat_3. Suppose the university only have 100 positions available, and such constraint can be modeled as the number of students in Cat_1 cannot exceed 100. Moreover, balancing gender in the selected students (100 students in total) can also be modeled as a constraint that the number of female students in Cat_1 should not be lower than 30. Adding such constraints to the selection process may result in rejecting some male students whose performances are better than those of the accepted female students. However, such portfolio is more satisfactory for the DMs in terms of gender balance. Modeling the DMs' preference as constraints eliminates some portfolios which don't satisfy their requirements on the whole portfolio.

3 Mathematical Program Formulation

3.1 Stating the Problem and Decision Variables

Given a set of alternatives A, a set of criteria indices J, evaluations of the alternatives $g_j(a), \forall a \in A, j \in J$, a set of category indices $K = \{1, 2, ..., k\}$, a set of profiles $b_h, 1 \leq h \leq k-1$, the goal of the program is to determine the performances of profiles $g_j(b_h), \forall j \in J, 1 \leq h \leq k-1$, weights w_j and majority threshold λ, satisfying all the constraints given by the DMs in the form of assignment examples and portfolio constraints. The MIP also defines additional variables involved in the way ELECTRE TRI assigns alternatives to categories. The binary variables $C_j(a, b_h), \forall a \in A, j \in J, 1 \leq h \leq k-1$ represent the partial concordance indices such that $C_j(a, b_h) = 1$ if and only if the performance of the alternative a on the criterion j is at least as good as the performance of the profile b_h. The continuous variables $\sigma_j(a, b_h)$ represent the weighted partial concordance indices, they are such that $\sigma_j(a, b_h) = w_j$ if and only if $C_j(a, b_h) = 1$. Finally, binary variables $n(a, h), \forall a \in A, h \in K$ are defined so that $n(a, h) = 1$ if and only if alternative a is assigned to category h. A slack variable s is used in the objective function which appreciates the ability of the ELECTRE TRI model to reproduce the assignment examples in a robust way.

The constraint $\sum_{j \in J} w_j = 1$ is posed, and the following constraints are used to ensure a correct ordering of the profiles defining the categories: $\forall j \in J, 2 \leq h \leq k-1 : g_j(b_{h-1}) \leq g_j(b_h)$.

3.2 Constraints Stemming from Preferences at Individual Level

The set of assignment examples E is the set of pairs $(a, h) \in A \times K$ specifying that alternative a is assigned to Cat_h. Recall that satisfying an assignment example (a, h) amounts to satisfy both $\sum_{j \in J: g_j(a) \geq g_j(b_{h-1})} w_j \geq \lambda$ and $\sum_{j \in J: g_j(a) \geq g_j(b_h)} w_j < \lambda$.

The sum of support in favor of the outranking of an alternative a over a profile b_h, $\sum_{j \in J: g_j(a) \geq g_j(b_h)} w_j$, can also be written $\sum_{j \in J} C_j(a, b_h) w_j$ with $C_j(a, b_h)$ equal to one iff $g_j(a) \geq g_j(b_h)$. Constraints (1) define the binary variables $C_j(a, b_h)$, $\forall j \in J, a \in A, 1 \leq h \leq k-1$, where ε is an arbitrary small positive value, and M is an arbitrary large value. See also Fig. 1.

$$\frac{1}{M}((g_j(a) - g_j(b_h)) + \varepsilon) \leq C_j(a, b_h) \leq \frac{1}{M}(g_j(a) - g_j(b_h)) + 1 \ . \qquad (1)$$

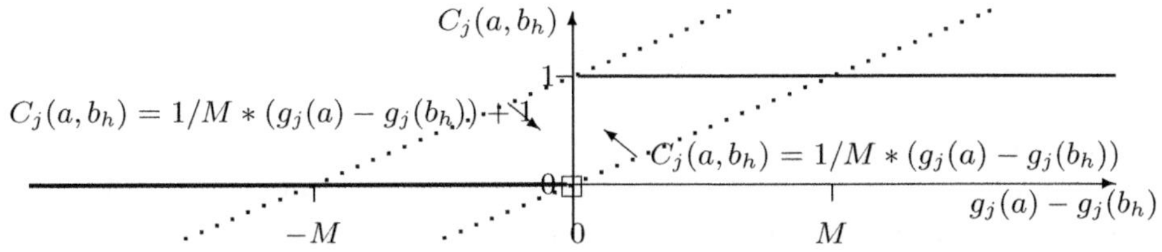

Fig. 1. constraining $C_j(a, b_h)$ to the appropriate value

The following constraints define the variables $\sigma_j(a, b_h)$ representing the sum of the support in favor of the assertion "a is at least as good as b_h" while avoiding the non-linear expression $\sigma_j(a, b_h) = C_j(a, b_h)w_j$ [19]. See also Fig. 2.

$$\forall j \in J, a \in A, 1 \leq h \leq k-1 : \begin{cases} \sigma_j(a, b_h) \leq w_j \\ \sigma_j(a, b_h) \geq 0 \\ \sigma_j(a, b_h) \leq C_j(a, b_h) \\ \sigma_j(a, b_h) \geq C_j(a, b_h) + w_j - 1 \ . \end{cases} \tag{2}$$

We also define, for simplicity of use in the next constraints, $\forall j \in J, a \in A$: $\sigma_j(a, b_0) = w_j$ and $\sigma_j(a, b_k) = 0$.

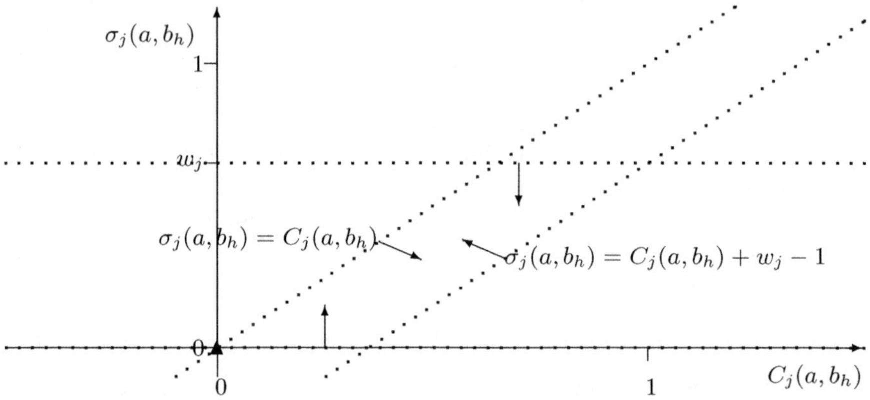

Fig. 2. constraining $\sigma_j(a, b_h)$ to the appropriate value

Finally, we need to ensure that each assignment example is assigned to the class specified by the DMs. The variable s is a slack variable used in the objective function.

$$\forall (a, h) \in E : \sum_{j \in J} \sigma_j(a, b_h) + s + \varepsilon \leq \lambda \leq \sum_{j \in J} \sigma_j(a, b_{h-1}) - s \ . \tag{3}$$

3.3 Constraints Stemming from Preferences at Portfolio Level

Suppose the DMs want to impose, in a student selection problem, that at least 30 students in the best category (i.e. Cat_k) are females. To model this, we define a function *Gender* on the set of alternatives that equals one if the student a is a female student and zero otherwise, and set as a constraint that the sum of *Gender*(a) on each alternative a assigned to Cat_k should be at least 30 ($\sum_{a \to \text{Cat}_k} Gender(a) \geqq 30$). In a project selection problem, suppose the DMs want to make sure that the sum of the costs of the selected projects (say, the projects in the best category) do not exceed the available budget x. A function *Cost* would be defined on the set of alternatives representing their cost attribute, and a constraint is added to ensure that the sum of $Cost(a)$ on alternatives a assigned to the best category should be no greater than the budget ($\sum_{a \to \mathrm{Cat}_k} Cost(a) \leq x$).

More generally, portfolio preferences are represented as a set N of tuples $\langle h, \underline{n_h}, \overline{n_h}, P \rangle, 1 \leq h \leq k, \underline{n_h}, \overline{n_h} \in \mathbb{R}, P$ a function from A to $\mathbb{R}$, representing the constraint that the preferential model inferred by the program should be such that the number of alternatives from A assigned to Cat_h weighted by their attribute P should be at least $\underline{n_h}$ and at most $\overline{n_h}$: $\underline{n_h} \leq \sum_{a \to \mathrm{Cat}_h} P(a) \leq \overline{n_h}$.

The following constraints define the binary variables $n(a,h)$, $\forall a \in A, 1 \leq h \leq k$, so that $n(a,h)$ equals one iff a is assigned to category Cat_h, that is, $\sum_{j \in J} \sigma_j(a, b_{h-1}) \geq \lambda$ and $\sum_{j \in J} \sigma_j(a, b_h) < \lambda$. The first constraints force that $n(a,h) = 1$ requires that a goes to category h, and the last ones force that exactly one $n(a,h)$ among all h equals one.

$$\forall a \in A, 1 \leq h \leq k : \begin{cases} n(a,h) \leq 1 + \sum_{j \in J} \sigma_j(a, b_{h-1}) - \lambda \ , \\ n(a,h) \leq 1 + \lambda - \sum_{j \in J} \sigma_j(a, b_h) - \varepsilon \ . \end{cases} \tag{4}$$

$$\forall a \in A : \sum_{1 \leq h \leq k} n(a,h) = 1. \tag{5}$$

These variables permit to guarantee the desired group sizes.

$$\forall \langle h, \underline{n_h}, \overline{n_h}, P \rangle \in N : \underline{n_h} \leq \sum_{a \in A} n(a,h) P(a) \leq \overline{n_h} \ . \tag{6}$$

3.4 Objective Function and Resolution Issues

In order to maximize the separation between the sum of support and the majority threshold, the objective of the MIP is set to maximize the slack variable s as defined in Constraints (3). The slack variable evaluates the ability of the Electre Tri model to "reproduce" the assignment examples in a robust way.

However the preference information of the DMs does not lead univocally to a single compatible portfolio. The optimization procedure finds out one of the compatible portfolios. In an interactive perspective, the DMs can provide further preference information considering the results of the MIP, and the information can be added to the optimization procedure to get a more satisfactory portfolio. The decision aiding process can proceed with several interactions until the DMs are content with the selected portfolio.

4 Illustrative Example

Let us illustrate the method with the following hypothetical decision situation. A government board has the responsibility to choose which research projects to finance among a list of 100 research proposals. The selection process involves sorting these proposals into three categories: projects that are considered very good and should be funded (category *Good*); projects that are good and should be funded if supplementary budget can be found (category *Average*); projects that are of insufficient quality and should not be funded (category *Bad*). To sort these projects in these three categories, the board agrees to use the following six criteria.

- sq The project's scientific quality, evaluated on a 5 points ordinal scale.
- rq The proposal's writing quality, evaluated on a 5 points ordinal scale.
- a The proposal's adequacy with respect to the government priorities, evaluated on a 3 points ordinal scale.
- te The experience of the researcher teams submitting the project, evaluated on a 5 points ordinal scale.
- ic Whether the proposal includes international collaboration, a binary assessment.
- ps The researchers' publication score evaluated by an aggregate measure of the total quality of publications of the researchers involved in the proposal (evaluated on a [0,100] scale).

The scales on all criteria are defined such that a greater value corresponds to a better evaluation.

Supplementary to these six criteria, the 100 projects to be evaluated are described by three attributes: the research domain to which the project belongs (Operational Research (OR), Artificial Intelligence (AI) or Statistics); the budget the project asks funding for; the originating country. Table 1 shows the data about the first 7 projects in the list (complete data lists for the whole example are available at `http://www.lgi.ecp.fr/~mousseau/ADT2011/`). In order to determine an appropriate preference model, the board gives as a first stage 30 examples of past research proposals whose performances on the six criteria and final quality evaluation are known. A part of this data is shown in Table 2.

Table 1. Some of the research projects to be evaluated. The budget is in tens of K€.

	evaluations criteria						descriptive attributes		
Project	rq	ps	a	sq	te	ic	budget	domain	country
Pr001	2	47	2	3	1	0	27	Stat.	Germany
Pr002	2	3	2	4	4	0	29	Stat.	France
Pr003	5	63	1	5	1	0	20	Stat.	Italy
Pr004	1	92	3	5	5	1	34	AI	Germany
Pr005	4	13	2	4	2	0	32	Stat.	Germany
Pr006	5	5	3	5	1	0	22	Stat.	Netherlands
Pr007	1	27	3	2	5	1	34	OR	Germany
⋮									

Table 2. Some research project examples and their respective assignments

Project	rq	ps	a	sq	te	ic	Cat
Ex01	4	50	2	3	3	0	Average
Ex02	4	85	3	1	5	1	Good
Ex03	3	95	1	2	5	1	Average
Ex04	5	91	2	2	5	1	Good
Ex05	5	89	1	5	3	0	Good
Ex06	3	5	3	2	2	1	Average
⋮							

The inference program is run with these assignment examples, and without supplementary portfolio constraints. Table 3 lists the resulting profiles and weights. Note that the profiles performances values in all our tables have been rounded up. Because each alternative used in this example has integer performance values on all criteria, doing so does not impact the way each alternative compares to these profiles. The resulting preference model is used to evaluate the 100 research projects, which leads to 22 projects being evaluated as good projects. The board is not satisfied with this set of projects because accepting these projects induces a total funding cost of 718 which exceeds the available budget (400). The program is thus run again with a supplementary constraint on the sum of the budget of the projects being assigned to the Good category to ensure that it stays below the available budget.

Table 3. Profiles, weights and majority threshold inferred during the first stage

	rq	ps	a	sq	te	ic	λ
b1	2	73	4	1	2	1	
b2	4	96	4	5	3	1	
w	0.2	0.2	0	0.2	0.2	0.2	0.5

This second stage inference yields other profiles and weights, given in Table 4, and a new list of assignments of which a part is displayed in Table 5. At this stage 11 projects are assigned to category Good and therefore are to be financed, leading to a total cost below 400. However the board is not fully satisfied yet because one domain is largely favored by this result, as the AI domain has 7 projects selected whereas only 1 project in the OR domain is to be financed. In a third stage, the inference program is thus run again with a new constraint requiring that the domain OR has at least 2 projects in the category Good. The final assignment results, shown partly in Table 6, are considered satisfactory.

The process could have continued had the board wished a better balance among the originating countries, or had they wished to consider more closely also the Average category. In case an infeasible problem had been reached at some point during the process, some constraints would have had to be relaxed

Table 4. Profiles, weights and majority threshold inferred with supplementary budget constraint

	rq	ps	a	sq	te	ic	λ
b1	2	2	2	1	2	1	
b2	3	84	2	4	3	2	
w	0.143	0.143	0.143	0.143	0.286	0.143	0.643

Table 5. A part of the assignment of the research projects with the preference model inferred during the second stage

Project	rq	ps	a	sq	te	ic	budget	domain	country	Cat
Pr001	2	47	2	3	1	0	27	Stat.	Germany	Bad
Pr002	2	3	2	4	4	0	29	Stat.	France	Average
Pr003	5	63	1	5	1	0	20	Stat.	Italy	Bad
Pr004	1	92	3	5	5	1	34	AI	Germany	Good
Pr005	4	13	2	4	2	0	32	Stat.	Germany	Average
Pr006	5	5	3	5	1	0	22	Stat.	Netherlands	Bad
Pr007	1	27	3	2	5	1	34	OR	Germany	Average
⋮										

Table 6. A part of the assignment of the research projects with the preference model inferred during the third stage

Project	rq	ps	a	sq	te	ic	budget	domain	country	Cat
Pr001	2	47	2	3	1	0	27	Stat.	Germany	Average
Pr002	2	3	2	4	4	0	29	Stat.	France	Average
Pr003	5	63	1	5	1	0	20	Stat.	Italy	Bad
Pr004	1	92	3	5	5	1	34	AI	Germany	Good
Pr005	4	13	2	4	2	0	32	Stat.	Germany	Average
Pr006	5	5	3	5	1	0	22	Stat.	Netherlands	Average
Pr007	1	27	3	2	5	1	34	OR	Germany	Average
⋮										

or deleted. The reader will find in Mousseau et al. [20] algorithms on how to proceed for constraints relaxation.

The proposed approach has been implemented as a free, open source Java library [5,6]. The mathematical programs implementation relies on the JLP software package, a free and open source Java Linear Programming wrapper on top of CPLEX (the solver we used) and other commercial and free solvers. Solving the problems used in this illustrative example takes less than one minute. A study of the solving time of a related problem is available for the interested reader [4]. That study does not take into account the portfolio constraints and thus examines a simpler problem. It shows that small to medium size problems (consisting of less than eight criteria, of three categories and of less than one hundred alternatives) are solvable within ninety minutes, which is a reasonable time provided that this kind of approach is primarily used in an off-line mode. Analysis of the solving time of the problem studied here, thus with the added portfolio constraints, is left as a future work.

5 Conclusion

The method applies constrained ELECTRE TRI model to portfolio selection problems in order to select a satisfactory portfolio considering DMs' preferences both at individual and portfolio level. Using a sorting model, the alternatives are evaluated by their intrinsic performances on criteria. Unsatisfactory portfolios which do not meet the DMs' requirements on portfolios as a whole are screened out by adding category size constraints to ELECTRE TRI model. Because of such category size constraints, the assignment of an alternative is dependent on its evaluation but also on other alternatives.

Our formalization permits to tackle the challenges the DMs may face during the decision of portfolio selection. (1) At individual level, an alternative is evaluated on multiple criteria which can be qualitative or quantitative criteria. Moreover, the DMs easily express their preferences on alternatives by assignment examples. (2) At portfolio level, the best alternatives do not necessarily compose the best portfolio. Our method takes into account the overall portfolio performance by modeling the DMs' preference on portfolio as constraints. (3) The preference information at the two levels (individual classification of alternatives and preference at the portfolio level) can be elicited from different stakeholders. (4) The proposed method involves the DMs deeply by asking them preference in an intuitive way.

The proposed method can be widely used in portfolio selection situations where the decision should be made taking into account the individual alternative and portfolio performance simultaneously. The proposed syntax of category size constraints has a broad descriptive ability for portfolio decision modeling. The method can be extended by providing robust recommendation to the DMs as a result of incomplete preference information. Moreover, the preference on portfolio level can be modeled as objectives rather than constraints of the optimization procedure, which would lead to a multiobjective problem.

References

1. Archer, N.P., Ghasemzadeh, F.: An integrated framework for project portfolio selection. International Journal of Project Management 17(4), 207–216 (1999)
2. Bouyssou, D., Marchant, T.: An axiomatic approach to noncompensatory sorting methods in MCDM, I: The case of two categories. European Journal of Operational Research 178(1), 217–245 (2007)

3. Bouyssou, D., Marchant, T.: An axiomatic approach to noncompensatory sorting methods in MCDM, II: more than two categories. European Journal of Operational Research 178(1), 246–276 (2007)
4. Cailloux, O., Meyer, P., Mousseau, V.: Eliciting ELECTRE TRI category limits for a group of decision makers. Tech. Rep. 2011-09, Laboratoire Génie Industriel, Ecole Centrale Paris (June 2011), `http://www.lgi.ecp.fr/Biblio/PDF/CR-LGI-2011-09.pdf` cahiers de recherche (2011-09)
5. Cailloux, O.: ELECTRE and PROMETHEE MCDA methods as reusable software components. In: Proceedings of the 25th Mini-EURO Conference on Uncertainty and Robustness in Planning and Decision Making (URPDM 2010). University of Coimbra, Portugal (2010)
6. Cailloux, O.: J-MCDA: free/libre java libraries for MCDA (2011), `http://sourceforge.net/projects/j-mcda/`
7. Cardinal, J.L., Mousseau, V., Zheng, J.: Multiple criteria sorting: An application to student selection. In: Salo, A., Keisler, J., Morton, A. (eds.) Portfolio Decision Analysis. Springer-Verlag New York Inc., Secaucus (2011)
8. Chien, C.: A portfolio-evaluation framework for selecting R&D projects. R and D Management 32(4), 359–368 (2002)
9. Coldrick, S., Longhurst, P., Ivey, P., Hannis, J.: An R&D options selection model for investment decisions. Technovation 25(3), 185–193 (2005)
10. Dias, L., Mousseau, V.: IRIS: a DSS for multiple criteria sorting problems. Journal of Multi-Criteria Decision Analysis 12, 285–298 (2003)
11. Dias, L., Mousseau, V., Figueira, J., Clímaco, J.: An aggregation/disaggregation approach to obtain robust conclusions with ELECTRE TRI. European Journal of Operational Research 138(2), 332–348 (2002)
12. Doumpos, M., Marinakis, Y., Marinaki, M., Zopounidis, C.: An evolutionary approach to construction of outranking models for multicriteria classification: The case of the ELECTRE TRI method. European Journal of Operational Research 199(2), 496–505 (2009)
13. Duarte, B.P., Reis, A.: Developing a projects evaluation system based on multiple attribute value theory. Computers & Operations Research 33(5), 1488–1504 (2006)
14. Farquhar, P.H., Rao, V.R.: A balance model for evaluating subsets of multiattributed items. Management Science 22(5), 528–539 (1976)
15. Golabi, K., Kirkwood, C.W., Sicherman, A.: Selecting a portfolio of solar energy projects using multiattribute preference theory. Management Science 27(2), 174–189 (1981)
16. Hall, N.G., Hershey, J.C., Kessler, L.G., Stotts, R.C.: A model for making project funding decisions at the national cancer institute. Operations Research 40(6), 1040–1052 (1992)
17. Liesiö, J., Mild, P., Salo, A.: Preference programming for robust portfolio modeling and project selection. European Journal of Operational Research 181(3), 1488–1505 (2007)
18. Liesiö, J., Mild, P., Salo, A.: Robust portfolio modeling with incomplete cost information and project interdependencies. European Journal of Operational Research 190(3), 679–695 (2008)
19. Meyer, P., Marichal, J., Bisdorff, R.: Disaggregation of bipolar-valued outranking relations. In: Le Thi, H.A., Bouvry, P., Pham Dinh, T. (eds.) Proc. of MCO 2008 Conference, pp. 204–213. Springer, Metz (2008)
20. Mousseau, V., Dias, L., Figueira, J.: Dealing with inconsistent judgments in multiple criteria sorting models. 4OR 4(3), 145–158 (2006)

21. Mousseau, V., Figueira, J., Naux, J.: Using assignment examples to infer weights for ELECTRE TRI method: Some experimental results. European Journal of Operational Research 130(2), 263–275 (2001)
22. Mousseau, V., Słowiński, R.: Inferring an ELECTRE TRI model from assignment examples. Journal of Global Optimization 12(2), 157–174 (1998)
23. Mousseau, V., Słowiński, R., Zielniewicz, P.: A user-oriented implementation of the ELECTRE TRI method integrating preference elicitation support. Computers & Operations Research 27(7-8), 757–777 (2000)
24. Phillips, L.D., Bana e Costa, C.A.: Transparent prioritisation, budgeting and resource allocation with multi-criteria decision analysis and decision conferencing. Annals of Operations Research 154(1), 51–68 (2007)
25. Rao, V.R., Mahajan, V., Varaiya, N.P.: A balance model for evaluating firms for acquisition. Management Science 37(3), 331–349 (1991)
26. Roy, B.: The outranking approach and the foundations of ELECTRE methods. Theory and Decision 31, 49–73 (1991)
27. Roy, B.: Multicriteria Methodology for Decision Aiding. Kluwer Academic, Dordrecht (1996)
28. Salo, A., Keisler, J., Morton, A.: Portfolio Decision Analysis. Springer-Verlag New York Inc., Secaucus (2011)

Author Index

GPSR Compliance

The European Union's (EU) General Product Safety Regulation (GPSR) is a set of rules that requires consumer products to be safe and our obligations to ensure this.

If you have any concerns about our products, you can contact us on ProductSafety@springernature.com

In case Publisher is established outside the EU, the EU authorized representative is:

Springer Nature Customer Service Center GmbH
Europaplatz 3
69115 Heidelberg, Germany

Batch number: 10844619

Printed by Printforce, the Netherlands